CPHC/BCS Distinguished Dissertations

Series edited by C.J. van Rijsbergen

Springer
London
Berlin
Heidelberg
New York
Barcelona
Budapest
Hong Kong
Milan
Paris
Santa Clara
Singapore
Tokyo

Martin Hofmann

Extensional Constructs in Intensional Type Theory

Springer

Martin Hofmann, PhD
Fachbereich Mathematik
Technische Hochschule Darmstadt
Schlossgartenstrasse 7
D-64289 Darmstadt
Germany

ISBN-13:978-1-4471-1243-3

British Library Cataloguing in Publication Data
Hofmann, Martin
Extensionalconstructs in intensional type theory.
(CPHC/BCS distinguished dissertation series)
1.Type theory
I.Title
511.3
ISBN-13:978-1-4471-1243-3

Library of Congress Cataloging-in-Publication Data
Hofmann, Martin, 1965-
Extensional constructs in intensional type theory / Martin Hofmann.
p. cm. -- (CPHC/BCS distinguished dissertation series)
Includes index.
ISBN-13:978-1-4471-1243-3 e-ISBN-13:978-1-4471-0963-1
DOI:10.1007/978-1-4471-0963-1

1. Automatic theorem proving. 2. Functional programming (Computer science)
3. Type theory. I.Title II. Series
QA76.9.A96H64 1997
511.3--dc21 96-37978

Typesetting: Camera ready by author

34/3830-543210 Printed on acid-free paper

Preface

Theories of dependent types have been proposed as a foundation of constructive mathematics and as a framework in which to construct certified programs. In these applications an important role is played by identity types which internalise equality and therefore are essential for accommodating proofs and programs in the same formal system.

This thesis attempts to reconcile the two different ways that type theories deal with identity types. In extensional type theory the propositional equality induced by the identity types is identified with definitional equality, i.e. conversion. This renders type-checking and well-formedness of propositions undecidable and leads to non-termination in the presence of universes. In intensional type theory propositional equality is coarser than definitional equality, the latter being confined to definitional expansion and normalisation. Then type-checking and well-formedness are decidable, and this variant is therefore adopted by most implementations.

However, the identity type in intensional type theory is not powerful enough for formalisation of mathematics and program development. Notably, it does not identify pointwise equal functions (functional extensionality) and provides no means of redefining equality on a type as a given relation, i.e. quotient types. We call such capabilities *extensional constructs*. Other extensional constructs of interest are uniqueness of proofs and more specifically of equality proofs, subset types, and propositional extensionality—the identification of equivalent propositions. In this work we investigate to what extent these extensional constructs may be added to intensional type theory without sacrificing decidability and existence of canonical forms. The method we use is the translation of identity types into equivalence relations defined by induction on the type structure. In this way type theory with extensional constructs can be understood as a high-level language for working with equivalence relations instead of equality. Such translations of type theory into itself turn out to be best described using categorical models of type theory.

We thus begin with a thorough treatment of categorical models with particular emphasis on the interpretation of type-theoretic syntax in such models. We then show how pairs of types and predicates can be organised into a model of type theory in which subset types are available and in which any two proofs of a proposition are equal. This model has applications in the areas of program extraction from proofs and modules for functional programs. For us its main purpose is to clarify the idea of syntactic translations via categorical model constructions.

The main result of the thesis consists of the construction of two models in which functional extensionality and quotient types are available. In the first one types are modelled by types together with proposition-valued partial equivalence relations. This model is rather simple and in addition provides subset types and propositional extensionality. However, it does not furnish proper dependent types such as vectors or matrices. We try to overcome this disadvantage by using another model based on families of type-valued equivalence relations which is however much more complicated and validates certain conversion rules only up to propositional equality.

We illustrate the use of these models by several small examples taken from both formalised mathematics and program development.

We also establish various syntactic properties of propositional equality including a proof of the undecidability of typing in extensional type theory and a correspondence between derivations in extensional type theory and terms in intensional type theory with extensional constructs added. Furthermore we settle affirmatively the hitherto open question of the independence of unicity of equality proofs in intensional type theory which implies that the addition of pattern matching to intensional type theory does not yield a conservative extension.

Preface to the present edition

This book contains the author's PhD thesis written under the supervision of Don Sannella and Gordon Plotkin at the LFCS Edinburgh. Except for corrections of minor mistakes and the addition of an index the text agrees with the examined version of the thesis.

Some parts of the book have been published elsewhere in [44] (Parts of Chapter 3), [48, 50, 46] (Parts of Chapter 5). In retrospect, the notion of categorical model described in Chapter 2.1 appears in parts unnecessarily complicated. A better notion, due to Peter Dybjer, has been described by the author in [45].

Acknowledgements

I would like thank my supervisor Don Sannella for his invaluable help and advice during the writing of this thesis. My second supervisor, Gordon Plotkin, has provided valuable guidance in discussions and in particular has suggested and made possible a visit to Gothenburg during which the idea for this thesis work arose.

I am grateful to Rod Burstall and Andy Pitts for agreeing to be examiners of this thesis as well as for detailed and helpful comments on the first version.

My friend and colleague Thorsten Altenkirch provided invaluable input to my research and moral support in many (sometimes lively) discussions. He carefully proof-read an earlier draft and suggested major improvements.

Thanks for stimulating conversations and suggestions on the subject of this thesis go to David Aspinall, Adriana Compagnoni, Thierry Coquand, Wolfgang Degen, Peter Dybjer, Bart Jacobs, Per Martin-Löf, Zhaohui Luo, James McKinna, Uwe Nestmann, Benjamin Pierce, Alex Simpson, Martin Steffen, Thomas Streicher, Terry Stroup, Makoto Takeyama, and the members of the Edinburgh Lego Club.

I wish to acknowledge financial support by Edinburgh University, the DAAD, and the European Union HCM programme.

Table of Contents

1. Introduction

Theories of dependent types have been proposed as a foundation of constructive mathematics [75, 20] and as a framework in which to construct certified programs [67, 20, 112] and to extract programs from proofs [28, 90]. Using implementations of such type theories, substantial pieces of constructive mathematics have been formalised and medium scale program developments and verifications have been carried out.

One key feature of dependent type theories making them so apt for these applications is that to a certain extent they allow for the internalisation of meta-theorems; they can "speak about themselves". For instance, it is possible to use such a type theory as a programming language and at the same time for reasoning about these programs. Then, the proof that e.g. a program meets its specification is an element of a certain type and the fact that this is so can be checked automatically.

To a great extent this is made possible through the internalisation of equality by *propositional equality* and more generally through the interplay between propositional equality and (intensional) *definitional equality*.

The main subject of this thesis is to study this interplay and in particular extensions to propositional equality which maintain the desirable properties of the definitional equality.

1.1 Definitional and propositional equality

In extensional type theory and axiomatic set theory the constant zero function and the function defined by

$$f(n) = \begin{cases} 0, & \text{if there is no counterexample to Fermat's} \\ & \text{conjecture of size less than } n \\ 1, & \text{otherwise} \end{cases}$$

are (probably) equal, whereas intensionally they are quite distinct, because their definitions are entirely different. On the other hand, even in an intensional setting there may be a proof that the *extensions* of the two functions agree. In the presence of such a proof it should be the case that whenever some definable property holds for the constant zero function then it also holds for f.[1]

However, certain equalities do not require a proof. For instance, in the above example the function f equals by definition the function which searches for a counterexample to Fermat's conjecture. Similarly, we have

$$\textit{if true then } e \textit{ else } e' = e \tag{1.1}$$

just by appealing to the *definition* of the if-then-else construct. One therefore speaks in the former case of a *propositional equality* whereas in the latter we have an example of a *definitional equality*. The precise borderline between propositional and definitional equality is fluid, but it appears to be commonly agreed that two objects are definitionally equal if after certain computation steps they evaluate to identical results.[2] Traditionally, definitional equality is therefore meant to be induced by a strongly normalising confluent rewriting system; in an extended sense we might simply require definitional equality to be decidable and furthermore that the decision procedure be induced by some notion of terminating computation on terms. Indeed, we shall consider notions of definitional equality defined indirectly via interpretation in some syntactically constructed model.

It is important that definitional equality is *pervasive* so if M and N are definitionally equal then $P(M)$ is definitionally equal to $P(N)$ whatever P is. In particular the *proposition* stating that M is propositionally equal to N is definitionally equal to the proposition that M is propositionally equal to itself. Since the latter proposition is always true, definitional equality always entails propositional equality, but certainly not vice versa.

This distinction between "obvious" identities which do not require a proof and "meaningful" identities which need to be verified is the key difference between intensional and extensional theories. In an extensional theory every identity needs a proof, whereas in an intensional theory a notion of computation or definitional expansion is built in.

An illustrative example of definitional equality arises in category theory. For conceptual reasons one often does not want to have an equality on objects. Nonetheless, such an equality is needed to state when a composition $f \circ g$ is

1 A simpler example of propositionally equal yet intensionally different functions is given by the identity on the natural numbers and the function $\lambda x{:}\,\mathrm{N}.Suc^x(0)$ sending x to the x-fold iteration of the successor function on zero. The example with Fermat's conjecture is however more convincing since one could imagine an intensional theory in which the former two functions become identified by some "η-rule" for natural numbers, whereas in no reasonable intensional or algorithmic setting would one identify two functions the propositional equality of which is equivalent to Fermat's conjecture.

2 Luo [70] confines definitional equality to actual definitional expansion and considers 1.1 as an instance of "computational equality". However, since computational and definitional equality behave identically syntactically this distinction is merely a philosophical one. On the other hand, Martin-Löf [74] explains the use of the term "definitional" for equalities like 1.1 on the grounds that the "non-canonical" term former if-then-else is *defined* by 1.1 and a corresponding clause for *false*.

defined, namely when the domain of f and the codomain of g agree. A careful analysis of common practice shows that the word "agree" here refers in most cases to definitional equality of objects. For example if $G(X) = A \times X$ one would be allowed to compose a morphism with codomain $G(X)$ with a projection, but no category theorist would write "let's prove that the domain and codomain of the following two morphisms are equal".

A formal framework in which both propositional and definitional equality are present is Martin-Löf's intensional type theory [88]. Definitional equality is expressed by the equality judgement $\Gamma \vdash M = N : \sigma$ meaning that M and N are definitionally equal terms of type σ in context (type assignment) Γ. There is also a notion of definitional equality on types: $\Gamma \vdash \sigma = \tau$. Propositional equality of terms M and N is expressed by the judgement $\Gamma \vdash P : Id_\sigma(M, N)$ meaning that P is a proof that M and N are propositionally equal terms of type σ in context Γ. Here "*Id*" is Martin-Löf's identity type which is defined as an inductive family with the single constructor $Refl(M) : Id_\sigma(M, M)$.

In the presence of a type of propositions "*Prop*" one can also *define* propositional equality by Leibniz's principle. According to this principle a proof that M is propositionally equal to N is a proof that whenever a property $P : \sigma \to Prop$ holds for M then it holds for N.

These formulations demonstrate an important aspect of propositional equality in intensional type theory—the fact that it internalises proofs of equality. Propositional equality is identified with the type of its proofs and it is decidable whether a term belongs to this type or not. In contrast, in set theory one may define an "identity type" by

$$Id^{\text{set}}_\sigma(M, N) = \begin{cases} \{\star\} & \text{if } M = N \\ \emptyset & \text{otherwise} \end{cases}$$

Then $Id^{\text{set}}_\sigma(M, N)$ is nonempty iff M and N are equal, but membership in $Id^{\text{set}}_\sigma(M, N)$ is not in general decidable, so $\star$ cannot be considered a proof that M equals N.

A similar situation occurs in *extensional Martin-Löf type theory* [75, 16] where propositional and definitional equality are forcefully identified by the *equality reflection rule*

$$\frac{\Gamma \vdash P : Id_\sigma(M, N)}{\Gamma \vdash M = N : \sigma} \quad \text{ID-DEFEQ}$$

This rule makes definitional equality extensional[3] and thus undecidable (see Section 3.2.2). Moreover, type checking becomes undecidable because $Refl(M) : Id_\sigma(M, N)$ holds iff M and N are definitionally equal. Also, the term $Refl(M)$ can no longer be considered a proof that M and N are equal as is the case

[3] Strictly speaking, definitional equality does not become extensional here, but the equality judgement no longer expresses definitional equality. To avoid the introduction of a third notion of equality, ***judgemental equality***, we shall by a slight abuse of language identify definitional equality with the relation expressed by the equality judgement.

in of the above "identity type" in set theory. For these and other reasons (see Section 3.2.6.2) interest in extensional type theory seems to have recently decreased. We shall however explain in Chapter 3 that the equality reflection rule can be seen as a shortcut for longer derivations in intensional type theory.

Unfortunately, the abovementioned formalisations of propositional equality (identity type and Leibniz equality) suffer from serious limitations when they are used for types which are not inductive. For instance, from a proof P : $\Pi x{:}\sigma.Id_\tau(f\ x, g\ x)$ that f and g are point-wise propositionally equal there is in general no way of constructing an element of $Id_{\sigma\to\tau}(f,g)$ although in the presence of P the functions f and g are clearly propositionally equal in the abstract sense. This has led to the common misunderstanding that the identity type was intensional. This is not so because if $M(x)$ and $N(x)$ are expressions containing a free variable of type σ then in certain cases an element of $Id_\tau(M(x), N(x))$ may be constructed by induction on x even if $M(x)$ and $N(x)$ are not definitionally (i.e. intensionally) equal.[4]

Summing up, we can say that definitional equality is intensional—it identifies objects which become identical upon definitional expansion and/or computation; and that (idealised) propositional equality is extensional, but that it is not adequately captured by the identity type or Leibniz equality. This thesis is an attempt at overcoming this mismatch.

1.2 Extensional constructs

We use the term "extensional construct" to refer to a desirable feature of propositional equality which is not present in its traditional formulation. In this thesis several such extensional constructs are of interest, notably the following:

i. Functional extensionality: Two functions which are point-wise propositionally equal are propositionally equal.
ii. Uniqueness of identity: Any two proofs of a propositional equality are propositionally equal.
iii. Proof irrelevance: Any two proofs of a proposition are propositionally equal.
iv. Subset types: The presence of a type former which permits the formation of a type of elements of a given type that satisfy a certain predicate.
v. Propositional extensionality: Two propositions which imply each other are propositionally equal.
vi. Quotient types: The presence of a type former which permits arbitrary definition of propositional equality on some underlying type.

These extensional constructs are not independent of each other. In particular, proof irrelevance gives rise to subset types and in the presence of quotient

[4] Luo [70] therefore calls the identity type "weakly intensional". A way of justifying the predicate "intensional" for the identity type consists of decreeing that intensional equality on open terms refers to definitional equality of all closed instances.

types one can derive functional extensionality. Also, propositional extensionality implies proof irrelevance if one assumes a distinguished proposition with exactly one proof.

Proof irrelevance and propositional extensionality only make sense if there is a difference between propositions and arbitrary types. In this thesis this difference is expressed by the assumption of a type of propositions *Prop* and a type $Prf(P)$ for each $P : Prop$. The propositions are then the types of the form $Prf(P)$ and the elements of these types are proofs. If for conceptual reasons one does not want to have a type of all propositions one can also introduce a new judgement $\Gamma \vdash \sigma\ Prop$ to mean that the type P is a proposition. We only hint at this alternative in Remarks 4.6.1 & 5.1.11.

The construct of uniqueness of identity is of course a special case of proof irrelevance, so it only makes sense in the absence of the latter, for example in a theory with propositions and types identified.

In extensional type theory, these extensional constructs are either present or can be added easily, see e.g. [16]. Here we take on the more difficult task of adding them to intensional type theory. Apart from the advantages of intensional type theory in general this has the desirable effect that we can add a choice principle for quotient types 5.1.7 which would be unsound in any extensional theory.

Extensional constructs could simply be added axiomatically as constants and justified semantically, say by a set-theoretic interpretation. For example, in order to achieve functional extensionality we could simply assume a family of constants

$$Ext_{\sigma,\tau} : \Pi f, g{:}\sigma \to \tau.(\Pi x{:}\sigma.Id_\tau(f\ x, g\ x)) \to Id_{\sigma\to\tau}(f,g)$$

However, in this case an important property, called N-canonicity (Definition 2.1.9), of definitional equality will be violated, namely that every closed term of the natural numbers is definitionally equal to a numeral.[5] The reason is that a term containing an instance of *Ext* cannot be definitionally equal to a numeral because there are no rules governing the behaviour of *Ext* with respect to definitional equality. This work can—to some extent—be seen as a quest for extensions of definitional equality so as to include rules for these constants.

An exception to this is uniqueness of identity. As noticed by Streicher [104] it can indeed by added axiomatically and the axiom can be endowed with a reduction rule so that no non-canonical elements arise. For us the concept is of relevance because it is needed to establish equivalence of extensional and intensional type theory with extensional constructs added (3.2.5).

1.3 Method

The main tool we use in order to achieve the goals set out above consists of defining interpretations of the type theory with extensional constructs inside

[5] The use of the natural numbers is arbitrary here. One can easily show that a non-canonical element in some other type induces one in the type of natural numbers.

pure type theory. For example, in order to achieve functional extensionality and quotient types, one interprets every type as a type together with an (internal) equivalence relation and propositional equality at a particular type as this equivalence relation. Since this approach is inspired by Bishop's definition of a set by its members and its equality [9] we call this translation the *setoid interpretation*. Similarly, for subset types and proof irrelevance we use a translation of types into types with unary predicates and of terms into terms plus proofs that these predicates are respected. Following Burstall and McKinna [13] we call this translation the *deliverables interpretation*.

Given such an interpretation we can consider two objects *definitionally equal* if they receive definitionally equal interpretations and in this way ensure N-canonicity.

1.3.1 The use of categorical models

It turns out that due to type dependency the verification that a certain translation indeed validates all the rules of type theory is quite difficult. It has therefore proven useful to insert the intermediate step of abstract categorical models. One then defines a sound interpretation function from the syntax to the abstract model once and for all and the task of proving that a certain translation is sound can then be reduced to the task of verifying that one has an instance of the abstract model.

This may be compared to the situation in the simply typed lambda calculus. Instead of directly giving an interpretation of typed lambda terms in some mathematical structure one can alternatively show that this structure forms a cartesian closed category and then appeal to the general interpretation function mapping lambda terms to morphisms in an arbitrary cartesian closed category [52]. In our situation the role of cartesian closed categories is played by *syntactic categories with attributes* —an equational presentation of Cartmell's notion of categories with attributes [15, 92]. Such an equational presentation has the advantage that the property of being a model can be checked using term rewriting (normalisation). In this way we have encoded all the syntactic models within the Lego proof checker [71] and advantageously used its normalisation features to verify them.

1.3.2 Syntactic models

So the task of interpreting type theory with extensional constructs can be rephrased as that of exhibiting a categorical model in which the desired extensional constructs are valid. The aims set out in Section 1.3 above place certain restrictions on such models.

Firstly, in order that (semantically defined) definitional equality be decidable, equality in the model must be decidable, which presupposes that the semantic objects come with some effective description.

Second, the model must have the property that the type of natural numbers consists of numerals only so that we obtain the desired property that even in the presence of an extensional construct every closed term of the natural numbers is definitionally equal to a numeral (N-canonicity).

Now any model satisfying these two requirements would meet the specification set out so far. However, in addition to this we want that the model explains the extensional constructs in terms of more basic ideas. "More basic" in this context means in terms of type theory without extensional constructs added. We thus seek a *syntactic model* the objects of which are syntactical and the equality of which is induced by the definitional equality of the underlying type theory.

The interpretation in the model then not only induces a coarser notion of definitional equality, but actually computes syntactic expressions which no longer contain the extensional constructs. Extensional constructs may therefore be seen as abbreviations or "macros" for longer derivations in the theory without them.

In particular, working in the syntactic model for functional extensionality and quotients can be viewed as using a high-level language for working with equivalence relations instead of equality and functions together with proofs that they respect these relations instead of mere functions. Similarly, the syntactic model for subset types and proof irrelevance can be understood as a high-level language for working with "deliverables" in the sense of [13], i.e. functions together with proofs that they respect certain predicates.

We also encounter a model which cannot be called syntactic—the *groupoid model*. It is similar in spirit to the syntactic models we give and serves similar purposes. The groupoid model can be seen as a refinement of the setoid interpretation in which proofs are constrained by equations. For example the proof that a function respects equality must respect symmetry. This allows for a more fine-grained definition of (semantic) propositional equality on the identity type itself which both shows the underivability of uniqueness of equality proofs and permits a view of propositional equality of types as isomorphism.

It turns out that the syntactic models for the various extensional constructs discriminate between the features present in the type theory under consideration. For instance, the simplest model for quotient types does not support universes and other more involved forms of type dependency like the definition of a family of types indexed over Booleans by case distinction. It is possible to extend this model to cover universes, but then the extensional constructs are available only or types in the universe and not for the universe itself. Definition of types by cases is still not possible. On the other hand, in a more complicated model in which these type dependencies are available, certain expected definitional equalities only hold propositionally, for example certain instances of the computation rules associated to elimination of natural numbers. Some of these limitations have a natural explanation in terms of the intended meaning of the particular extensional construct in question; others, like the lack of certain definitional equalities, are rather arbitrary but nevertheless seem unavoidable.

Of course here the question arises whether the method of syntactic models is not too restrictive. Certainly, any approach to extensional constructs would (modulo some coding) be induced by a model which meets the first two requirements on a syntactic model, viz. semantic equality would have to be decidable and the interpretation of the natural numbers would have to consist of numerals only. Given that from the model one also expects some insight about the nature of the extensional constructs, not much choice is left. We cannot exclude the existence of such non-syntactic models, but have not encountered any in the course of the research reported here.

1.4 Applications

The applications of extensional constructs are numerous. Functional extensionality is required when one wants to reason about higher-order functions and also when coding coinductive types and greatest fixpoints using function types [63]. Quotient types are of pivotal use in formalisations of proofs in constructive algebra, since the most elementary algebraic constructions involve quotients. Quotient types also provide bisimulation principles to reason about infinite types. This may facilitate semantics-based verification of concurrent programs. Subset types can be used for modular specification. Here the additional computational principle that a function on a subset type should embody an algorithm defined on the whole type is important (Section 4.6.2). Proof irrelevance and propositional extensionality are of more theoretical interest. For example, propositional extensionality implies that quotient types are *effective*, i.e. that if two equivalence classes are equal in a quotient type then their representatives are related (Section 5.1.6.4). We work out several of these applications in some detail in Chapter 6.

Most of the applications also go through if the extensional constructs are merely assumed axiomatically without supporting them by a syntactic model. Indeed, we establish a conservativity property (Theorem 3.2.5) of which this fact is a general consequence. The benefit of the syntactic models in applications is three-fold. First, they show that using (partial) equivalence relations (called "book-equalities" in the Automath tradition [26]) is essentially equivalent to using propositional equality with extensional constructs added. Even if one does insist on "book-equalities", the syntactic models may guide their use, which is particularly important in the presence of type dependency. Second, they allow us to gauge the differences in expressiveness between various approaches to "book-equalities", e.g. equivalence relations vs. partial equivalence relations, Prop-valued vs. type-valued relations, *et cetera*. Finally, two of the syntactic models we look at give rise to a complete separation between proofs and computations. Definitional equality in the model only compares the computational parts, so that the decision procedure for equality and type checking becomes more efficient.

1.4.1 Application to machine-assisted theorem proving

Theorem proving in type theories with or without extensional constructs is in principle possible by writing down derivations according to the rules. For realistic applications these rules are, however, unmanageable and one either has to move to informal notation or use machine support. For ordinary type theory without extensional constructs such machine support exists in the form of proof assistants like Lego [71]. There are in principle two possibilities as to how a proof assistant can be extended so as to provide extensional constructs.

As said above, most of the applications go through with axiomatically added extensional constructs. The Lego system allows for the addition of axioms or constants and even for the addition of certain definitional equalities as rewrite rules. In Appendix Appendix A we give Lego code which approximates in this sense one of the syntactic models for extensionality and quotient types (a setoid interpretation). The main disadvantage of this simple-minded approach is that terms of the natural numbers may contain instances of these axioms and thus not reduce to a canonical form. We also encounter certain non-standard operations on subset types and quotient types which cannot be cast into the form of Lego constants.

A deeper application of the work described here consists of changing the implementation of the proof assistant itself so that it would compute interpretations in a syntactic model of all types and terms occurring during a proof session and use these interpretations to decide definitional equality and thus type checking. This would restore the ability of internal computation on terms and would also support the non-standard operations. Additionally, the interpretations could be made accessible to the user for subsequent processing. For instance, in the case of the syntactic model for proof irrelevance the computational part of the interpretation can be seen as a program corresponding to the computational content of a proof.

Unfortunately, such a change to a proof assistant involves substantial practical effort and thus falls beyond the scope of this thesis.

1.5 Overview

Chapter 2 recapitulates the syntax of dependent type theory including universes and impredicative quantification. We use a judgement-oriented presentation without a logical framework. In order to compensate for the lack of a logical framework we develop a high-level syntax for substitutions and free variables. In the second part of Chapter 2 we introduce our notion of categorical model for dependent type theory, so-called syntactic categories with attributes, and we describe the interpretation of syntax therein. The definition of the interpretation function and the soundness proof closely follows [103].

The third chapter is mainly devoted to a comparison between intensional type theory with extensional constructs added and extensional type theory. The main result is that the latter is conservative over the former. Together with the undecidability of extensional type theory we consider this result the main theoretical justification for the use of intensional type theory with extensional constructs added. We also describe some basic constructions with the identity type that are required later, such as the definition of Martin-Löf's elimination rule in terms of a Leibniz principle.

Chapter 4 contains the first syntactic model construction, a "deliverables interpretation" of the Calculus of Constructions, which models types as types together with a predicate. The model supports subset types and proof irrelevance, and it interprets a non-standard rule for subset types which under certain conditions allows one to lift a function defined on a subset to a function on the whole set. The interpretation establishes a correspondence between two methodologies for program development in type theory, the *deliverables approach* and the *refinement approach*. We also sketch an application of the interpretation to typing of higher-order modules.

Our main purpose in presenting the deliverables interpretation is, however, to introduce the methodology of describing translations of type theory into itself using categorical models, and to indicate how the verification of such a categorical model can be mechanised using Lego.

Chapter 5 forms the kernel of this thesis. It contains three model constructions centred around extensionality and quotient types. Section 5.1 describes an interpretation under which types are modelled as types together with a *Prop*-valued partial equivalence relation ("setoids"). The interpretation is rather simple because type dependency is only allowed at the level of the relations, not for the types themselves. In this model functional and propositional extensionality, subset types, and quotient types can be interpreted. For the quotient types a choice operator may be defined which under certain conditions allows for a representative to be recovered from an equivalence class. The model does not contain any properly dependent types such as matrices and universes. To alleviate this we look at an extension with universes obtained by amalgamating the setoid model with the deliverables interpretation. In this way a universe is supported, but extensional constructs become restricted to types inside the universe.

Section 5.3 describes an attempt to overcome the lack of type dependency in the previous model. We try to answer the question of what the right definition of a family of setoids indexed over a setoid might be, under the hypothesis that the underlying types of such a family should depend on the underlying type of the indexing setoid. The criterion for the value of such a definition is of course that the families must form a syntactic category with attributes and support as many type formers as possible as well as functional extensionality. The answer turns out to be rather unsatisfactory; the model is quite complicated and the verification even of the type of natural numbers is extremely complicated.

In Section 5.2 the—in our opinion—most natural solution to the above question of dependent setoids is described: the groupoid interpretation of type theory. It is, however, not a syntactic model, but can only be defined in extensional type theory or set theory. We show how the groupoid interpretation answers the question of independence of uniqueness of identity and how it permits one to interpret propositional equality on a universe as isomorphism. It also motivates the setup of the dependent setoid model in Section 5.3, which is why we describe it first.

Chapter 6 is devoted to applications of the extensional constructs and of the conservativity of extensional type theory over intensional type theory with extensional constructs added. Many of these applications have been formally developed within the Lego proof checker.

Section 6.1 studies a formalisation of Tarski's generalised fixpoint theorem using subset types. The main lemma—a property of posets—is instantiated in the course of the proof with various sub-posets of a given one, whence the need for subset types. Our development is based on an earlier formalisation by Pollack, which can be viewed as the result of translating our development to intensional type theory using the interpretation in the deliverables model.

Section 6.2 studies an important application of functional extensionality—an encoding of infinite lists ("streams") as functions. We derive a coinduction principle for these streams and define a fixpoint combinator the verification of which makes heavy use of coinduction.

In Section 6.3 we describe a simple-minded encoding of category theory in type theory with extensional constructs (propositional equality is used to compare both objects and morphisms). We investigate how this encoding gets translated under the various interpretations in terms of setoids. The resulting encodings in pure intensional type theory resemble encodings proposed independently by other authors.

Section 6.4 serves a similar purpose. We show how an encoding of disjoint-union types in extensional type theory due to Troelstra gets translated under the setoid interpretations. In particular we show in some detail how functional extensionality gets eliminated.

Section 6.5 is devoted to an examination of our syntax for quotient types by redoing certain constructions with quotient *sets* collected by Bourbaki [10] in order to become independent of their encoding as sets of equivalence classes. All but one construction go through; we explain why this one (isomorphism between the two definitions of the image of a function as a subset and as a quotient) must fail under a constructive understanding of quotienting as a redefinition of (propositional) equality. The material in this section explains some basic techniques for the use of quotient types such as the definition of binary functions on a quotient type.

Finally, Section 6.6 gives an application of the conservativity of extensional type theory over intensional type theory. We give an extensional type-theoretic variant of a theorem due to Mendler which involves indexed families of sets. Our formalisation is such that the conservativity theorem implies the existence of a proof of this theorem in intensional type theory with functional extensionality and uniqueness of identity. However, we were not able to come up explicitly with such a proof. This "application" suggests that in certain situations extensional type theory might be useful as a high-level language for type theory with extensional constructs.

Each of the four theoretical chapters ends with a section discussing possible extensions and related work. General conclusions and some directions for future research are summarised in Chapter 7.

The appendix contains a summary of the type-theoretic rules used throughout the thesis, a glossary of type theories appearing in the thesis, and an index of symbols. We also include a translation into Lego syntax of some of the rules for extensional constructs.

2. Syntax and semantics of dependent types

In this chapter we fix a particular syntax for a dependently typed calculus and define an abstract notion of model as well as a general interpretation function mapping syntactical objects to entities in a model. This interpretation function is shown to be sound with respect to the syntax.

Γ	$::=$	$\diamond$	empty context
	$\mid$	$\Gamma, x{:}\,\sigma$	context extension
σ, τ	$::=$	$\mathbf{N}$	natural numbers
	$\mid$	$\Pi x{:}\,\sigma.\tau$	dependent product
	$\mid$	$Id_\sigma(M, N)$	identity type
M, N, O, P	$::=$	x	variable
	$\mid$	0	$\mathbf{N}$ introduction (zero)
	$\mid$	$Suc(M)$	$\mathbf{N}$ introduction (successor)
	$\mid$	$R^{\mathbf{N}}_\sigma(M, N, O)$	$\mathbf{N}$ elimination (induction)
	$\mid$	$\lambda x{:}\,\sigma.M^\tau$	typed abstraction
	$\mid$	$App_{\sigma,\tau}(M, N)$	typed application
	$\mid$	$Refl_\sigma(M)$	identity introduction
	$\mid$	$J_{\sigma,\tau}(M, N, O, P)$	identity elimination

Fig. 2.1. Syntax of pre-constructions

2.1 Syntax for a core calculus

We start by giving a syntactic description of intensional Martin-Löf type theory with Π-types, natural numbers, and the identity type [74]. On top of this we can add further inductive types, Σ-types, universes, impredicative quantification and so forth, in order to accommodate richer systems like the Calculus of Constructions [20] or Lego/ECC [68, 70]. The syntax we give does not make use of a "logical framework" as e.g. in [88] or [104]. We could, however, view the core calculus as a framework and make all further definitions of types inside some universe. So we do not really commit ourselves to a particular style of presentation.

2.1.1 Raw syntax

The raw syntax is formed out of three categories: pre-contexts, pre-types, and pre-terms. These are defined by the grammar in Fig. 2.1.

A few remarks concerning this presentation are in order. First, note that we include type information in application and abstraction. This makes the definition of the interpretation function for models easier, and it appears somewhat arbitrary to leave the type information out in this particular place. Indeed much more type information could be inferred automatically and for an actual implementation one would set up some sort of front end which would generate as much type information as possible so that the syntax we give here would only appear as an intermediate language. See also Streicher [103] where situations are indicated in which the typing information in an application can *not* be inferred. To increase readability we shall, however, often suppress type information informally and in particular we abbreviate $App_{\sigma,\tau}(M,N)$ by $(M\ N)$ and $\lambda x{:}\sigma.M^{\tau}$ by $\lambda x{:}\sigma.M$ or even $\lambda x.M$ where appropriate. See Coquand's survey article [17, Section 2.3] for a deeper discussion of such *abus de langage*.

Second, we are a bit sloppy about variable binding in our presentation. For example, as will emerge from the typing rules, the type σ in of $R^{\mathbf{N}}_{\sigma}$ contains a free variable of type $\mathbf{N}$ which becomes bound in $R^{\mathbf{N}}_{\sigma}$. To be consistent with the usual notation we make this binding explicit in the case of Π-types, but otherwise not, since it would lead to overly clumsy terms.

We identify all constructions up to renaming of both free and bound variables and ensure by a suitable renaming policy that no unwanted variable captures occur. This could be made more precise by using de Bruijn indices instead of named variables, but the price to pay is that syntactic weakening would no longer be the identity and thus notation would become less readable.

The union of pre-terms, -types, and -contexts is called the set of *pre-constructions*. The prefix "pre" indicates that these need not "typecheck" in any sense. The actual terms, types, and contexts (constructions) are those which occur inside the typing judgements to be defined in the next section. To avoid repetition we use the synonyms "element", "function", "operation" for "terms", where the latter two are mostly used for terms with distinguished free variables. We also sometimes say "set" instead of "type".

2.1.2 Judgements

There are six kinds of judgements.

$\Gamma \vdash$	Γ is a well-formed context of variable declarations
$\Gamma \vdash \sigma$	σ is a type in context Γ
$\Gamma \vdash M : \sigma$	M is a term of type σ in context Γ
$\vdash \Gamma = \Delta$	The contexts Γ and Δ are equal
$\Gamma \vdash \sigma = \tau$	The types σ and τ both in context Γ are equal
$\Gamma \vdash M = N : \sigma$	The terms M and N of type σ are equal

The last three kinds of judgements state *definitional* equality of contexts, types, and terms. The valid judgements are defined inductively by the following clauses.

2.1.2.1 Context rules. Valid contexts are generated from the empty context by successive extensions.

$$\frac{}{\diamond \vdash}\ \textsc{Empty} \qquad\qquad \frac{\Gamma \vdash \sigma}{\Gamma, x : \sigma \vdash}\ \textsc{Compr}$$

It follows from our conventions on variable names that in rule Compr the variable x does not occur in Γ.

2.1.2.2 Type formation rules. For each of the type formers **N**, *Id*, Π there is a formation rule stating the conditions under which a pre-type having that type former as outermost constructor is well-formed. In particular the type formation rule for Π specifies its binding behaviour.

$$\frac{\Gamma, x : \sigma \vdash \tau}{\Gamma \vdash \Pi x : \sigma.\tau}\ \Pi\textsc{-Form} \qquad \frac{\Gamma \vdash M : \sigma \quad \Gamma \vdash N : \sigma}{\Gamma \vdash Id_\sigma(M, N)}\ \textsc{Id-Form} \qquad \frac{\Gamma \vdash}{\Gamma \vdash \mathbf{N}}\ \textsc{N-Form}$$

2.1.2.3 Term formation rules. First, there are two "structural rules" which define the typing of variables and type conversion along type equalities.

$$\frac{\Gamma, x{:}\sigma, \Delta \vdash}{\Gamma, x{:}\sigma, \Delta \vdash x : \sigma}\ \textsc{Var} \qquad\qquad \frac{\Gamma \vdash \sigma = \tau \quad \Gamma \vdash M : \sigma}{\Gamma \vdash M : \tau}\ \textsc{Conv}$$

The following rules each correspond to a constructor on pre-terms and define its well-formedness and typing. In particular they define the binding behaviour of the term formers λ, $R^{\mathbf{N}}_\sigma$, and $J_{\sigma,\tau}$. The rules can be grouped in pairs according to whether a certain type appears in the conclusion (introduction) or in a premise (elimination) of the rule. The rules corresponding to the Π-type are

$$\frac{\Gamma, x : \sigma \vdash M : \tau}{\Gamma \vdash \lambda x{:}\sigma.M^\tau : \Pi x{:}\sigma.\tau}\ \Pi\textsc{-Intro}$$

$$\frac{\Gamma \vdash M : \Pi x{:}\sigma.\tau \qquad \Gamma \vdash N : \sigma}{\Gamma \vdash App_{\sigma,\tau}(M, N) : \tau[x := N]}\ \Pi\textsc{-Elim}$$

The introduction rules for the natural numbers define the constructors 0 and *Suc* whereas the elimination rule defines mathematical induction.

$$\frac{\Gamma \vdash}{\Gamma \vdash 0 : \mathbf{N}}\ \textsc{N-Intro-0} \qquad\qquad \frac{\Gamma \vdash M : \mathbf{N}}{\Gamma \vdash Suc(M) : \mathbf{N}}\ \textsc{N-Intro-Suc}$$

$$\frac{\begin{array}{l}\Gamma\,,\,x{:}\mathbf{N} \vdash \sigma \\ \Gamma \vdash M_z : \sigma[x := 0] \\ \Gamma\,,\,x{:}\mathbf{N}\,,\,p{:}\sigma \vdash M_s : \sigma[x := Suc(x)] \\ \Gamma \vdash N : \mathbf{N}\end{array}}{\Gamma \vdash R^{\mathbf{N}}_{\sigma}(M_z, M_s, N) : \sigma[x := N]} \quad \text{N-ELIM}$$

Notice here that the free variables x in σ as well as x and p in M_s are bound by $R^{\mathbf{N}}$. We could have chosen a notation like

$$R^{\mathbf{N}}_{x:\mathbf{N}.\sigma}(M_z,\, x{:}\mathbf{N}.p{:}\sigma.M_s,\, N)$$

to emphasise this. Instead we define later in Section 2.2 a high-level syntax for terms and types with free variables.

Finally, for the identity type $Id_\sigma(M, N)$ the introduction rule corresponds to reflexivity, whereas the elimination rule is a generalised Leibniz principle which we study in more detail in Chapter 3.

$$\frac{\Gamma \vdash M : \sigma}{\Gamma \vdash \mathit{Refl}_\sigma(M) : Id_\sigma(M, M)} \quad \text{ID-INTRO}$$

$$\frac{\begin{array}{l}\Gamma\,,\,x{:}\sigma, y{:}\sigma\,,\,p{:}\,Id_\sigma(x, y) \vdash \tau \\ \Gamma\,,\,x{:}\sigma \vdash M : \tau[x := x][y := x][p := \mathit{Refl}_\sigma(x)] \\ \Gamma \vdash N_1 : \sigma \\ \Gamma \vdash N_2 : \sigma \\ \Gamma \vdash P : Id_\sigma(N_1, N_2)\end{array}}{\Gamma \vdash J_{\sigma,\tau}(M, N_1, N_2, P) : \tau[x := N_1][y := N_2][p := P]} \quad \text{ID-ELIM-J}$$

2.1.2.4 Definitional equality. Definitional equality of types $\Gamma \vdash \sigma = \tau$, terms $\Gamma \vdash M = N : \sigma$, and contexts $\Gamma = \Delta$ is defined as the least congruence with respect to all type and term formers closed under "definitional expansion". This means that definitional equality is reflexive, symmetric, and transitive:

$$\frac{\Gamma \vdash \sigma}{\Gamma \vdash \sigma = \sigma} \quad \text{T-REFL}$$

$$\frac{\Gamma \vdash \sigma = \tau}{\Gamma \vdash \tau = \sigma} \quad \text{T-SYM}$$

$$\frac{\Gamma \vdash \sigma = \tau \qquad \Gamma \vdash \tau = \rho}{\Gamma \vdash \sigma = \rho} \quad \text{T-TRANS}$$

with analogous rules for definitional equality on contexts and terms. Furthermore there are congruence rules for every type and term former; for example the congruence rule for the successor and for the Π-type are:

$$\frac{\Gamma \vdash M = M' : \mathbf{N}}{\Gamma \vdash Suc(M) = Suc(M') : \mathbf{N}} \quad \text{C-SUC} \qquad \frac{\Gamma \vdash \sigma = \sigma' \qquad \Gamma, x{:}\sigma \vdash \tau = \tau'}{\Gamma \vdash \Pi x{:}\sigma.\tau = \Pi x{:}\sigma'.\tau'} \quad \text{C-}\Pi$$

Notice also the congruence rule for context extension.

$$\frac{\vdash \Gamma = \Delta \qquad \Gamma \vdash \sigma = \tau}{\vdash \Gamma, x{:}\sigma = \Delta, y{:}\tau} \quad \text{C-COMPR}$$

Special attention is drawn to the congruence rule for abstraction ("ξ-rule") because it is left out in some presentations of type theory. See [74] and Section 3.2.1.

$$\frac{\Gamma \vdash \Pi x{:}\sigma.\tau = \Pi x{:}\sigma'.\tau' \qquad \Gamma, x{:}\sigma \vdash M = M' : \tau}{\Gamma \vdash \lambda x{:}\sigma.M^{\tau} = \lambda x{:}\sigma'.M'^{\tau'} : \Pi x{:}\sigma.\tau} \quad \text{C-ABSTR}$$

Finally, closure under definitional expansion is formalised by the following rules which are all understood under the premise that the respective left and right hand sides both have the indicated types.

$$\frac{}{\Gamma \vdash App_{\sigma,\tau}(\lambda x{:}\sigma.M^{\tau}, N) = M[x := N] : \tau[x := N]} \quad \Pi\text{-BETA}$$

$$\frac{}{\Gamma \vdash R^{\mathbf{N}}_{\sigma}(M_z, M_s, 0) = M_z : \sigma[x := 0]} \quad \text{NAT-COMP-ZERO}$$

$$\frac{}{\begin{array}{l}\Gamma \vdash R^{\mathbf{N}}_{\sigma}(M_z, M_s, Suc(N)) = \\ \qquad M_s[x := N][p := R^{\mathbf{N}}_{\sigma}(M_z, M_s, N)] : \sigma[x := Suc(N)]\end{array}} \quad \text{NAT-COMP-SUC}$$

$$\frac{}{\begin{array}{l}\Gamma \vdash J_{\sigma,\tau}(M, N, N, Refl_{\sigma}(N)) = \\ \qquad M[x := N] : \tau[x := N][y := N][p := Refl_{\sigma}(N)]\end{array}} \quad \text{ID-COMP}$$

Notice that, unlike propositional equality given by Id the definitional equality is not annotated by any justification or proof whatsoever. Thus, $\Gamma \vdash M = M' : \sigma$ states that M and M' are definitionally equal terms of type σ whereas $\Gamma \vdash P : Id_{\sigma}(M, M')$ states that P is a proof that M and M' are propositionally equal terms of type σ in context Γ. Also a definitional equality can never appear as an assumption in a context, whereas propositional equality obviously can. This stands in contrast to the system presented in [34] in the framework of so-called *labelled deductive systems*.

2.1.3 Notation

The empty context is usually left out in judgements; so we write $\vdash \mathbf{N}$ instead of $\diamond \vdash \mathbf{N}$. We sometimes write $\Gamma \vdash J$ for an indeterminate judgement where J may be of the form Γ, $\Gamma = \Delta$, σ, $\sigma = \tau$, $M : \sigma$, $M = N : \sigma$. In the running text the ambient context Γ and the derivation symbol $\vdash$ are often left out and informal equality reasoning may be used to justify equality judgements thereby implicitly using reflexivity, symmetry, transitivity, and the congruence rules.

We shall in the sequel sometimes use an abbreviated notation for inference rules whereby we omit the context Γ carried along previously. So a rule $\frac{\Delta \vdash J}{\Theta \vdash J'}$ is understood as $\frac{\Gamma, \Delta \vdash J}{\Gamma, \Theta \vdash J'}$. This abbreviated notation appears also in [6] and [108]. We also use this abbreviated notation in the running text, so phrases like "let $M : \sigma$ be ..." are understood in arbitrary context Γ. If $\Gamma \vdash \sigma$ then we sometimes write $\Gamma \vdash \sigma$ *true* to mean that there exists M with $\Gamma \vdash M : \sigma$.

The implicit typing premises to equality rules mentioned above (Section 2.1.2.4) are also required in all subsequent equality rules. So an equality rule of the form

$$\frac{\mathcal{P}}{\Gamma \vdash M = N : \sigma}$$

is understood as

$$\frac{\mathcal{P}, \Gamma \vdash M : \sigma, \Gamma \vdash N : \sigma}{\Gamma \vdash M = N : \sigma}$$

We use the symbol $\equiv$ for syntactic identity including α-conversion. So $\lambda x{:}\mathbf{N}.x \equiv \lambda y{:}\mathbf{N}.y$, but $(\lambda x{:}\mathbf{N}.x)\ 0 \not\equiv 0$.

The set of set-theoretic natural numbers is denoted ω. If $n \in \omega$ we sometimes write n also for the corresponding numeral in $\mathbf{N}$, for example we write abbreviate $\mathit{Suc}(\mathit{Suc}(\mathit{Suc}(0)))$ by 3.

2.1.4 Derived rules and meta-theoretic properties

The meta-theory of systems like the above has been the subject of extensive study [2, 103, 108, 20, 70, 38]. For our development we need the following meta-theoretic properties whose proofs are standard and may be found e.g. in [70].

Proposition 2.1.1 (Weakening). *Let $\Gamma, \Delta \vdash$ and $\Gamma \vdash \sigma$. If the judgement $\Gamma, \Delta \vdash J$ is derivable then so is $\Gamma, x{:}\sigma, \Delta \vdash J$.*

Proposition 2.1.2 (Substitution). *Let $\Gamma, x{:}\sigma, \Delta \vdash$ and $\Gamma \vdash M : \sigma$. If the judgement $\Gamma, x{:}\sigma, \Delta \vdash J$ is derivable then so is $\Gamma, \Delta[x := M] \vdash J[x := M]$*

Proposition 2.1.3. *If $\vdash \Gamma = \Delta$ then $\Gamma \vdash$. If $\Gamma \vdash \sigma = \tau$ then $\Gamma \vdash \sigma$. If $\Gamma \vdash M = N : \sigma$ then $\Gamma \vdash M : \sigma$.*

Proposition 2.1.4 (Strong normalisation). *The rewrite system obtained by directing the non-logical equality rules (Π-*Beta*, *Id-Comp*, *Nat-Comp-Zero*, *Nat-Comp-Suc*) from left to right is strongly normalising. This induces a decision procedure for definitional equality.*

Proposition 2.1.5 (Decidability). *It is decidable whether a judgement $\Gamma \vdash J$ is derivable.*

Remark 2.1.6. The strong normalisation property implies in particular that every term in the empty context is definitionally equal to a canonical one, i.e. a term having *Refl*, 0, *Suc*, or λ as outermost term former.

Proposition 2.1.7 (Unicity of typing). *If $\Gamma \vdash M : \sigma$ and $\Gamma \vdash M : \tau$ then $\Gamma \vdash \sigma = \tau$.*

Proposition 2.1.8 (Unicity of type formation). *If $\Gamma \vdash \Pi x{:}\sigma.\tau = \Pi x{:}\sigma'.\tau'$ then $\Gamma \vdash \sigma = \sigma'$ and $\Gamma, x{:}\sigma \vdash \tau = \tau'$. If $\Gamma \vdash \mathit{Id}_\sigma(M, N) = \mathit{Id}'_\sigma(M', N')$ then $\Gamma \vdash \sigma = \sigma'$ and $\Gamma \vdash M = M' : \sigma$ and $\Gamma \vdash N = N' : \sigma$.*

These meta-properties continue to hold *mutatis mutandis* for the various extensions to the core type theory we are going to make.

2.1.4.1 N-canonicity. We are now in the position to make precise what we mean by the property that in the empty context terms of the natural numbers are canonical which we hinted at in the Introduction.

Definition 2.1.9 (N-canonicity). A type theory with natural numbers is *N-canonical* if whenever $\diamond \vdash M : \mathbf{N}$ then $\diamond \vdash M = \underbrace{Suc(\ldots Suc(0)\ldots)}_{n \text{ times}}$ for some $n \in \omega$.

It follows from Proposition 2.1.5 and Remark 2.1.6 above that the core type theory described so far is N-canonical. We remark that if $\Gamma \vdash M : \mathbf{N}$ is a *non-canonical* element of the natural numbers and $\Gamma \vdash L : \sigma$ is *any* term of type σ then $\Gamma \vdash R^{\mathbf{N}}_{\sigma}(L, [x{:}\mathbf{N}, y{:}\sigma]L, M) : \sigma$ is a non-canonical element of type σ so that the role of $\mathbf{N}$ in the definition of N-canonicity is arbitrary.

2.2 High-level syntax

In the syntactic model constructions we are interested in we need to define various operations on contexts and will require the notion of relative contexts or "telescopes" [27]. We thus introduce some high-level syntax and derived rules and judgements in order to deal with these. We also introduce some abbreviational machinery to deal with definitions and substitutions.

2.2.1 Telescopes

If Γ and $\Delta = x_1{:}\tau_1, \ldots, x_n{:}\tau_n$ are pre-contexts then we write $\Gamma \vdash \Delta$ as an abbreviation for the n judgements $\Gamma \vdash \tau_1$, $\Gamma, x_1{:}\tau_1 \vdash \tau_2$ up to $\Gamma, x_1{:}\tau_1, \ldots, x_{n-1}{:}\tau_{n-1} \vdash \tau_n$. In this case we say that Δ is a *context relative to* or a *telescope* with respect to Γ. It can be shown by induction that $\Gamma \vdash \Delta$ iff $\Gamma, \Delta \vdash$ where according to our convention the variables in Δ have been suitably renamed. The type equality judgements are extended accordingly to telescopes. That is to say, we write $\Gamma \vdash \Delta = \Delta'$ if Δ and Δ' have the same length and definitionally equal components.

2.2.2 Elements of telescopes and context morphisms

Assume $\Gamma \vdash \Delta$ with $\Delta = x_1 : \tau_1, \ldots, x_n : \tau_n$. If $f = (M_1, \ldots, M_n)$ is an n-tuple of pre-terms then we write $\Gamma \vdash f : \Delta$ as an abbreviation for the n judgements $\Gamma \vdash M_1 : \tau_1$, $\Gamma \vdash M_2 : \tau_2[x_1 := M_1]$, up to $\Gamma \vdash M_n : \tau_n[x_1 := M_1] \ldots [x_{n-1} := M_{n-1}]$. We then call f an *element of the telescope* Δ. In the special case where Δ is independent of Γ, i.e. we have $\Gamma \vdash$ and $\Delta \vdash$, we call f a *substitution* or a *context morphism* from Γ to Δ. In this situation we write $\Gamma \vdash f \Rightarrow \Delta$. Definitional equality is extended canonically to elements of telescopes, that is we write $\Gamma \vdash f = f' : \Delta$ if f and f' have the same length and definitionally equal components.

Syntactic substitution generalises to elements of telescopes. If $\Gamma \vdash f : \Delta$ and $\Gamma, \Delta \vdash \sigma$ then we write $\sigma[(x_1, \ldots, x_n) := f]$ for the parallel substitution of all Δ-variables x_1 through x_n in σ by their companions in the n-tuple f. We then have $\Gamma \vdash \sigma[(x_1, \ldots, x_n) := f]$. If the variables are clear from the (linguistic) context we also write $\sigma[f]$. An analogous notation applies to substitution inside terms and telescopes.

The *context morphisms* include the *identity* from Γ to Γ which consists of the tuple of variables in Γ, and they are closed under composition, i.e. if f is a context morphism from Γ to Δ and g is a context morphism from Δ to Θ then the parallel substitution of f into g gives a context morphism from Γ to Θ denoted $g \circ f$. In this way the contexts form a category, i.e. composition is associative and the identity is neutral.

We remark that the property of unicity of typing (Proposition 2.1.7) which holds for terms does not extend to elements of telescopes. For example the pair $(0, \mathit{Refl}_{\mathbf{N}}(0))$ can be viewed as a substitution from $\diamond$ to $x : \mathbf{N}, p : \mathit{Id}_{\mathbf{N}}(x, x)$ or to $x : \mathbf{N}, p : \mathit{Id}_{\mathbf{N}}(0, 0)$. Therefore, elements of telescopes ought to be annotated with typing information which we shall, however, usually omit.

Since we are going to describe constructions involving abstractly given contexts and telescopes we need a means of introducing variables for them. If Γ is a pre-context of length n and if $\gamma = (\gamma_1, \ldots \gamma_n)$ is an n-tuple of variables then $\gamma : \Gamma$ denotes Γ with its variables renamed to $\gamma_1 \ldots \gamma_n$. This allows us for example to write $\gamma : \Gamma \vdash \gamma : \Gamma$.

2.2.3 Definitions and substitution

Assume $\Gamma, \Delta \vdash \sigma$ for some type expression σ. By the declaration

$$\tau[\delta : \Delta] := \sigma$$

the meta-variable τ becomes an abbreviation for the expression σ. This notation emphasises that the variables from Δ are free in τ. For example we may define

$$\mathit{eqzero}[n : \mathbf{N}] := \mathit{Id}_{\mathbf{N}}(n, 0)$$

and in ambient context $\mathit{maxint} : \mathbf{N}$

$$\mathit{eqmax}[n : \mathbf{N}] := \mathit{Id}_{\mathbf{N}}(n, \mathit{maxint})$$

Explicit variable names in a substitution may now be omitted, for example the expression $\mathit{eqzero}[5]$ denotes $\mathit{Id}_{\mathbf{N}}(5, 0)$ and $\mathit{eqmax}[5]$ denotes $\mathit{Id}_{\mathbf{N}}(5, \mathit{maxint})$. We also use square brackets as a form of first-order abstraction, i.e. we may write

$$[\delta : \Delta]\sigma$$

to emphasise the Δ-variables in σ. The above declaration can thus equivalently be written as

$$\tau := [\delta : \Delta]\sigma$$

Whenever τ or $[\delta : \Delta]\sigma$ appears inside some term then by suitable renaming it is ensured that none of the Δ-variables are captured.

Analogous conventions apply to terms, telescopes, and elements. The main application of first-order abstraction is the instantiation of binding operators like $R^{\mathbf{N}}$. For example the following is a (somewhat contrived) proof that every natural number is equal to itself

$$
\begin{array}{ll}
n:\mathbf{N} \vdash & \\
\quad R^{\mathbf{N}}_{[n:\mathbf{N}]Id_{\mathbf{N}}(n,n)}(& \text{by ind. on } n \\
\qquad \mathit{Refl}_{\mathbf{N}}(0)\,, & \text{base case} \\
\qquad [n:\mathbf{N}, p: Id_{\mathbf{N}}(n,n)]\mathit{Refl}_{\mathbf{N}}(Suc(n))\,, & \text{ind. case} \\
\qquad n) : Id_{\mathbf{N}}(n,n) & \text{conclusion}
\end{array}
$$

Observe the scoping of variable names here. The n in the subscript to $R^{\mathbf{N}}$ is different from the one declared in the global context.

It is the hope of the author that with these conventions a reasonable compromise between readability and formal correctness has been found.

2.3 Further type formers

In this section we introduce the syntax of a few further type formers such as Σ-types and universes. We also describe how Coquand-Huet's Calculus of Constructions fits into our syntactic framework. It is understood that whenever a rule introduces new type or term formers then the raw syntax is suitably extended to account for these, and definitional equality is extended accordingly.

2.3.1 Unit type

We start with a unit type, which is sometimes handy for technical reasons. It introduces a type $\mathbf{1}$ with single canonical element $\star$. The unit type comes with an induction principle stating that $\star$ is the only inhabitant of $\mathbf{1}$ up to propositional equality. We shall later (Section 4.5.4) consider an extensional unit types which contains a single element $\star$ up to definitional equality.

$$\frac{}{\vdash \mathbf{1}}\ \textsc{Unit-Form} \qquad \frac{}{\star : \mathbf{1}}\ \textsc{Unit-Intro}$$

$$\frac{x{:}\mathbf{1} \vdash \sigma[x] \quad \vdash M : \sigma[\star] \quad \vdash N : \mathbf{1}}{\vdash R^{\mathbf{1}}_{\sigma}(M,N) : \sigma[N]}\ \textsc{Unit-Elim}$$

$$\frac{}{\vdash R^{\mathbf{1}}_{\sigma}(M,\star) = M : \sigma[\star]}\ \textsc{Unit-Comp}$$

2.3.2 Σ-types

Σ-types serve to internalise telescopes. They are given by the following rules.

$$\frac{\vdash \sigma \qquad x{:}\,\sigma \vdash \tau[x]}{\vdash \Sigma x{:}\,\sigma.\tau[x]}\ \Sigma\text{-FORM} \qquad \frac{\vdash M : \sigma \qquad \vdash N : \tau[M]}{\vdash pair_{\sigma,\tau}(M,N) : \Sigma x{:}\,\sigma.\tau[x]}\ \Sigma\text{-INTRO}$$

$$\frac{\begin{array}{l} p : \Sigma x{:}\,\sigma.\tau[x] \vdash \rho[p] \\ x{:}\,\sigma, y{:}\,\tau[x] \vdash M[x,y] : \rho[pair_{\sigma,\tau}(x,y)] \qquad \vdash P : \Sigma x{:}\,\sigma.\tau[x] \end{array}}{\vdash R^{\Sigma}_{\sigma,\tau,\rho}(M,P) : \rho[P]}\ \Sigma\text{-ELIM}$$

$$\frac{}{R^{\Sigma}_{\sigma,\tau,\rho}(M, pair_{\sigma,\tau}(N,O)) = M[N,O] : \rho[pair_{\sigma,\tau}(N,O)]}\ \Sigma\text{-COMP}$$

In case type information can be inferred from the context we abbreviate the pairing operator $pair_{\sigma,\tau}(M,N)$ by (M,N). By suitable instantiation of R^{Σ} we can define projections .1 and .2 which provide a left inverse to the pairing operator viewed as the context morphism $x{:}\,\sigma, y{:}\,\tau \vdash pair_{\sigma,\tau}(x,y) \Rightarrow \Sigma x{:}\,\sigma.\tau$:

$$\begin{array}{l} M.1 := R^{\Sigma}_{\sigma,\tau,\sigma}([x{:}\,\sigma, y{:}\,\tau]x\ ,\ M) : \sigma \\ M.2 := R^{\Sigma}_{\sigma,\tau,[p:\Sigma x:\sigma.\tau]\tau[p.1]}([x{:}\,\sigma, y{:}\,\tau]y\ ,\ M) : \tau[M.1] \end{array}$$

Now from Σ-COMP we get $pair(M,N).1 = M$ and $pair(M,N).2 = N$. But pairing is not surjective which is why Σ-types only approximate the summation obtained from telescopes. However, using R^{Σ} we can find a term of the type

$$p{:}\,\Sigma x{:}\,\sigma.\tau \vdash Id_{\Sigma x:\sigma.\tau}(p, \langle p.1, p.2\rangle)$$

namely $R^{\Sigma}(p\ ,\ [x{:}\,\sigma, y{:}\,\tau]Refl_{\Sigma x:\sigma.\tau}(pair(x,y)))$. Thus "surjective pairing holds propositionally".

2.3.3 Function and cartesian product types

Function spaces and cartesian products are special cases of dependent product and sum. Writing $\sigma \to \tau := \Pi x{:}\,\sigma.\tau$ and $\sigma \times \tau := \Sigma x{:}\,\sigma.\tau$ we have $\vdash \sigma \to \tau$ and $\vdash \sigma \times \tau$ provided $\vdash \sigma$ and $\vdash \tau$ and we obtain corresponding introduction and elimination rules.

Although Σ-types and unit type are not absolutely crucial we consider them from now on as part of the core type theory which we henceforth refer to by TT. The type theory TT is a sub-system of Martin-Löf's intensional type theory [88].

2.3.4 The Calculus of Constructions

Coquand-Huet's [20] Calculus of Constructions is a dependently typed calculus with products (Π-types) only and a universe *Prop* that is closed under impredicative quantification. The latter is described by the following rules.

$$\frac{}{\vdash Prop}\ \textsc{Prop-Form} \qquad \frac{\vdash S : Prop}{\vdash Prf(S)}\ \textsc{Prf-Form}$$

$$\frac{x{:}\,\sigma \vdash S[x] : Prop}{\vdash \forall x{:}\,\sigma.S[x] : Prop}\ \textsc{Prop-Intro}$$

$$\frac{}{\vdash Prf(\forall x{:}\,\sigma.S[x]) = \Pi x{:}\,\sigma.Prf(S[x])}\ \textsc{Prop-Eq}$$

The terms of *Prop* are called "propositions". If $M : Prop$ is a proposition then $Prf(M)$ is the type of its proofs. A type of the form $Prf(M)$ for some $M : Prop$ is also called a proposition. The term "propositions" for these terms (and types) can be misleading because in some applications of the Calculus of Constructions the universe *Prop* is used to represent both datatypes and propositions [90]. In the present work this will not be done.

The operator $\forall$ together with the equation PROP-EQ gives that the product over a family of propositions (indexed over σ) is a proposition again. In Coquand and Huet's original presentation, terms of type *Prop* and their corresponding *Prf*-types were syntactically identified. The presentation used here is due to Streicher [103]. The Calculus of Constructions has types in the empty context, for example the type of falsehood $\vdash \Pi s{:}\, Prop.Prf(s)$.

We use logical connectives $\Rightarrow, \wedge, \vee, tt, ff, \neg$ for their encodings in the Calculus of Constructions. If $P, Q : Prop$ then their implication $P \Rightarrow Q$ is defined as the proposition $\forall x{:}\, Prf(P).Q$, their conjunction is defined as the proposition $\forall c{:}\, Prop.(P \Rightarrow Q \Rightarrow c) \Rightarrow c$, and their disjunction $P \vee Q$ is given by $\forall c{:}\, Prop.(P \Rightarrow c) \Rightarrow (Q \Rightarrow c) \Rightarrow c$. The true proposition tt is defined as $\forall c{:}\, Prop.c \Rightarrow c$. The false proposition ff is defined as $\forall c{:}\, Prop.Prf(c)$. The negation $\neg P$ of proposition P is defined as $P \Rightarrow ff$. If $x{:}\,\sigma \vdash P : Prop$ then we define the existential quantification of P as $\exists x{:}\,\sigma.P := \forall c{:}\, Prop.(\forall x{:}\,\sigma.P \Rightarrow c) \Rightarrow c$. For $M, N{:}\,\sigma$ we define Leibniz equality $M \stackrel{L}{=} N$ as

$$\forall P{:}\,\sigma \to Prop.(P\ M) \Rightarrow (P\ N)$$

It is well-known that the proof rules for these connectives in higher-order intuitionistic logic are derivable [20]. Extending the notation from Section 2.1.3 we write $\Gamma \vdash P\ true$ for $\Gamma \vdash P : Prop$ to mean that there exists P with $\Gamma \vdash P : Prf(P)$.

The syntactic properties stated in Section 2.1.4 including decidability of equality continue to hold for the Calculus of Constructions, see e.g. [3].

2.3.5 Universes

A universe is a type U containing names or codes for types. This is achieved by an operator *El* which for each term $M : U$ gives a type $El(M)$. A universe can be closed under various type formers in which case there are functions similar to $\forall$ which perform the type formation on the level of codes. As an example we give the rules for a universe closed under Π and containing the natural numbers.

$$\frac{}{\vdash U}\ U\textsc{-Form} \qquad \frac{\vdash M : U}{\vdash El(M)}\ El\textsc{-Form}$$

$$\frac{\vdash S : U \qquad x : El(S) \vdash T : U}{\vdash \hat{\Pi}(S,T) : U} \; U\text{-INTRO-}\Pi$$

$$\frac{}{El(\hat{\Pi}(S,T)) = \Pi x{:}\, El(S).El(T)} \; U\text{-EQ-}\Pi$$

$$\frac{}{\vdash \hat{\mathbf{N}} : U} \; U\text{-INTRO-N} \qquad \frac{}{\vdash El(\hat{\mathbf{N}}) = \mathbf{N}} \; U\text{-EQ-N}$$

We do not consider "universe-induction" [88, p. 101ff.]. Universes can be used to represent modules [41] and also increase the strength of the type theory. For example in the presence of a universe containing an empty type $\mathbf{0}$ (which we haven't defined) it becomes possible to find an inhabitant of the type $\vdash \Pi p : Id_{\mathbf{N}}(0, Suc(0)).\mathbf{0}$ which corresponds to Peano's fourth axiom [100]. For a type theory without universes it may therefore be appropriate to introduce this as an axiom by the following rule:

$$\frac{\Gamma \vdash M : Id_{\mathbf{N}}(0, Suc(0)) \qquad \Gamma \vdash \sigma}{\Gamma \vdash Peano_{\sigma}(M) : \sigma} \; \text{PEANO-4}$$

This rule does not introduce non-canonical elements in the empty context, because with $\Gamma = \diamond$ the premises to PEANO-4 cannot be met. In [109] it is shown how a term $Peano_{\sigma}(M)$ can be *defined* by induction on the form of σ in the core type theory without empty types and universes. See also Section 6.4.

2.3.6 Quotient types

Quotient types permit the redefinition of the propositional equality on a type. According to the particular notion of propositional equality chosen we obtain different formulations of quotient types. In this thesis we study quotient types with respect to Leibniz equality (Section 5.1), the identity type (Sects. 3.2.6.1, 5.3), and (marginally) also with respect to extensional 'definitional' equality (Section 3.2.6.1). We defer the syntactic rules to these sections. Here we only remark that intensional formulations of quotient types must invariably introduce non-canonical elements of identity types which establish propositional equality of equivalence classes of related elements, and therefore the statement made in Remark 2.1.6 is no longer true in the presence of quotient types unless definitional equality is redefined using a syntactic model like the ones presented in Chapter 5.

2.4 Abstract semantics of type theory

As described in the Introduction we require a precise definition of what a model for a dependently typed calculus should be and also a generic interpretation function mapping syntactic objects to objects in the model. Many notions of model have been proposed in the literature; for a comprehensive treatment see e.g. [59]. We use a derivative of *categories with attributes* as introduced by Cartmell [15] and later used by Pitts [92] (dubbed *type categories* there) and Moggi [87]. This notion seems to offer the right level of generality for our purposes, and nevertheless does not introduce any unnecessary formal padding.

As opposed to the "display-map" approach [106] or "contextual categories" [15, 102], in categories with attributes a distinction is made between "families" (the denotations of types) and their corresponding context extensions. In particular, the map which assigns the corresponding context extension to a family need not be injective. This extra generality will be required in almost all the model constructions we shall study.

The original definition of categories with attributes involves conditional equations, since certain diagrams are required to be pullbacks. This is a disadvantage if one wants to construct models internally in some type theory and check the model equations using normalisation. We thus describe a notion of model without conditional equations inspired by Curien's categorical combinators [22, 23, 24]. This means that in addition to abstract contexts and families we introduce a third sort of *abstract sections* which are in one-one correspondence to sections (right inverses) to projections. Their separate introduction allows to quantify over them without using conditional equations. Ehrhard [31] and Ritter [95] achieves the same goal by introducing arbitrary morphisms between families as a new sort. For our purposes this appears overly general and the price to pay is that a formal unit type must be introduced and that the equational theory becomes more complicated.

Since a guideline in our description of the semantics of dependent types is to stay as close to the syntax as possible without losing the advantages which abstract semantics offers, we have chosen the name *syntactic categories with attributes* for the notion of model to be described here.

2.4.1 Syntactic categories with attributes

Definition 2.4.1. A *syntactic category with attributes* is given by the following data:

i. A category $\mathbf{C}$ with terminal object $\top$. The objects of $\mathbf{C}$ are called *contexts* (A, B[1], Γ, Δ, ...), the morphisms (f, g, h, ...) are called *context morphisms*. The unique morphism from the terminal object to Γ is denoted $!_\Gamma$.
ii. A functor $Fam : \mathbf{C}^{op} \to \mathbf{Sets}$. If $f \in \mathbf{C}(\Gamma, \Delta)$ is a context morphism and $\sigma \in Fam(\Delta)$ then $Fam(f)(\sigma) \in Fam(\Gamma)$ is abbreviated by $\sigma\{f\}$. The elements of $Fam(\Gamma)$ are called *types* or *families* (σ, τ, ρ,...) in context Γ.
iii. If σ is a family in context Γ then there is a context $\Gamma \cdot \sigma$ called the *comprehension* of σ, and a context morphism $p(\sigma) \in \mathbf{C}(\Gamma \cdot \sigma, \Gamma)$ called the *display map* or the *projection* of σ. Moreover, if in addition $f \in \mathbf{C}(B, \Gamma)$ then $q(f, \sigma) \in \mathbf{C}(B \cdot \sigma\{f\}, \Gamma \cdot \sigma)$ and

[1] Read "Alpha", "Beta".

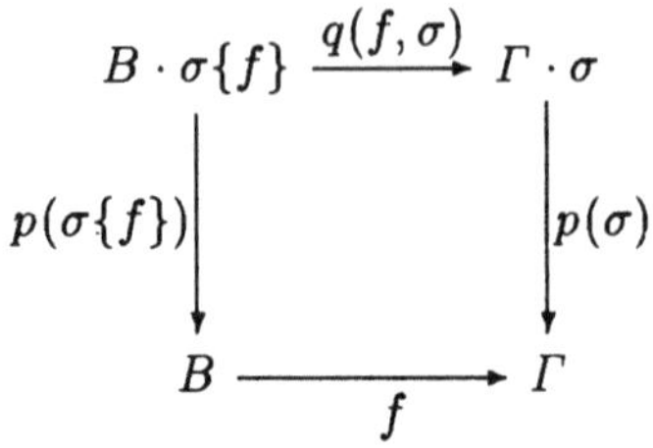

commutes.[2] The assignment $q(-,-)$ is functorial in the sense that

$$q(id_\Gamma, \sigma) = id_{\Gamma \cdot \sigma}$$

and for $g \in \mathbf{C}(A, B)$ also

$$q(f \circ g, \sigma) = q(f, \sigma) \circ q(g, \sigma\{f\})$$

Note that these equations "typecheck" by virtue of the functoriality of *Fam*.

iv. For each $\sigma \in Fam(\Gamma)$ there is a set $Sect(\sigma)$ of *sections* or *terms* (M, N, ...).

v. If $M \in Sect(\sigma)$ for $\sigma \in Fam(\Gamma)$ then $\overline{M} \in \mathbf{C}(\Gamma, \Gamma \cdot \sigma)$ and

$$p(\sigma) \circ \overline{M} = id_\Gamma \qquad \text{(SECT-CHAR)}$$

or diagrammatically

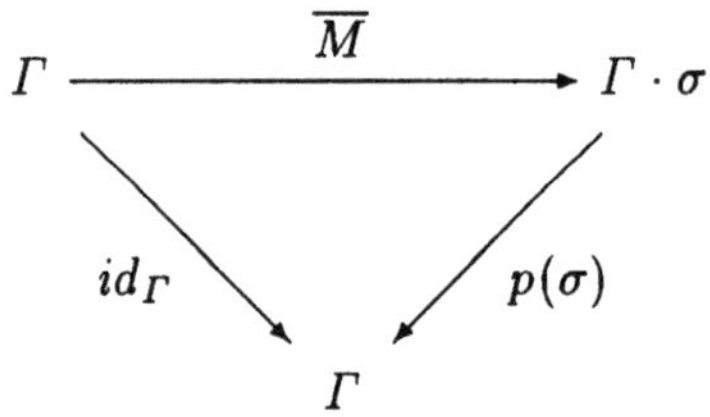

vi. If $\sigma \in Fam(\Delta)$ and $f \in \mathbf{C}(\Gamma, \Delta \cdot \sigma)$ then $Hd(f) \in Sect(\sigma\{p(\sigma) \circ f\})$ and

$$q(p(\sigma) \circ f, \sigma) \circ \overline{Hd(f)} = f \qquad \text{(MOR-INV)}$$

or diagrammatically

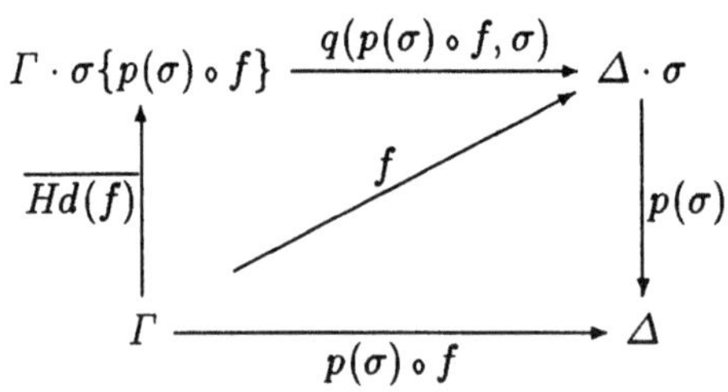

Moreover, if $f : \Gamma \to \Delta$ and $M \in Sect(\sigma\{f\})$ then

[2] Henceforth all diagrams are supposed to commute unless stated otherwise. The diagrams have been typeset using Paul Taylor's LaTeX diagram package.

$$M = Hd(q(f,\sigma) \circ \overline{M}) \qquad \text{(SECT-INV)}$$

which at the level of contexts becomes

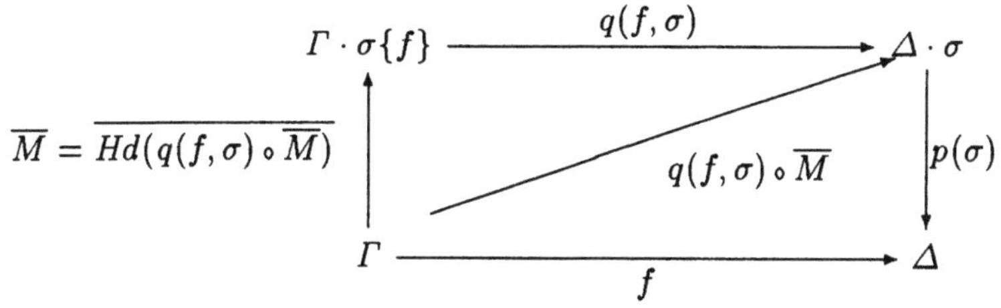

This definition is rather condensed and may need some explanation. **C** being a category means that there is an associative composition ($\circ$) for context morphisms which has a left- and right-neutral element, the identity id_Γ. *Fam* being a functor from $\mathbf{C}^{op}$ to the category **Sets** of sets and functions means that for each context Γ there is a set $Fam(\Gamma)$ of possible interpretations of types over Γ and if $f : B \to \Gamma$ then $Fam(f)$ is a function from $Fam(\Gamma)$ to $Fam(B)$ written $-\{f\}$ and satisfying $\sigma\{id_\Gamma\} = \sigma$ and $\sigma\{f \circ g\} = \sigma\{f\}\{g\}$ for $\sigma \in Fam(\Gamma)$ and $g : A \to B$.

The projection $p(\sigma)$ associates to each family $\sigma \in Fam(\Gamma)$ a forgetful context morphism $\Gamma \cdot \sigma \to \Gamma$. Its codomain $\Gamma \cdot \sigma$ corresponds to the extension of context Γ by type σ. Substitution along the morphism $p(\sigma)$ corresponds to weakening.

The morphism $q(f,\sigma) : \Gamma \cdot \sigma\{f\} \to \Delta \cdot \sigma$ can be seen as a "weakening" of f. On the Δ-part of $\Delta \cdot \sigma\{f\}$ it behaves like f (this expressed by the commutative square for q) and on the $\sigma\{f\}$-part it does nothing (this is indirectly expressed by the equation MOR-INV).

The mapping $M \mapsto \overline{M}$ associates a context morphism to a section M from Γ to $\Gamma \cdot \sigma$ which is the identity everywhere but on the last component. This is expressed by the equation SECT-CHAR. The equation MOR-INV is a kind of surjective pairing property for context morphisms. It expresses that (and how) $f : \Gamma \to \Delta \cdot \sigma$ can be recovered from its two components $Hd(f)$ and $p(\sigma) \circ f$.

The (at first sight strange) "typing" of *Hd* may be explained as follows: A context morphism $f : \Gamma \to \Delta \cdot \sigma$ should be thought of as containing two components: A context morphism from Γ to Δ (obtainable as $p(\sigma) \circ f$) and a section of σ submitted to substitution along this first component of f, i.e. an element of $Sect(\sigma\{p(\sigma) \circ f\})$. This is $Hd(f)$.

We continue by giving a few examples of syntactic categories with attributes which should clarify the definition further. In particular, the term model provides some more intuition and should be carefully gone through by the reader unfamiliar with categorical models of type theory.

Example 2.4.1 (Term model). Consider a calculus of dependent types $\mathcal{T}$ like the core calculus TT described in Section 2.1. We construct a syntactic category with attributes, which will later turn out to be the initial one. The underlying category of contexts **C** has as objects the well-formed contexts of $\mathcal{T}$ and as morphisms context morphisms as defined in Section 2.2.2 modulo definitional equality. This is easily seen to be a category in which the empty context is a terminal object. For a context Γ the set $Fam(\Gamma)$ is the set of well-formed types in this context, i.e. the set of valid judgements $\Gamma \vdash \sigma$. Context extension is inherited from the syntax, so $\Gamma \cdot (\Gamma \vdash \sigma) = \Gamma, x{:}\sigma$ and the display map $p(\Gamma \vdash \sigma)$ is the context morphism consisting of the variables in Γ. Substitution is syntactic substitution, i.e. if $(M_1, \ldots, M_n) : B \rightarrow \Gamma = (x_1 : \sigma_1, \ldots, x_n : \sigma_n)$ then $(\Gamma \vdash \sigma)\{(M_1, \ldots, M_n)\} = B \vdash \sigma[x_1 := M_1] \ldots [x_n := M_n]$. The "weakened" context morphism $q((M_1, \ldots, M_n), \sigma) : B, x{:}\sigma[x_1 := M_1] \ldots [x_n := M_n] \longrightarrow \Gamma, x{:}\sigma$ between the respective comprehensions then becomes $(M_1, \ldots, M_n, x)$ where $x : \sigma[x_1 := M_1] \ldots [x_n := M_n]$ is the last variable of the comprehension of the substituted type. If $\Gamma \vdash \sigma$ is a family then $Sect(\Gamma \vdash \sigma)$ is the set of terms $\Gamma \vdash M : \sigma$ again modulo definitional equality. To such a section we associate the context morphism $\overline{x_1 : \sigma_1, \ldots, x_n : \sigma_n \vdash M : \sigma} = (x_1, \ldots, x_n, M)$. Finally if $f = (M_1, \ldots, M_n)$ is a context morphism we put $Hd(f) := M_n$. The verification that this forms a syntactic category with attributes is straightforward. The term model of the Calculus of Constructions has been thoroughly studied by Streicher [103]; in particular a completeness result for abstract categorical semantics with respect to syntax is proved there.

The next example is taken from [58] with minor adjustments.

Example 2.4.2 (Families over sets). Let $\mathbf{F}$ be a small category with terminal object. We form a syntactic category with attributes over the category **Sets** of sets and functions as follows: If Γ is a set then $Fam(\Gamma)$ is the set of all functions from Γ to $Ob(\mathbf{F})$. If $\sigma \in Fam(\Gamma)$ then $\Gamma \cdot \sigma$ is the disjoint union $\{(\gamma, s) \mid \gamma \in \Gamma \text{ and } s \in \mathbf{F}(1, \sigma(\gamma))\}$. The map $p(\sigma)$ is the first projection. If furthermore $f : B \rightarrow \Gamma$ then $\sigma\{f\}$ is the composition $\sigma \circ f$ and $q(f, \sigma)$ sends (b, s) where $b \in B$ and $s \in \mathbf{F}(1, \sigma(f(b)))$ to $(f(b), s)$. Finally $Sect(\sigma)$ is the cartesian product $\prod_{\gamma \in \Gamma} \mathbf{F}(1, \sigma(\gamma))$. We leave the remaining definitions and the verifications to the reader.

In this example families carry an intensional structure, namely the morphisms in $\mathbf{F}$ which are not global sections and thus are not reflected into the underlying category.

Example 2.4.3 (Set-theoretic model). An important special case of the previous example arises when $\mathbf{F}$ is chosen to be the full subcategory of **Sets** consisting of small sets, i.e. with cardinality smaller than some inaccessible cardinal (see Section 5.2.2.6). We then obtain the familiar set-theoretic model [68, 17]. of type theory, which may e.g. be used to demonstrate equational consistency, i.e. the fact that the type $\vdash Id_{\mathbf{N}}(0, Suc(0))$ is not inhabited.

Example 2.4.4 (Families of ω-sets). An ω-set is a pair $X = (|X|, \Vdash_X)$ where $|X|$ is a set and $\Vdash_X$ is a surjective relation between $|X|$ and the set ω of natural numbers. If $n \Vdash_X x$ for some $n \in \omega$ and $x \in |X|$ then we say that n *realises* x or that n is a *realiser* for x. Surjectivity means that every element of $|X|$ has a realiser. A *morphism* between two ω-sets X and Y is a function f from $|X|$ to $|Y|$ for which there exists a natural number n (viewed as a code for a partial recursive function) such that for all $x \in |X|$ and realisers

$m \Vdash_X x$, the computation $\{n\}(m)$ terminates and is a realiser for $f(x)$. (We write $\{n\}(m)$ for the application of the partial recursive function coded by n to input m.) We say that f is *tracked* by n. The ω-sets with these morphisms form a category with terminal object given by the singleton set $\{\star\}$ equipped with the trivial realisability relation, i.e. every natural number realises $\star$ and the unique morphism is realised by any code for a total recursive function.

We construct a syntactic category with attributes over the category of ω-sets as follows: A family over an ω-set Γ is an assignment of a (small) ω-set to each element of $|\Gamma|$. If $\sigma = (\sigma_\gamma)_{\gamma \in |\Gamma|}$ is such a family then its comprehension $\Gamma \cdot \sigma$ is defined by $|\Gamma \cdot \sigma| := \{(\gamma, s) \mid \gamma \in |\Gamma| \text{ and } s \in |\sigma_\gamma|\}$ and $n \Vdash_{\Gamma \cdot \sigma} (\gamma, s)$ iff $\{p_1\}(n) \Vdash_\Gamma \gamma$ and $\{p_2\}(n) \Vdash_{\sigma_\gamma} s$ where p_1 and p_2 are codes for the components of some recursive bijection between ω and $\omega \times \omega$. The display map $p(\sigma)$ is the first projection tracked by p_1. If σ is such a family above Γ and f is a morphism from B to Γ tracked by $n \in \omega$ then we define the substitution $\sigma\{f\}$ simply by composition, viz. as the family $(\sigma_{f(\beta)})_{\beta \in |B|}$. The morphism $q(f, \sigma)$ sends (β, s) to $(f(\beta), s)$ and is tracked by the pairing of n and a code for the identity. If σ is as before then a section is an element M of the cartesian product $\prod_{\gamma \in |\Gamma|} |\sigma_\gamma|$ together with a realiser, i.e. some natural number n such that whenever $m \Vdash_\Gamma \gamma$ then $\{n\}(m) \Vdash_{\sigma_\gamma} M_\gamma$. We leave the definition of the remaining components to the reader.

Remark 2.4.2. One might try to define a syntactic category with attributes over **Sets**, say, by defining $Fam(\Gamma)$ as the set of all functions with codomain Γ and if $\sigma : \Sigma \longrightarrow \Gamma$ is such a function and furthermore $f : B \to \Gamma$ then to define substitution as the pullback

$$\begin{array}{ccc} \{(b, s) \in B \times \Sigma \mid f(b) = \sigma(s)\} & \longrightarrow & \Sigma \\ \big\downarrow & & \big\downarrow \sigma \\ B & \xrightarrow[f]{} & \Gamma \end{array}$$

This does not work because then for example $\sigma\{id_\Gamma\}$ is only isomorphic, but not equal to σ. Also the composition law is not strictly satisfied. In [47] we describe a canonical construction which turns a structure like the above into a syntactic category with attributes.

Some remarks on notation. We usually refer to a syntactic category with attributes by the name of its category of contexts. If **C** is any category then a syntactic category with attributes having **C** as underlying category of contexts will be called a *syntactic category with attributes* over **C**. Whenever we use one of the operator symbols which were part of the definition of a syntactic category with attributes it should be clear from the context to which syntactic category with attributes we are referring.

Although the formulation of the above definition might suggest something different, a syntactic category with attributes is meant to be a *structure*, i.e.

the various components referred to by the operator symbols are explicitly given and not merely required to exist. The same applies to further definitions in this style.

We begin our investigation of syntactic categories with attributes with a characterisation of the sections as special context morphisms.

Proposition 2.4.3. *Let $\sigma \in Fam(\Gamma)$. The assignment $M \mapsto \overline{M}$ is a bijection between $Sect(\sigma)$ and $\{f : \Gamma \to \Gamma \cdot \sigma \mid p(\sigma) \circ f = id_\Gamma\}$.*

Proof. One direction is part of the definition of a syntactic category with attributes. Conversely, if $f : \Gamma \to \Gamma \cdot \sigma$ has the specified property then $Hd(f)$ is a section of $\sigma\{id_\Gamma\} = \sigma$. It is immediate from the definition that these two maps are inverse to each other.

So up to bijective correspondence the set $Sect(\sigma)$ is the set of sections of the corresponding display map. By making these sections part of the structure we can quantify over sections without having to introduce conditional equations.

Our next step consists of establishing a correspondence to the usual notions of model which make use of pullbacks.

Proposition 2.4.4. *Every diagram of the form*

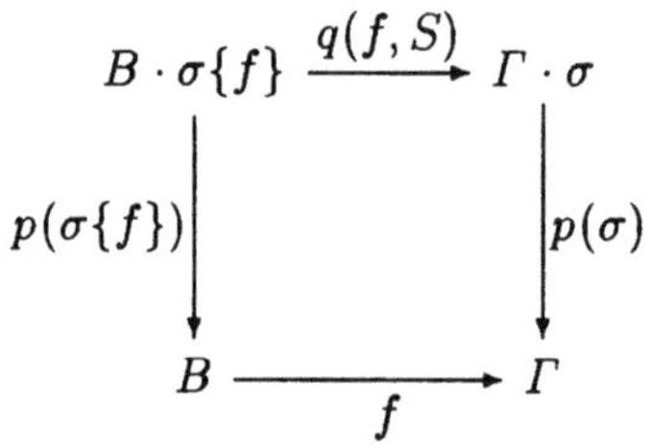

is a pullback.

Proof. Let $u : A \to B$ and $v : A \to \Gamma \cdot \sigma$ be context morphisms such that $f \circ u = p(\sigma) \circ v$. The unique mediating morphism from A to $B \cdot \sigma\{f\}$ is given by $q(u, \sigma\{f\}) \circ \overline{Hd(v)}$. Indeed

$$\begin{array}{ll} p(\sigma\{f\}) \circ q(u, \sigma\{f\}) \circ \overline{Hd(v)} = & \\ u \circ p(\sigma\{u \circ f\}) \circ \overline{Hd(v)} = & \text{since } Hd(v) \in Sect(\sigma\{u \circ f\}) \\ u & \end{array}$$

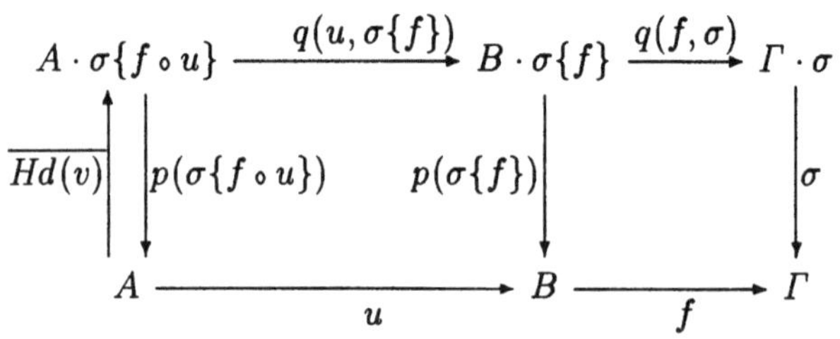

Similarly we have

$$\begin{array}{lll} & q(f,\sigma)\circ q(u,\sigma\{f\})\circ\overline{Hd(v)} & \\ = & q(f\circ u,\sigma)\circ\overline{Hd(v)} & \\ = & v & \text{by M}\textsc{or}\text{-I}\textsc{nv} \end{array}$$

For uniqueness let $w : A \to B\cdot\sigma\{f\}$ be such that $q(f,\sigma)\circ w = v$ and $p(\sigma\{f\})\circ w = u$. Then

$$\begin{array}{lll} & Hd(w) & \\ = & Hd(q(fu,\sigma)\circ\overline{Hd(w)}) & \text{by S}\textsc{ect}\text{-I}\textsc{nv} \\ = & Hd(q(f,\sigma)\circ q(u,\sigma\{f\})\circ\overline{Hd(w)}) & \\ = & Hd(q(f,\sigma)\circ w) & \text{by M}\textsc{or}\text{-I}\textsc{nv} \\ = & Hd(v) & \text{by assumption} \end{array}$$

and thus

$$\begin{array}{l} w = \\ q(u,\sigma\{f\})\circ\overline{Hd(w)} = \\ q(u,\sigma\{f\})\circ\overline{Hd(v)} \end{array}$$

as required.

This property has the consequence that if $f : B \to \Gamma\cdot\sigma$ then $Hd(f)$ is the *unique* section $M \in Sect(\sigma\{f\})$ with $q(p(\sigma)\circ f,\sigma)\circ\overline{M} = f$. This implies the following identities.

Lemma 2.4.5. *Let A, B, Γ be contexts and $\sigma \in Fam(\Gamma)$.*

i. $Hd(\underline{q(f,\sigma)}\circ g) = Hd(g)$ *for* $f : B \to \Gamma$ *and* $g : A \to B\cdot\sigma\{f\}$.
ii. $Hd(\overline{Hd(f)}\circ g) = Hd(f\circ g)$ *for* $f : B \to \Gamma\cdot\sigma$ *and* $g : A \to B$.

We can furthermore use the pullback property in order to characterise syntactic categories with attributes.

Proposition 2.4.6. *Let* $\mathbf{C}$ *be a category with terminal object and equipped with the following data:*

- *A functor* $Fam : \mathbf{C}^{op} \to \mathbf{Sets}$ *with morphism part written* $-\{-\}$ *as in Definition 2.4.1.*
- *For each* $\sigma \in Fam(\Gamma)$ *a* $\mathbf{C}$*-morphism with codomain* Γ *denoted* $\Gamma\cdot\sigma \xrightarrow{p(\sigma)} \Gamma$.
- *For* $f \in \mathbf{C}(B,\Gamma)$ *and* $\sigma \in Fam(\Gamma)$ *a* pullback square

$$\begin{array}{ccc} B\cdot\sigma\{f\} & \xrightarrow{q(f,\sigma)} & \Gamma\cdot\sigma \\ {\scriptstyle p(\sigma\{f\})}\downarrow & & \downarrow{\scriptstyle p(\sigma)} \\ B & \xrightarrow[f]{} & \Gamma \end{array}$$

- *For each* $\sigma \in Fam(\Gamma)$ *a set* $Sect(\sigma)$ *and a bijection*

$$\overline{-} : Sect(\sigma) \longrightarrow \{f \in \mathbf{C}(\Gamma,\Gamma\cdot\sigma) \mid p(\sigma)\circ f = id_\Gamma\}$$

Then there is a unique way of completing this structure to a syntactic category with attributes.

Proof. It only remains to define the $Hd(-)$ operation and to check the equations and uniqueness of the particular choice made. Let $f : B \to \Gamma \cdot \sigma$. The unique mediating morphism in

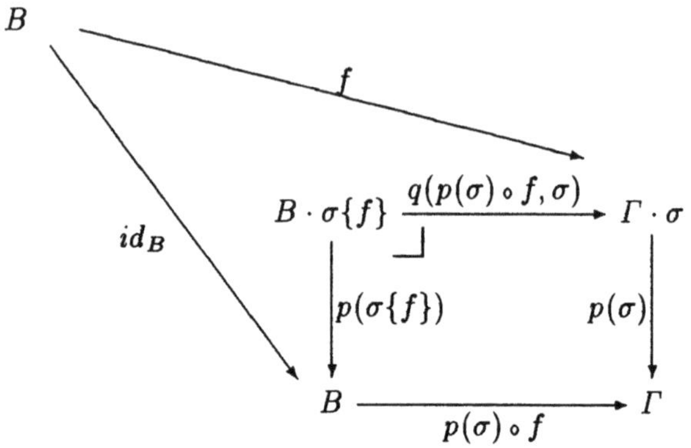

is a section of $\sigma\{p(\sigma) \circ f\}$. We let $Hd(f)$ be the corresponding section of $\sigma\{p(\sigma) \circ f\}$. Uniqueness of this choice follows from the universality of the pullback and the equations are routine calculation.

Remark 2.4.7. A special case of Proposition 2.4.6 arises when the bijection $\overline{-}$ is an identity, i.e. if $Sect(\sigma)$ equals the set of right inverses to $p(\sigma)$. Such a structure is called "category with attributes" by Cartmell and Moggi [15, 87] and a "type category" by Pitts [92]. This notion in turn is equivalent to Jacobs' notion of *full split comprehension category* [59], a notion of model based on fibrations.

2.4.1.1 Substitution on sections. Assume $f : B \to \Gamma$, $\sigma \in Fam(\Gamma)$, and $M \in Sect(\sigma)$. We construct a section $M\{f\} \in Sect(\sigma\{f\})$ as

$$M\{f\} = Hd(\overline{M} \circ f)$$

This reflects the intuition that substitution in terms is composition rather than induced by abstract universal properties. If in particular the morphism f happens to be a display map, i.e. of the form $p(\tau)$ for some $\tau \in Fam(\Gamma)$, then the "substitution" $M\{p(\tau)\}$ is the *weakening* of M. Alternatively we can identify the morphism $\overline{M\{f\}}$ as the unique mediating morphism in the diagram

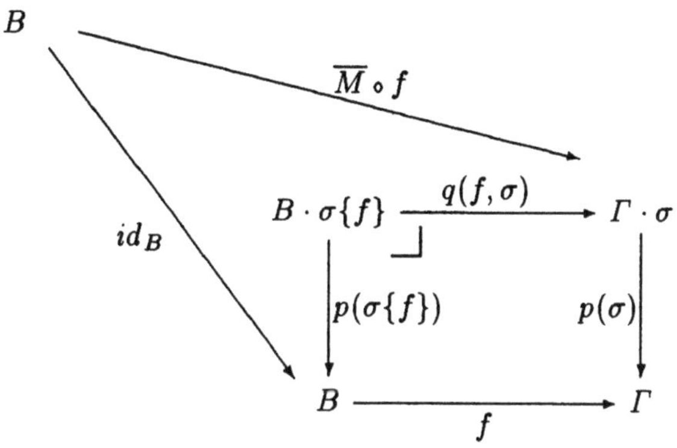

whereby we get that if $q(f,\sigma) \circ \overline{N} = \overline{M} \circ f$ for some $N \in Sect(\sigma\{f\})$ then $N = M\{f\}$.

2.4.1.2 Notation for weakening. If $\sigma, \tau \in Fam(\Gamma)$, $M \in Sect(\sigma)$, and $f : B \to \Gamma$ we introduce the following abbreviated notations for weakening.

$$\begin{array}{lcl} \sigma^+ & := & \sigma\{p(\tau)\} \\ M^+ & := & M\{p(\tau)\}(= Hd(\overline{M} \circ p(\tau))) \\ f^+ & := & q(f,\sigma) \end{array}$$

This notation being ambiguous (e.g. f^+ can mean both $q(f,\sigma)$ and $q(f,\tau)$) we use it only if the meaning is clear from the context. In the term model we have $(\Gamma \vdash \sigma)^+ = \Gamma, x{:}\tau \vdash \sigma$ and $(\Gamma \vdash M : \sigma)^+ = \Gamma, x{:}\tau \vdash M : \sigma$.

2.4.1.3 Handling of variables. If $\sigma \in Fam(\Gamma)$ then $Hd(id_{\Gamma \cdot \sigma})$ is a section of $\sigma\{p(\sigma)\}$. In the term model it corresponds to the term $\Gamma, x : \sigma \vdash x : \sigma$ where the second instance of σ is actually weakened. We abbreviate this section by v_σ. The thus defined "semantic variables" are stable under substitution in the following sense.

Lemma 2.4.8. *If $f : B \to \Gamma$ and $\sigma \in Fam(\Gamma)$ then*

$$v_\sigma\{q(f,\sigma)\} = v_{\sigma\{f\}}$$

We can also establish that the substitution of a section into a variable results in that section; this corresponds to the syntactic triviality $x[x := M] = M$.

Lemma 2.4.9. *If $\sigma \in Fam(\Gamma)$ and $M \in Sect(\sigma)$ then*

$$v_\sigma\{\overline{M}\} = M$$

Proof. Routine calculation.

We could have introduced v_σ as a primitive together with an operator *Cons* which turns a context morphism $f : B \to \Gamma$ and a section $M \in Sect(\sigma\{f\})$ into a context morphism $Cons(M, f) : B \to \Gamma \cdot \sigma$. The morphism $q(f,\sigma)$ then becomes definable as $Cons(v_{\sigma\{f\}}, f \circ p(\sigma\{f\}))$. We leave it as an exercise to work out the details of this alternative description.[3]

The variables other than the last one are obtained as successive weakenings of v_σ. For example if $\sigma \in Fam(\Gamma)$ and $\tau \in Fam(\Gamma \cdot \sigma)$ then $v_\tau \in Sect(\tau^+)$ and $v_\sigma{}^+ \in Sect(\sigma^{++})$.

2.4.2 Type constructors

We now turn to the description of the semantic companions to the type formers described in Section 2.1. These semantic operators will be defined in such a way that the term model always provides an instance. This precludes their definition by universal properties as is usually done in the literature (e.g. [103, 59]). Instead we must ask for stability under substitution of all type and term formers not only the constructors. This sometimes leads to more complicated formulations, but it is the only way to identify as models the various syntactic constructions to be defined later.

2.4.2.1 Dependent Products.

Definition 2.4.10. Assume a syntactic category with attributes **C**. We say that **C** *is equipped with dependent products* if the following hold.

- For every context Γ and families $\sigma \in Fam(\Gamma)$, $\tau \in Fam(\Gamma \cdot \sigma)$ there is a distinguished family $\Pi(\sigma, \tau) \in Fam(\Gamma)$.
- If $M \in Sect(\tau)$ then there is a distinguished section $\lambda_{\sigma,\tau}(M) \in Sect(\Pi(\sigma, \tau))$.
- If $M \in Sect(\Pi(\sigma, \tau))$ and $N \in Sect(\sigma)$ then there is a distinguished section $App_{\sigma,\tau}(M, N) \in Sect(\tau\{\overline{N}\})$.
- If $f : B \to \Gamma$ then

  $$\Pi(\sigma, \tau)\{f\} = \Pi(\sigma\{f\}, \tau\{q(f, \sigma)\})$$

  and for $M \in Sect(\Pi(\sigma, \tau))$, $N \in Sect(\sigma)$

  $$App_{\sigma,\tau}(M, N)\{f\} = App_{\sigma\{f\},\tau\{q(f,\sigma)\}}(M\{f\}, N\{f\})$$

  and for $M \in Sect(\tau)$

  $$\lambda_{\sigma,\tau}(M)\{f\} = \lambda_{\sigma\{f\},\tau\{q(f,\sigma)\}}(M\{q(f, \sigma)\})$$

- If $M \in Sect(\tau)$ and $N \in Sect(\sigma)$ then $App_{\sigma,\tau}(\lambda_{\sigma,\tau}(M), N) = M\{\overline{N}\}$.

This definition of dependent products almost verbally follows the syntactic definition of Π-types. The requirements on compatibility of reindexing with the dependent product constructor corresponds to the syntactic definition of substitution as distributive over type- and term constructors.

---

[3] In the meantime this "exercise" has been (independently) worked out by P. Dybjer and it turns out that the resulting notion of model without conditional equations is much more natural and intuitive than ours. If this thesis would be written again, certainly this notion would be used instead.

Remark 2.4.11. We could require an "η-rule" too, i.e. for $M \in Sect(\Pi(\sigma,\tau))$

$$\lambda_{\sigma,\tau}(App_{\sigma\{p(\sigma)\},\tau\{q(p(\sigma),\sigma)\}}(M^+, v_\sigma)) = M$$

Then the compatibility condition on $\lambda_{\sigma,\tau}(-)$ could be dropped and in fact Π would enjoy a universal property. However then the term model would not have dependent products unless an η-rule were present in the syntax.

Proposition 2.4.12. *The term model from Example 2.4.1 is equipped with dependent products if the underlying syntax has Π-types.*

Proof. If $\sigma \in Fam(\Gamma)$ and $\tau \in Fam(\Gamma\cdot\sigma)$ then Γ is a valid context and $\Gamma \vdash \sigma$ and $\Gamma, x:\sigma \vdash \tau$ by definition of the term model. We then put $\Pi(\sigma,\tau) := \Pi x : \sigma.\tau \in Fam(\Gamma)$. If $f : B \to \Gamma$ is a (syntactic) context morphism then clearly

$$(\Pi x : \sigma.\tau)[f] = \Pi x : \sigma[f].\tau[(f,x)]$$

by definition of syntactic substitution. Next if $M \in Sect(\tau)$ then $\Gamma, x : \sigma \vdash M : \tau$ so we put $\lambda_{\sigma,\tau}(M) := \lambda x{:}\sigma.M \in Sect(\Pi(\sigma,\tau))$. Conversely, if $M \in Sect(\Pi(\sigma,\tau))$ and $N \in Sect(\sigma)$ then $\Gamma \vdash M : \Pi x : \sigma.\tau$ and $\Gamma \vdash N : \sigma$ so we put $App_{\sigma,\tau}(M,N) := (M\ N)$. Again by definition of syntactic substitution these choices meet the coherence requirements and moreover by PI-BETA the defining equation is also satisfied. We also see that the "η-rule" mentioned in the above remark corresponds to $\lambda x : \sigma.M\ x = M$ which does not necessarily hold.

We now give another characterisation of dependent products which is sometimes easier to check.

Proposition 2.4.13. *A syntactic category with attributes has dependent products iff for each family $\sigma \in Fam(\Gamma)$ and $\tau \in Fam(\Gamma\cdot\sigma)$ there is a distinguished family $\Pi(\sigma,\tau) \in Fam(\Gamma)$ and a morphism*

$$ev_{\sigma,\tau} : \Gamma\cdot\sigma\cdot\Pi(\sigma,\tau)\{p(\sigma)\} \to \Gamma\cdot\sigma\cdot\tau$$

with $p(\tau)\circ ev_{\sigma,\tau} = p(\Pi(\sigma,\tau)\{p(\sigma)\})$ in such a way that for every section $M \in Sect(\tau)$ there exists a distinguished section $\lambda_{\sigma,\tau}(M) \in Sect(\Pi(\sigma,\tau))$ such that $ev_{\sigma,\tau}\circ\overline{\lambda_{\sigma,\tau}(M)\{p(\sigma)\}} = \overline{M}$, and moreover the following three coherence conditions are satisfied for any morphism $f : B \to \Gamma$:

– $\Pi(\sigma,\tau)\{f\} = \Pi(\sigma\{f\},\tau\{q(f,\sigma)\})$

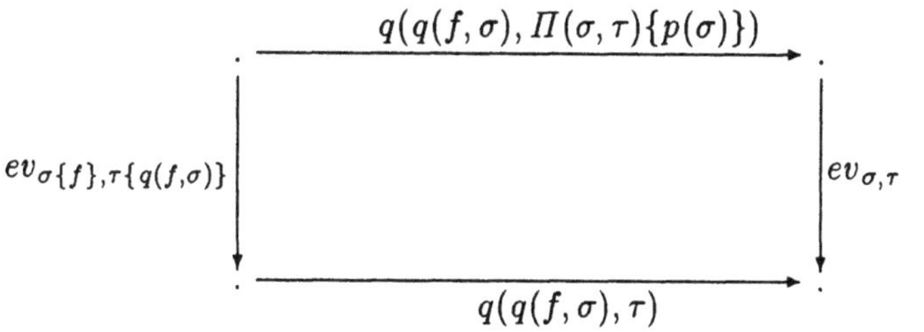

– $\lambda_{\sigma,\tau}(M)\{f\} = \lambda_{\sigma\{f\},\tau\{q(f,\sigma)\}}(M\{q(f,\sigma)\})$

Proof. We give the proof for the term model. The general case is an exercise in categorical notation. In the situation $\gamma : \Gamma \vdash \sigma$ and $\gamma : \Gamma, x : \sigma \vdash \tau$ we define the context morphism $ev_{\sigma,\tau}$ as

$$\gamma{:}\, \Gamma, s{:}\, \sigma, f{:}\, \Pi x{:}\, \sigma.\tau \vdash (\gamma, s, f\ s)\ :\ \gamma{:}\, \Gamma, s{:}\, \sigma, t{:}\, \tau$$

Conversely if we are given a context morphism $ev_{\sigma,\tau}$ with the above source and target and satisfying the commutativity requirement then we must have

$$ev_{\sigma,\tau} = [\gamma{:}\, \Gamma, s{:}\, \sigma, f{:}\, \Pi x{:}\, \sigma.\tau](\gamma, s, A[s, f])$$

for some term

$$\gamma{:}\, \Gamma, s{:}\, \sigma, f{:}\, \Pi x{:}\, \sigma.\tau \vdash A[s, f] : \tau$$

We then define for $\Gamma \vdash M : \Pi x{:}\, \sigma.\tau$ and $\Gamma \vdash N : \sigma$

$$(M\ N) := A[N, M]$$

The verifications are straightforward.

From the example of the dependent product it should have become clear how in principle every syntactic type or term former can be translated into its semantic equivalent. For every type forming rule one introduces an operator on families and for every term forming rule an operator on sections where in each case suitable instances of weakening have to be inserted. Then for every equation one introduces an equation between the corresponding semantic entities and finally one imposes equations which ensure the compatibility of all semantic operators with substitution.

In most cases, however, this approach leads to very clumsy formulations of the semantic type formers. Indeed, it is often possible to replace a term operator by a single context morphism (or sometimes by a single section) and a type operator by a single family. An example for the former is provided by the alternative characterisation of dependent products in Proposition 2.4.13 where we have replaced the application operator by the evaluation morphisms. An example for the latter arises from the identity type we define in Section 2.4.2.4 below.

We often follow this strategy in the definition of further semantic type formers without explicitly stating and proving correspondences like Proposition 2.4.13 above.

In most cases we relate our definition to stronger ones which make use of universal properties and thus are more elegant, but are usually not met by syntactic models like the term model.

2.4.2.2 Unit types. The semantic definition corresponding to the unit type defined in Section 2.3.1 is as follows:

Definition 2.4.14. A syntactic category with attributes is *equipped with unit types*, if for every context Γ there is a distinguished family $\mathbf{1}_\Gamma \in \mathit{Fam}(\Gamma)$ and a distinguished section $\star_\Gamma \in \mathit{Sect}(\mathbf{1}_\Gamma)$ such that for every $\sigma \in \mathit{Fam}(\Gamma \cdot \mathbf{1}_\Gamma)$ and $M \in \mathit{Sect}(\sigma\{\overline{\star_\Gamma}\})$ there is a distinguished section $R^{\mathbf{1}_\Gamma}_\sigma(M) \in \mathit{Sect}(\sigma)$ with

$$R^{\mathbf{1}_\Gamma}_\sigma(M)\{\overline{\star_\Gamma}\} = M$$

in such a way that the following coherence equations are satisfied for every morphism $f : B \to \Gamma$, and $M \in \mathit{Sect}(\sigma\{\star_\Gamma\})$:

- $\mathbf{1}_\Gamma\{f\} = \mathbf{1}_B$
- $\star_\Gamma\{f\} = \star_B$
- $(R^{\mathbf{1}_\Gamma}_\sigma(M))\{f^+\} = R^{\mathbf{1}_B}_{\sigma\{f^+\}}(M\{f^+\})$

In view of the first coherence condition one might require only $\mathbf{1}_\top$ and obtain the other units by substitution along the unique morphism into the terminal object. The operator $R^{\mathbf{1}_\Gamma}$ would, however, still have to be required for arbitrary contexts.

The correspondence to the syntax is as follows: If $\sigma \in \mathit{Fam}(\Gamma \cdot \mathbf{1}_\Gamma)$ and $M \in \mathit{Sect}(\sigma\{\overline{\star_\Gamma}\})$ and $N \in \mathit{Sect}(\mathbf{1}_\Gamma)$, then $R^{\mathbf{1}}(M,N) := R^{\mathbf{1}_\Gamma}_\sigma(M)\{\overline{N}\} \in \mathit{Sect}(\sigma\{\overline{N}\})$.

In many models unit types exist by virtue of the following proposition.

Proposition 2.4.15. *A syntactic category with attributes* $\mathbf{C}$ *can be equipped with unit types if there exists a family* $\mathbf{1} \in \mathit{Fam}(\top)$ *such that* $p(\mathbf{1})$ *is an isomorphism.*.

Proof. We put $\mathbf{1}_\Gamma := \mathbf{1}\{!_\Gamma\}$. Since

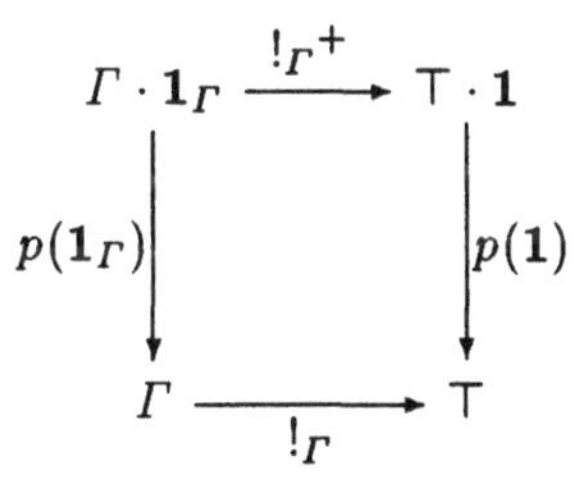

is a pullback, $p(\mathbf{1}_\Gamma)$ is an isomorphism, too. Let φ be its inverse. We put $\star_\Gamma := \mathit{Hd}(\varphi)$ and for $\sigma \in \mathit{Fam}(\Gamma \cdot \mathbf{1}_\Gamma)$ and $M \in \mathit{Sect}(\sigma\{\varphi\})$ we put $R^{\mathbf{1}_\Gamma}_\sigma(M) := M\{p(\mathbf{1}_\Gamma)\} \in \mathit{Sect}(\sigma\{p(\mathbf{1}_\Gamma) \circ \varphi\}) = \mathit{Sect}(\sigma)$. Stability under substitution follows since all the operators are given by substitutions themselves.

2.4.2.3 Natural numbers.

Definition 2.4.16. A syntactic category with attributes $\mathbf{C}$ *is equipped with natural numbers* if for every context Γ there are the following data:

- a distinguished family $\mathbf{N}_\Gamma \in \mathit{Fam}(\Gamma)$,

- a distinguished section $0_\Gamma \in Sect(\mathbf{N}_\Gamma)$ and a distinguished morphism $Suc_\Gamma : \Gamma \cdot \mathbf{N}_\Gamma \to \Gamma \cdot \mathbf{N}_\Gamma$ with $p(\mathbf{N}_\Gamma) \circ Suc_\Gamma = p(\mathbf{N}_\Gamma)$,
- for every $\sigma \in Fam(\Gamma \cdot \mathbf{N}_\Gamma)$, $M_z \in Sect(\sigma\{\overline{0}\})$, and $M_s \in Sect(\sigma\{Suc_\Gamma \circ p(\sigma)\})$ a distinguished section $R^{\mathbf{N}_\Gamma}(M_z, M_s) \in Sect(\sigma)$ with $R^{\mathbf{N}_\Gamma}(M_z, M_s)\{\overline{0_\Gamma}\} = M_z$ and $R^{\mathbf{N}_\Gamma}(M_z, M_s)\{Suc_\Gamma\} = M_s\{\overline{R^{\mathbf{N}_\Gamma}(M_z, M_s)}\}$

in such a way that all these data are stable under substitution, i.e. for $f : B \to \Gamma$, $\sigma \in Fam(\Gamma \cdot \mathbf{N}_\Gamma)$, $M_z \in Sect(\sigma\{\overline{0}\})$, $M_s \in Sect(\sigma\{Suc_\Gamma \circ p(\sigma)\})$, we have:

- $\mathbf{N}_\Gamma\{f\} = \mathbf{N}_B$
- $0_\Gamma\{f\} = 0_B$
- $f^+ \circ Suc_B = Suc_\Gamma \circ f^+$
- $R^{\mathbf{N}_\Gamma}_\sigma(M_z, M_s)\{f^+\} = R^{\mathbf{N}_B}_\sigma(M_z\{f\}, M_s\{f^+\})$

By straightforward calculation we obtain:

Proposition 2.4.17. *The term model described in 2.4.1 is equipped with natural numbers given by*

$$\begin{array}{lcl} \mathbf{N}_\Gamma & := & \Gamma \vdash \mathbf{N} \\ 0_\Gamma & := & \Gamma \vdash 0 : \mathbf{N} \\ Suc_\Gamma[\gamma{:}\,\Gamma, n{:}\,\mathbf{N}] & := & (\gamma, Suc(n)) \; : \; \Gamma, n{:}\,\mathbf{N} \\ R^{\mathbf{N}_\Gamma}_\sigma(M_z, M_s) & := & \Gamma, x{:}\,\mathbf{N} \vdash R^{\mathbf{N}}(M_z, M_s, x) : \sigma \end{array}$$

Again, we relate our notion of natural numbers to a stronger one often offered in the literature, e.g. [81].

Definition 2.4.18. Let $\mathbf{B}$ be a cartesian category (with terminal object and products). A *parametrised natural numbers object* in $\mathbf{B}$ is a diagram $1 \xrightarrow{0} \mathbf{N} \xrightarrow{Suc} \mathbf{N}$ such that for any diagram $G \xrightarrow{z} S \xrightarrow{s} S$ there exists a unique morphism $R(z, s) : G \times \mathbf{N} \to S$ such that

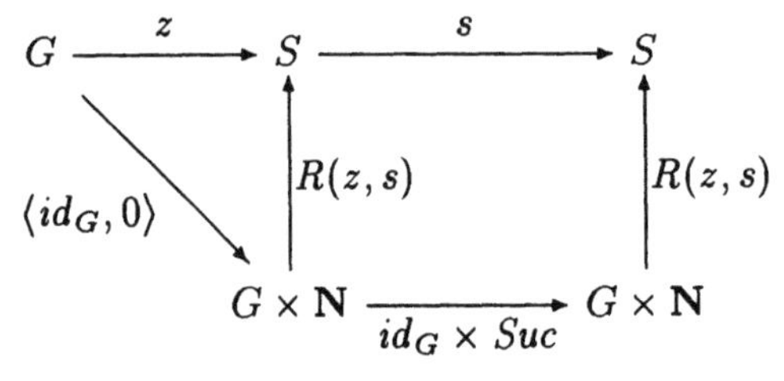

Proposition 2.4.19. *Let* $\mathbf{C}$ *be a syntactic category with attributes. If there exists a family* $\mathbf{N} \in Fam(\top)$ *such that* $\top \cdot \mathbf{N}$ *is the object part of a parametrised natural numbers object in* $\mathbf{C}$ *then* $\mathbf{C}$ *can be equipped with natural numbers in the sense of Definition 2.4.16.*

Proof. We define

$$\begin{array}{lcl} \mathbf{N}_\Gamma & := & \mathbf{N}\{!_\Gamma\} \\ 0_\Gamma & := & Hd(0)\{!_\Gamma\} \end{array}$$

Next we notice that $\Gamma \cdot \mathbf{N}_\Gamma$ is isomorphic to the cartesian product $\Gamma \times \top \cdot \mathbf{N}$ by virtue of Proposition 2.4.4. Without loss of generality we can identify $\Gamma \cdot \mathbf{N}_\Gamma$

with this product. We then put $Suc_\Gamma := id_\Gamma \times Suc$. We also have that $p(\mathbf{N}_\Gamma)$ is a product projection and $\overline{0_\Gamma} = \langle id_\Gamma, 0\rangle$.

Next we note that for $\sigma \in Fam(\Gamma \cdot \mathbf{N}_\Gamma)$, sections of $\sigma\{Suc_\Gamma\}\{p(\sigma)\}$ (this is the "type" of M_s) are in bijective correspondence to the set of endomorphisms h of $\Gamma \cdot \mathbf{N}_\Gamma \cdot \sigma$ with $p(\sigma) \circ h = Suc_\Gamma \circ p(\sigma)$. On the other hand elements of $Sect(\sigma\{\overline{0_\Gamma}\})$ (the "type" of M_z) are in bijective correspondence to morphisms $g : \Gamma \to \Gamma \cdot \mathbf{N}_\Gamma \cdot \sigma$ with $p(\sigma) \circ g = \langle id_\Gamma, 0\rangle$.

Now given such g and h arising from appropriate sections M_z and M_s then $R(g,h) : \Gamma \cdot \mathbf{N}_\Gamma \to \Gamma \cdot \mathbf{N}_\Gamma \cdot \sigma$ and $p(\sigma) \circ R(g,h) = id_{\Gamma \cdot \mathbf{N}_\Gamma}$ by the uniqueness property of the natural numbers object. So $R(g,h)$ induces a section of σ which is the desired $R_\sigma^{\mathbf{N}_\Gamma}(M_z, M_s)$. The commutativity of the diagram defining $R(g,h)$ gives the required equation for $R_\sigma^{\mathbf{N}_\Gamma}$. The coherence laws for $\mathbf{N}_\Gamma$, 0_Γ, and Suc_Γ follow from their definition by substitution (pullback) and the coherence of $R_\sigma^{\mathbf{N}_\Gamma}$ is again a consequence of the uniqueness property.

One may compare this to the proof of mathematical induction for natural numbers objects in a topos, see e.g. [40].

2.4.2.4 Identity types. For the definition of the identity type we first observe that for $\sigma \in Fam(\Gamma)$ the morphism

$$\overline{v_\sigma} : \Gamma \cdot \sigma \to \Gamma \cdot \sigma \cdot \sigma^+$$

is the "diagonal" which in the term model is given by

$$\gamma{:}\, \Gamma, x{:}\,\sigma \vdash (\gamma, x, x) \ : \ \Gamma, x{:}\,\sigma, y{:}\,\sigma$$

It has the property

$$p(\sigma^+) \circ \overline{v_\sigma} = p(\sigma)^+ \circ \overline{v_\sigma} = id_{\Gamma \cdot \sigma}$$

where $p(\sigma^+)$ and $p(\sigma)^+$ are the two projections $\Gamma \cdot \sigma \cdot \sigma^+ \to \Gamma \cdot \sigma$.

Definition 2.4.20. A syntactic category with attributes $\mathbf{C}$ *is equipped with identity types* if for every context Γ and $\sigma \in Fam(\Gamma)$ there is

- a distinguished family $Id(\sigma) \in Fam(\Gamma \cdot \sigma \cdot \sigma^+)$,
- a distinguished morphism $Refl_\sigma : \Gamma{\cdot}\sigma \to \Gamma \cdot \sigma \cdot \sigma^+{\cdot} Id_\sigma$ with $p(Id_\sigma) \circ Refl_\sigma = \overline{v_\sigma}$,
- for every family $\rho \in Fam(\Gamma \cdot \sigma \cdot \sigma^+ \cdot Id_\sigma)$ and $M \in Sect(\rho\{Refl_\sigma\})$ a distinguished section $J_{\sigma,\rho}(M) \in Sect(\rho)$ with $J_{\sigma,\rho}(M)\{Refl_\sigma\} = M$

in such a way that all these data are stable under substitution, i.e. for $f : B \to \Gamma$, σ, ρ as above, and $M \in Sect(\rho\{Refl_\sigma\})$ we have:

- $Id(\sigma)\{f^{++}\} = Id(\sigma\{f\})$
- $Refl_\sigma \circ f^+ = f^{+++} \circ Refl_{\sigma\{f\}}$
- $J_{\sigma,\rho}(M)\{f^{+++}\} = J_{\sigma\{f\},\rho\{f^{+++}\}}(M\{f^+\})$

Again in many "semantic" models we can give a simpler definition as follows:

Proposition 2.4.21. *A syntactic category with attributes can be equipped with identity types if for every $\sigma \in Fam(\Gamma)$ there is a distinguished family $Id(\sigma) \in Fam(\Gamma \cdot \sigma \cdot \sigma^+)$ stable under substitution such that*

$$\begin{array}{ccc} \Gamma \cdot \sigma \cdot \sigma^+ \cdot Id(\sigma) & \xrightarrow{p(Id(\sigma))} & \Gamma \cdot \sigma \cdot \sigma^+ \\ {\scriptstyle p(Id(\sigma))}\downarrow & & \downarrow{\scriptstyle p(\sigma)^+} \\ \Gamma \cdot \sigma \cdot \sigma^+ & \xrightarrow[p(\sigma^+)]{} & \Gamma \cdot \sigma \end{array}$$

(commutes and) is a pullback.

Proof. The universal property of the pullback applied to

$$\Gamma \cdot \sigma \cdot \sigma^+ \xleftarrow{\overline{v_\sigma}} \Gamma \cdot \sigma \xrightarrow{\overline{v_\sigma}} \Gamma \cdot \sigma \cdot \sigma^+$$

gives the morphism

$$Refl_\sigma : \Gamma \cdot \sigma \to \Gamma \cdot \sigma \cdot \sigma^+ \cdot Id(\sigma)$$

with $p(Id(\sigma)) \circ Refl_\sigma = \overline{v_\sigma}$ as required. Now $Refl_\sigma$ is actually an isomorphism because

$$\Gamma \cdot \sigma \cdot \sigma^+ \xleftarrow{\overline{v_\sigma}} \Gamma \cdot \sigma \xrightarrow{\overline{v_\sigma}} \Gamma \cdot \sigma \cdot \sigma^+$$

is also the pullback of the two projections

$$p(\sigma)^+, p(\sigma^+) : \Gamma \cdot \sigma \cdot \sigma^+ \to \Gamma \cdot \sigma$$

For the inverse of $Refl_\sigma$ we may for example take $p(\sigma^+) \circ p(Id(\sigma))$. So substituting along this inverse defines the desired function $J_{\sigma,\rho} : Sect(\rho\{Refl_\sigma\}) \to Sect(\rho)$. All the data except $Id(\sigma)$ are defined by universal properties and thus are stable under substitution.

2.4.2.5 Σ-types.

Definition 2.4.22. A syntactic category with attributes **C** *is equipped with Σ-types* if for every context Γ and families $\sigma \in Fam(\Gamma)$, $\tau \in Fam(\Gamma \cdot \sigma)$ there is:

- a distinguished family $\Sigma(\sigma, \tau) \in Fam(\Gamma)$,
- a distinguished morphism

$$pair_{\sigma,\tau} : \Gamma \cdot \sigma \cdot \tau \to \Gamma \cdot \Sigma(\sigma, \tau)$$

  with

$$p(\Sigma(\sigma, \tau)) \circ pair_{\sigma,\tau} = p(\sigma) \circ p(\tau)$$

- for each family $\rho \in Fam(\Gamma \cdot \Sigma(\sigma, \tau))$ and $M \in Sect(\rho\{pair_{\sigma,\tau}\})$ a distinguished section $R^{\Sigma}_{\sigma,\tau}(M) : Sect(\rho)$ with $R^{\Sigma}_{\sigma,\tau}(M)\{pair_{\sigma,\tau}\} = M$

in such a way that all these data are stable under substitution, i.e. for every situation $f : B \to \Gamma$, $\sigma \in Fam(\Gamma)$, $\tau \in Fam(\Gamma \cdot \sigma)$, $\rho \in Fam(\Gamma \cdot \Sigma(\sigma, \tau))$, $M \in Fam(\rho\{pair_{\sigma,\tau}\})$ we have:

- $\Sigma(\sigma,\tau)\{f\} = \Sigma(\sigma\{f\}, \tau\{f^+\})$
- $f^+ \circ pair_{\sigma,\tau} = pair_{\sigma\{f\},\tau\{f^+\}} \circ f^{++}$
- $R^{\Sigma}_{\sigma,\tau}(M)\{f^+\} = R^{\Sigma}_{\sigma\{f\},\tau\{f^+\}}(M\{f^{++}\})$

The term model is equipped with Σ-types with the obvious choice. A more specialised but simpler definition of Σ-types is the following:

Proposition 2.4.23. *A syntactic category with attributes can be equipped with Σ-types if for every $\sigma \in Fam(\Gamma)$ and $\tau \in Fam(\Gamma \cdot \sigma)$ there is a distinguished family $\Sigma(\sigma, \tau) \in Fam(\Gamma)$ and an* isomorphism *$pair_{\sigma,\tau} : \Gamma \cdot \sigma \cdot \tau \to \Gamma \cdot \Sigma(\sigma, \tau)$ with $p(\Sigma(\sigma, \tau)) \circ pair_{\sigma,\tau} = p(\sigma) \circ p(\tau)$ both stable under substitution in the sense of Definition 2.4.22.*

Proof. The elimination operation $R^{\Sigma_{\sigma,\tau}}$ is defined by substitution along the inverse of $pair_{\sigma,\tau}$.

The Σ-types obtained by virtue of the above proposition are called *extensional* for they have the property that any $P \in Sect(\Sigma(\sigma, \tau))$ is of the form $Hd(pair_{\sigma,\tau} \circ \overline{P.1}^+ \circ \overline{P.2})$ for uniquely determined sections $P.1 \in Fam(\sigma)$ and $P.2 \in Fam(\tau\{\overline{P.1}\})$.

2.4.2.6 Universes. In this section we define a general semantic framework which allows us to interpret various kinds of universes including the universe *Prop* in the Calculus of Constructions. We first present a very concise definition of semantic universes which, however, turns out to be too restrictive for many applications. Therefore we define a looser notion and give a canonical construction which allows to pass from one to the other. The Calculus of Constructions will play the role of a "running example". We give the explicit constructions for this particular case and appeal to the imagination of the reader for the generalisation to other instances of universes.

Throughout this section we assume a fixed syntactic category with attributes **C** equipped with dependent products.

Definition 2.4.24. A *full submodel* of **C** consists of a subset $Fam'(\Gamma)$ of $Fam(\Gamma)$ for each context Γ such that whenever $f : B \to \Gamma$ and $\sigma \in Fam'(\Gamma)$ then $\sigma\{f\} \in Fam'(B)$. In other words a full submodel is a subfunctor of $Fam : \mathbf{C}^{op} \to \mathbf{Sets}$.

Definition 2.4.25. A full submodel Fam' is *closed* under some type former if whenever all family arguments to the type former are in Fam' then so is the newly formed family.

For example Fam' is closed under dependent products if for $\sigma \in Fam'(\Gamma)$ and $\tau \in Fam'(\Gamma \cdot \sigma)$ the product $\Pi(\sigma, \tau)$ is in $Fam'(\Gamma)$ (instead of just in $Fam(\Gamma)$). In case the type former takes no arguments, i.e. it is a constant like **N**, then closure means that this constant is in Fam'.

Definition 2.4.26. A full submodel Fam' is closed under *impredicative quantification* if whenever $\sigma \in Fam(\Gamma)$ and $\tau \in Fam'(\Gamma{\cdot}\sigma)$ then $\Pi(\sigma, \tau) \in Fam'(\Gamma)$.

So impredicative quantification means that the product of a family in Fam' over an arbitrary family is in Fam' again.

Definition 2.4.27. A pair (U, El) is a *generic family* for a full submodel Fam' if $U \in Fam(\top)$ and $El \in Fam'(\top \cdot U)$ and for every family $\sigma \in Fam'(\Gamma)$ there exists a unique morphism $s : \Gamma \to \top \cdot U$ such that $\sigma = El\{s\}$.

Example 2.4.5. If $U \in Fam(\top)$ and $El \in Fam(\top \cdot U)$ are any two families with the property that whenever $El\{s\} = El\{s'\}$ for some $s, s' : \Gamma \to \top \cdot U$ then the assignment $Fam'(\Gamma) := \{\sigma \in Fam(\Gamma) \mid \exists s : \Gamma \to \top \cdot U \,.\, \sigma = El\{s\}\}$ defines a full submodel of $\mathbf{C}$ with generic family (U, El).

In fact by definition every full submodel with generic family is of the form given in Example 2.4.5. The reason for introducing the notion of full submodel is that it allows to state closure properties in a shorter way as becomes clear in the next definition.

Definition 2.4.28. The syntactic category with attributes $\mathbf{C}$ is a *model of the Calculus of Constructions* if it has a full submodel closed under impredicative quantification with a distinguished generic family, denoted $(\mathbf{Prop}, \mathbf{Prf})$.

Proposition 2.4.29. *Let $\mathbf{C}$ be a model of the Calculus of Constructions. Then for any context Γ, family $\sigma \in Fam(\Gamma)$ and $s : \Gamma \cdot \sigma \to \top \cdot \mathbf{Prop}$ there is a distinguished morphism $\forall_\sigma(s) : \Gamma \to \top \cdot \mathbf{Prop}$ with*

$$\mathbf{Prf}\{\forall_\sigma(s)\} = \Pi(\sigma, \mathbf{Prf}\{s\})$$

Moreover, these morphisms are stable under substitution, that is for every $f : B \to \Gamma$ and σ, s as above we have

$$\forall_\sigma(s) \circ f = \forall_{\sigma\{f\}}(s \circ f^+)$$

Proof. If $s : \Gamma \cdot \sigma \to \top \cdot \mathbf{Prop}$ then $\tau := \mathbf{Prf}\{s\}$ lies in the full submodel. Since the latter is closed under impredicative quantification the product $\Pi(\sigma, \tau)$ does too. We thus define $\forall_\sigma(s)$ as the unique morphism with $\mathbf{Prf}\{\forall_\sigma(s)\} = \Pi(\sigma, \tau)$. For stability under substitution we observe that substituting either side into $\mathbf{Prf}$ gives rise to the same family. So the equation follows by the definition of generic family.

Proposition 2.4.30. *The term model of the Calculus of Constructions is a model for same.*

Proof. The generic family is given by the pair $(\vdash Prop\,,\, x{:}Prop \vdash Prf(x))$. We must show that this is indeed a generic family, i.e. that $\Gamma \vdash Prf(M) = Prf(N)$ implies $\Gamma \vdash M = N$. We prove this by induction along derivations together with the statement that $\Gamma \vdash Prf(M) = \Pi x{:}\sigma.Prf(N)$ implies $\Gamma \vdash M = \forall x{:}\sigma.N$. So assume $\Gamma \vdash Prf(M) = Prf(N)$. If this is an instance of reflexivity then $\Gamma \vdash M = N$ by reflexivity, too. If it is an instance of the congruence rule for

Prf we must have $\Gamma \vdash M = N$ as a premise. In all other cases we can find τ such that $\Gamma \vdash \mathit{Prf}(M) = \tau$ and $\Gamma \vdash \tau = \mathit{Prf}(N)$ such that both these judgements have shorter derivations than $\Gamma \vdash \mathit{Prf}(M) = \mathit{Prf}(N)$ and moreover the last instance in the proof of $\Gamma \vdash \mathit{Prf}(M) = \tau$ is neither reflexivity, nor transitivity, and if it is symmetry then the derivation of its premise does not end with either of these three. In the first case either τ is $\mathit{Prf}(M')$ and $\Gamma \vdash M = M'$ or M is $\forall x{:}\sigma.M'$ and τ is $\Pi x{:}\sigma.\mathit{Prf}(M')$; in either case the result follows from the inductive hypothesis. All the other cases are similar.

We thus have shown that the types of the form $\mathit{Prf}(M)$ form a full submodel with generic family. They are closed under impredicative quantification by virtue of the $\forall$-operator.

2.4.2.7 Other universes. In order to define semantic counterparts to other sorts of universes we proceed in a similar way. We ask for a full submodel with a generic family closed under the desired type-forming operations. For example in order to interpret the particular universe used in Section 2.3.5 we ask for closure under dependent products and natural numbers. We can now define operators similar to $\forall_\sigma(s)$ witnessing the closure on the level of morphisms. Again using uniqueness we can show that these operators are stable under substitution. Finally, by suitably extending the argument in the proof of Proposition 2.4.30 we can show that the term model provides an instance.

The set-theoretic model admits a cumulative hierarchy of universes all closed under all the standard type formers using inaccessible cardinals. This is worked out e.g. in [68]. The ω-set model admits such a chain of universes all inside the impredicative universe of partial equivalence relations hinted at in Example 2.4.6 below even without using inaccessible cardinals. This is worked out in [1] and [6].

2.4.2.8 Loose models for universes. Unfortunately the definitions above are too restrictive to account for many natural models like the set-theoretic model and the ω-set model. For example if in the set-theoretic model we put $\mathbf{Prop} := \{\mathit{tt}, \mathit{ff}\}$ and $\mathbf{Prf}(\mathit{tt}) = \{\star\}$ and $\mathbf{Prf}(\mathit{ff}) = \emptyset$ then the induced full submodel, i.e. those families $\{\sigma_\gamma\}_{\gamma \in \Gamma}$ with $\sigma_\gamma \in \{\{\star\}, \emptyset\}$, are not closed under impredicative quantification because if $\sigma_\gamma = \{\star\}$ for all $\gamma \in \Gamma$ then the product over the family is a singleton, but not the chosen singleton $\{\star\}$. We present a laxer notion of model for the Calculus of Constructions of which the set-theoretic model forms an instance (see Example 2.4.6) together with a canonical construction which turns such a model into an actual model in the sense of the definitions given above. The notion as well as the construction generalise to other universes in a straightforward way.

Definition 2.4.31. A syntactic category with attributes $\mathbf{C}$ equipped with dependent products is a *loose model for the Calculus of Constructions* if the following are given:

- distinguished families $\mathbf{Prop} \in \mathit{Fam}(\top)$ and $\mathbf{Prf} \in \mathit{Fam}(\top \cdot \mathbf{Prop})$,
- for each $\sigma \in \mathit{Fam}(\Gamma)$ and $s : \Gamma \cdot \sigma \to \top \cdot \mathbf{Prop}$ a distinguished morphism $\forall_\sigma(s) : \Gamma \to \top \cdot \mathbf{Prop}$ and a distinguished morphism

$$ev_{\sigma,s} : \Gamma \cdot \sigma \cdot \mathbf{Prf}\{\forall_\sigma(s) \circ p(\sigma)\} \to \Gamma \cdot \sigma \cdot \mathbf{Prf}\{s\}$$

with

$$p(\mathbf{Prf}\{s\}) \circ ev_{\sigma,s} = p(\mathbf{Prf}\{\forall_\sigma(s) \circ p(\sigma)\})$$

– for each section $M \in Sect(\mathbf{Prf}\{s\})$ a distinguished section

$$\lambda_{\sigma,s}(M) \in Sect(\mathbf{Prf}\{\forall_\sigma(s)\})$$

with

$$ev_{\sigma,s} \circ \overline{\lambda_{\sigma,s}(M)\{p(\sigma)\}} = \overline{M}$$

in such a way that all these data are stable under substitution, that is for $f : B \to \Gamma$, $\sigma \in Fam(\Gamma)$, $s : \Gamma \cdot \sigma \to \top \cdot \mathbf{Prop}$, $M \in Sect(\mathbf{Prf}\{s\})$ we have:

- $\forall_\sigma(s) \circ f = \forall_{\sigma\{f\}}(s \circ f^+)$
- $ev_{\sigma,s} \circ f^{++} = f^{++} \circ ev_{\sigma\{f\},s \circ f^+}$
- $\lambda_{\sigma,s}(M)\{f\} = \lambda_{\sigma\{f\},s \circ f^+}(M\{f^+\})$

Notation. To simplify notation we shall in the sequel denote by "**Prop**" the family in $Fam(\top)$, its substitutions $\mathbf{Prop}\{!_\Gamma\} \in Fam(\Gamma)$, and its comprehension $\top \cdot \mathbf{Prop}$.

Example 2.4.6. The set-theoretic model can be turned into a loose model of the Calculus of Constructions as follows: We define **Prop** and **Prf** as in the erroneous attempt above, i.e. $\mathbf{Prop} = \{tt, ff\}$ and $\mathbf{Prf}(tt) = \{\star\}$ and $\mathbf{Prf}(ff) = \emptyset$, and for $s : \Gamma \cdot \sigma \to \mathbf{Prop}$ we define $\forall_\sigma(s)(\gamma) := tt$ if $s(\gamma, x) = tt$ for all $x \in \sigma_\gamma$ and ff otherwise. Then we can define $ev_{\sigma,s}$ as the function sending $(x, \star)$ to $(x, \star)$ in the former case and as the empty function otherwise. The other components are defined accordingly and the verifications are straightforward.

This loose model is trivial in the sense that for any two $M, N \in Sect(\mathbf{Prf}\{s\})$ for some $s : \Gamma \to \mathbf{Prop}$ we must have $M = N$. It is known that the ω-set model can be turned into a nontrivial loose model by letting **Prop** be the set of partial equivalence relations on ω endowed with the trivial realisability structure [103, 2, 58].

Proposition 2.4.32. *There exists a canonical construction which for a given loose model* **C** *for the Calculus of Constructions produces a model in the sense of Definition 2.4.28 over the same category of contexts. Moreover the new model supports whatever type former the original one did.*

Proof. The families of the new model are defined as the disjoint union

$$Fam_{new}(\Gamma) := Fam(\Gamma) \dot\cup \mathbf{C}(\Gamma, \mathbf{Prop})$$

If $f : B \to \Gamma$ then substitution on families in the left summand is inherited from the original model whereas on the right summand it is given by composition.

For $\sigma \in Fam_{new}(\Gamma)$ let $\hat\sigma$ be σ if σ is already in $Fam(\Gamma)$ and $\mathbf{Prf}\{\sigma\}$ if σ is a morphism from Γ to **Prop**. All the other components of the new model except Π are inherited from the original model by pre-composition with "$\hat{-}$".

If $\sigma \in Fam_{new}(\Gamma)$ and $\tau \in Fam(\Gamma \cdot \sigma)$ then we define the product $\Pi(\sigma, \tau)$ in the new model as $\Pi(\hat{\sigma}, \tau)$. On the other hand, if τ is a morphism from $\Gamma \cdot \sigma$ to **Prop** then the product $\Pi(\sigma, \tau)$ in the new model is defined as the morphism $\forall_{\hat{\sigma}}(\tau) : \Gamma \rightarrow \mathbf{Prop}$. For evaluation and abstraction we use either the operations from the original model or the ones which are part of the definition of a loose model. Both satisfy the required equations and are stable under substitution.

The families in the right summand, viz. the morphisms into **Prop**, thus form a full submodel closed under impredicative quantification. It has the generic family $(\mathbf{Prop}, id_{\mathbf{Prop}})$. Indeed if $s : \Gamma \rightarrow \mathbf{Prop} \in Fam_{new}(\Gamma)$ then there is a unique morphism — namely s itself — such that s arises by substitution from the generic family.

The above proposition is a bit vague as stated, because for example the promised construction might produce the same (trivial) result for every loose model. Had we defined a notion of morphism between models we could identify the construction as an equivalence.

In case the loose model we start with has the property that the families of the form $\mathbf{Prf}\{f\}$ form a full submodel with generic object, i.e. the functions $\mathbf{Prf}\{-\}$ are injective, then we can choose Fam_{new} to be Fam and define $\Pi(\sigma, \tau)$ by case distinction according to whether $\tau = \mathbf{Prf}\{t\}$ for some (then unique) t or not. A special case of this procedure appears in [103] for the particular case of the ω-set model and in [2] for a particular class of models.

2.5 Interpreting the syntax

In this section we define a semantic function which maps well-formed terms, types, and contexts of the dependently typed calculus defined in Section 2.1 to entities in a syntactic category with attributes which supports the type formers and other features present in the syntax. We proceed by first defining a partial interpretation function on pre-constructions which is then proven total on well-formed constructions by induction on derivations. This method, which circumvents the necessity to prove independence of the interpretation of a particular derivation, is due to Streicher [103]. Our treatment here differs from *loc. cit.* in that our notion of model is more general, in particular we do not require extensional Π-types, and that in *loc. cit.* only the particular example of the Calculus of Constructions is considered.

2.5.1 Partial interpretation

Assume a syntactic category with attributes **C** equipped with dependent products, natural numbers, and identity types. An *a priori partial* interpretation function $[\![-]\!]$ is defined which maps:

- pre-contexts to objects of **C**
- pairs $\Gamma \mid \sigma$, where Γ is a pre-context and σ is a pre-type, to families in $Fam([\![\Gamma]\!])$

- pairs $\Gamma \mid M$, where Γ is a pre-context and M is a pre-term to sections in $\mathit{Sect}(\sigma)$ for some $\sigma \in \mathit{Fam}(\Gamma)$.

We show below in Theorem 2.5.6 that this semantic function is defined on all contexts, types, and terms.

The semantic clauses are the following, where in order to reduce the number of cases we adopt the convention that an expression containing an undefined subexpression is itself undefined. We also adopt the convention that expressions which do not "typecheck", like $\Gamma \cdot \sigma$ if $\sigma \notin \mathit{Fam}(\Gamma)$, are undefined.

- $[\![\diamond]\!] = \top$
- $[\![\Gamma, x{:}\sigma]\!] = [\![\Gamma]\!] \cdot [\![\Gamma \mid \sigma]\!]$
- $[\![\Gamma \mid \Pi x{:}\sigma.\tau]\!] = \Pi([\![\Gamma \mid \sigma]\!], [\![\Gamma, x{:}\sigma \mid \tau]\!])$
- $[\![\Gamma \mid \mathbf{N}]\!] = \mathbf{N}_{[\Gamma]}$
- $[\![\Gamma \mid \mathit{Id}_\sigma(M,N)]\!] = \mathit{Id}_{[\Gamma \mid \sigma]}\{\overline{[\![\Gamma \mid M]\!]^+}\}\{\overline{[\![\Gamma \mid N]\!]}\}$
- $[\![\Gamma \mid x]\!] = v_{[\Gamma' \mid \sigma]}\underbrace{{}^{+\cdots+}}_{n \text{ times}}$ if $\Gamma = \Gamma', x{:}\sigma, \Gamma''$ where Γ'' has length n and x does not occur in Γ', Γ''. Undefined otherwise.
- $[\![\Gamma \mid \mathit{App}_{\sigma,[x:\sigma]\tau}(M,N)]\!] = \mathit{App}_{[\Gamma \mid \sigma],[\Gamma,x:\sigma \mid \tau]}([\![\Gamma \mid M]\!], [\![\Gamma \mid N]\!])$
- $[\![\Gamma \mid \lambda x{:}\sigma.M^\tau]\!] = \lambda_{[\Gamma \mid \sigma],[\Gamma,x:\sigma \mid \tau]}([\![\Gamma, x{:}\sigma \mid M]\!])$
- $[\![\Gamma \mid 0]\!] = 0_{[\Gamma]}$
- $[\![\Gamma \mid \mathit{Suc}(M)]\!] = \mathit{Hd}(\mathit{Suc}_{[\Gamma]} \circ \overline{[\![\Gamma \mid M]\!]})$
- $[\![\Gamma \mid R^{\mathbf{N}}(M_z, M_s, N)]\!] = R^{\mathbf{N}_{[\Gamma]}}(\overline{[\![\Gamma \mid M_z]\!]}, [\![\Gamma, x{:}\mathbf{N}, p{:}\sigma \mid M_s]\!])\{\overline{[\![\Gamma \mid N]\!]}\}$
- $[\![\Gamma \mid \mathit{Refl}_\sigma(M)]\!] = \mathit{Refl}_{[\Gamma \mid \sigma]}\{\overline{[\![\Gamma \mid M]\!]}\}$
- $[\![\Gamma \mid J_{\sigma,[xy:\sigma,p:\mathit{Id}_\sigma(x,y)]\tau}(M, N_1, N_2, P)]\!] =$
  $J_{[\Gamma \mid \sigma],[\Gamma,xy:\sigma,p:\mathit{Id}_\sigma(x,y) \mid \tau]}([\![\Gamma, x{:}\sigma \mid M]\!])\{\overline{[\![\Gamma \mid N_1]\!]^{++}}\}\{\overline{[\![\Gamma \mid N_2]\!]^+}\}\{\overline{[\![\Gamma \mid P]\!]}\}$

The n-fold weakening in the clause for variables is understood along the semantics of the n types in Γ'' in their respective contexts. Notice that the semantic combinators like $R^{\mathbf{N}}$ and J take *sections* as arguments, whereas substitution $(-\{-\})$ takes morphisms as arguments, so that when arguments are supplied via substitution, like the three last arguments to J we must use $\overline{-}$ to transform sections into morphisms.

2.5.2 Soundness of the interpretation

The aim of this section is to provide a series of Definitions and Lemmas leading up to the statement and proof of a soundness theorem for the interpretation of dependent type theory in syntactic categories with attributes. The casual reader can skip this section and proceed directly to the soundness theorem itself (Theorem 2.5.6)

Before we state and prove a soundness theorem for this interpretation we establish correspondences between syntactic and semantic substitution and weakening. In order to state these we need to define a semantic equivalent of telescopes.

Definition 2.5.1. Let **C** be a model and $\Gamma \in Ob(\mathbf{C})$. A *(semantic) telescope* over Γ is a list of families $(\sigma_1, \dots, \sigma_n)$ with $\sigma_1 \in Fam(\Gamma)$, $\sigma_2 \in Fam(\Gamma \cdot \sigma_1)$, $\dots$, $\sigma_n \in Fam(\Gamma \cdot \sigma_1 \cdot \dots \cdot \sigma_{n-1})$. The set of semantic telescopes is denoted by $Tel(\Gamma)$. The empty telescope over Γ is written $()_\Gamma$. The operations of comprehension and substitution are extended to telescopes by the following recursive definition.

- $\Gamma \cdot ()_\Gamma = \Gamma$
- $p(()_\Gamma) = id_\Gamma$
- $()_\Gamma\{f\} = ()_B$ for $f : B \to \Gamma$.
- $q(f, ()_\Gamma) = f$
- $\Gamma \cdot (\Delta, \sigma) = \Gamma \cdot \Delta \cdot \sigma$ where (Δ, σ) is the telescope Δ extended by family σ.
- $p((\Delta, \sigma)) = p(\Delta) \circ p(\sigma)$
- $(\Delta, \sigma)\{f\} = (\Delta\{f\}, \sigma\{q(f, \Delta)\})$
- $q(f, (\Delta, \sigma)) = q(q(f, \Delta), \sigma)$

By a straightforward induction on the length of telescopes we now obtain

Proposition 2.5.2. *The telescopes together with the operations defined above satisfy the premises of Proposition 2.4.6 and thus define a syntactic category with attributes where* $Sect(\Delta) =$ *"the set of sections of* $p(\Delta)$*".*

The partial interpretation function can be extended to telescopes by defining for pre-contexts Γ and $\Delta = x_1 : \tau_1, \dots, x_n : \tau_n$

$$[\![\Gamma \mid \Delta]\!] := ([\![\Gamma \mid \tau_1]\!], [\![\Gamma, x_1 : \tau_1 \mid \tau_2]\!], \dots, [\![\Gamma, x_1 : \tau_1, \dots, x_{n-1} : \tau_{n-1} \mid \tau_n]\!])$$

Clearly, if $[\![\Gamma \mid \Delta]\!]$ is defined then so is $[\![\Gamma]\!]$ and the former is a semantic telescope over the latter. Moreover, the following two properties of the extension are obvious from the definitions of $p(-)$ and $q(-,-)$ on telescopes:

$$p([\![\Gamma \mid \Delta, \Theta]\!]) = p([\![\Gamma \mid \Delta]\!]) \circ p([\![\Gamma, \Delta \mid \Theta]\!])$$

$$q(f, [\![\Gamma \mid \Delta, \Theta]\!]) = q(q(f, [\![\Gamma \mid \Delta]\!]), [\![\Gamma, \Delta \mid \Theta]\!])$$

Lemma 2.5.3 (Weakening). *Let Γ, Δ be pre-contexts, σ, τ pre-types, M a pre-term and x a fresh variable. The following equations hold if either side is defined:*

$$[\![\Gamma, x : \sigma, \Delta]\!] = [\![\Gamma, x : \sigma]\!] \cdot [\![\Gamma \mid \Delta]\!]\{p([\![\Gamma \mid \sigma]\!])\}$$

$$[\![\Gamma, x : \sigma, \Delta \mid \tau]\!] = [\![\Gamma, \Delta \mid \tau]\!]\{q(p([\![\Gamma \mid \sigma]\!]), [\![\Gamma \mid \Delta]\!])\}$$

$$[\![\Gamma, x : \sigma, \Delta \mid M]\!] = [\![\Gamma, \Delta \mid M]\!]\{q(p([\![\Gamma \mid \sigma]\!]), [\![\Gamma \mid \Delta]\!])\}$$

Proof. An *instance* of this lemma is one of the left hand sides of the above equations. The *weight* of such an instance is the number of symbols required to write it down including variables and punctuation symbols. Thus for example the weight of $[\![\Gamma, x : \sigma, \Delta, y : \tau]\!]$ is greater than the weight of $[\![\Gamma, x : \sigma \mid \tau]\!]$ because y is not contained in the latter. We also need the notion of *length* of a pre-context Δ denoted $|\Delta|$ which is the number of variable declarations it contains. We proceed by induction on the weight and give a representative selection of the many cases required.

First consider the case where $\Delta = \diamond$. If either side of the instance is defined then $p(\llbracket \Gamma \mid \sigma \rrbracket)$ must be defined, too. Now $\llbracket \Gamma, x{:}\,\sigma, \Delta \rrbracket = \llbracket \Gamma, x{:}\,\sigma \rrbracket = \llbracket \Gamma, x{:}\,\sigma \rrbracket \cdot \llbracket \Gamma \mid \Delta \rrbracket\{p(\llbracket \Gamma \mid \sigma \rrbracket)\}$ because $\llbracket \Gamma \mid \Delta \rrbracket = ()_{\llbracket \Gamma \rrbracket}$ by definition.

If $\Delta = \Delta', y{:}\,\tau$ then

$$
\begin{array}{ll}
 & \llbracket \Gamma, x{:}\,\sigma, \Delta \rrbracket \\
= & \llbracket \Gamma, x{:}\,\sigma, \Delta' \rrbracket \cdot \llbracket \Gamma, x{:}\,\sigma, \Delta' \mid \tau \rrbracket \qquad \text{by definition} \\
= & \llbracket \Gamma, x{:}\,\sigma \rrbracket \cdot \llbracket \Gamma \mid \Delta' \rrbracket\{p(\llbracket \Gamma \mid \sigma \rrbracket)\} \cdot \llbracket \Gamma, \Delta' \mid \tau \rrbracket\{q(p(\llbracket \Gamma \mid \sigma \rrbracket), \llbracket \Gamma \mid \Delta' \rrbracket)\} \\
= & \llbracket \Gamma, x{:}\,\sigma \rrbracket \cdot (\, \llbracket \Gamma \mid \Delta' \rrbracket\{p(\llbracket \Gamma \mid \sigma \rrbracket)\}\, ,\, \llbracket \Gamma, \Delta' \mid \tau \rrbracket\{q(p(\llbracket \Gamma \mid \sigma \rrbracket), \llbracket \Gamma \mid \Delta' \rrbracket)\}\,) \\
= & \llbracket \Gamma, x{:}\,\sigma \rrbracket \cdot \llbracket \Gamma \mid \Delta \rrbracket\{p(\llbracket \Gamma \mid \sigma \rrbracket)\}
\end{array}
$$

The last two steps follow from the definition of substitution and comprehension on semantic telescopes. The inductive hypothesis has been used in the third step on the instances $\llbracket \Gamma, x{:}\,\sigma, \Delta' \mid \tau \rrbracket$ and $\llbracket \Gamma, x{:}\,\sigma, \Delta' \rrbracket$ which both have smaller weight.

Next we consider the instance $\llbracket \Gamma, x{:}\,\sigma, \Delta \mid \mathbf{N} \rrbracket$. We have

$$
\begin{array}{ll}
 & \llbracket \Gamma, x{:}\,\sigma, \Delta \mid \mathbf{N} \rrbracket \\
= & \mathbf{N}_{\llbracket \Gamma, x:\sigma, \Delta \rrbracket} \\
= & \mathbf{N}_{\llbracket \Gamma, x:\sigma \rrbracket \cdot \llbracket \Gamma \mid \Delta \rrbracket\{p(\llbracket \Gamma \mid \sigma \rrbracket)\}} \qquad \text{by IH} \\
= & \mathbf{N}_{\llbracket \Gamma, \Delta \rrbracket}\{q(p(\llbracket \Gamma \mid \sigma \rrbracket), \llbracket \Gamma \mid \Delta \rrbracket)\} \\
= & \llbracket \Gamma, \Delta \mid \mathbf{N} \rrbracket\{q(p(\llbracket \Gamma \mid \sigma \rrbracket), \llbracket \Gamma \mid \Delta \rrbracket)\}
\end{array}
$$

For the penultimate step we have used that $q(p(\llbracket \Gamma \mid \sigma \rrbracket), \llbracket \Gamma \mid \Delta \rrbracket)$ is a morphism from $\llbracket \Gamma, x{:}\,\sigma \rrbracket \cdot \llbracket \Gamma \mid \Delta \rrbracket\{p(\llbracket \Gamma \mid \sigma \rrbracket)\}$ to $\llbracket \Gamma \rrbracket \cdot \llbracket \Gamma \mid \Delta \rrbracket = \llbracket \Gamma, \Delta \rrbracket$ and the fact that $\mathbf{N}_\Gamma\{f\} = \mathbf{N}_B$ for *any* morphism $f : B \to \Gamma$.

Now we consider the instance $\llbracket \Gamma, x{:}\,\sigma, \Delta \mid \Pi y{:}\,\tau.\rho \rrbracket$. We have

$$
\begin{array}{ll}
 & \llbracket \Gamma, x{:}\,\sigma, \Delta \mid \Pi y{:}\,\tau.\rho \rrbracket \\
= & \Pi(\, \llbracket \Gamma, x{:}\,\sigma, \Delta \mid \tau \rrbracket\, ,\, \llbracket \Gamma, x{:}\,\sigma, \Delta, y{:}\,\tau \mid \rho \rrbracket\,) \\
= & \Pi(\;\; \llbracket \Gamma, \Delta \mid \tau \rrbracket\{q(p(\llbracket \Gamma \mid \sigma \rrbracket), \llbracket \Gamma \mid \Delta \rrbracket)\}, \\
 & \qquad \llbracket \Gamma, \Delta, y{:}\,\tau \mid \rho \rrbracket\{q(p(\llbracket \Gamma \mid \sigma \rrbracket), \llbracket \Gamma \mid \Delta, y{:}\,\tau \rrbracket)\}) \\
= & \Pi(\;\; \llbracket \Gamma, \Delta \mid \tau \rrbracket\{q(p(\llbracket \Gamma \mid \sigma \rrbracket), \llbracket \Gamma \mid \Delta \rrbracket)\}, \\
 & \qquad \llbracket \Gamma, \Delta, y{:}\,\tau \mid \rho \rrbracket\{q(q(p(\llbracket \Gamma \mid \sigma \rrbracket), \llbracket \Gamma \mid \Delta \rrbracket), \llbracket \Gamma, \Delta \mid y{:}\,\tau \rrbracket)\}) \\
= & \Pi(\llbracket \Gamma, \Delta \mid \tau \rrbracket, \llbracket \Gamma, \Delta, y{:}\,\tau \mid \rho \rrbracket)\{q(p(\llbracket \Gamma \mid \sigma \rrbracket), \llbracket \Gamma \mid \Delta \rrbracket)\} \\
= & \llbracket \Gamma, \Delta \mid \Pi y{:}\,\tau.\rho \rrbracket\{q(p(\llbracket \Gamma \mid \sigma \rrbracket), \llbracket \Gamma \mid \Delta \rrbracket)\}
\end{array}
$$

In the second step we have used the IH on the instances $\llbracket \Gamma, \Delta \mid \tau \rrbracket$ and $\llbracket \Gamma, \Delta, y{:}\, \tau \mid \rho \rrbracket$ both of which have smaller weight than the original instance. We see here the need for the rather involved structure of the inductive argument. The third step uses the definition of the $q(-,-)$ operation on semantic telescopes and the fourth step is a consequence of the stability of Π under substitution.

The instance $\llbracket \Gamma, x{:}\,\sigma, \Delta \mid Id_\tau(M, N) \rrbracket$ is similar. We expand the definition and use the IH on the instances $\llbracket \Gamma, x{:}\,\sigma, \Delta \mid \tau \rrbracket$, $\llbracket \Gamma, x{:}\,\sigma, \Delta \mid M \rrbracket$, and $\llbracket \Gamma, x{:}\,\sigma, \Delta \mid N \rrbracket$ all of which have smaller weight.

Let us now consider the instance $\llbracket \Gamma, x{:}\,\sigma, \Delta \mid y \rrbracket$ where y is a variable. We observe that using our notation for telescopes and functoriality of substitution we can rewrite the iterated weakening in the semantic clause for variables as follows:

$$\llbracket\Gamma', y{:}\,\tau, \Gamma'' \mid y\rrbracket = \upsilon_{\llbracket\Gamma'|\tau\rrbracket}\{p(\llbracket\Gamma', y{:}\,\tau \mid \Gamma''\rrbracket)\}$$

where y does not occur in Γ', Γ''. Suppose now that $\llbracket\Gamma, x{:}\,\sigma, \Delta \mid y\rrbracket$ is defined. Then either $\Gamma = \Gamma', y{:}\,\tau, \Gamma''$ for some type τ and y does not occur in $\Gamma', \Gamma'', \Delta$ or $\Delta = \Delta', y{:}\,\tau, \Delta''$ and y does not not occur in $\Gamma, \Delta', \Delta''$. In the former case we calculate as follows:

$$\begin{aligned}
& \llbracket\Gamma', y{:}\,\tau, \Gamma'', x{:}\,\sigma, \Delta \mid y\rrbracket \\
= \;& \upsilon_{\llbracket\Gamma'|\tau\rrbracket}\{p(\llbracket\Gamma', y{:}\,\tau \mid \Gamma'', x{:}\,\sigma, \Delta\rrbracket)\} \\
= \;& \upsilon_{\llbracket\Gamma'|\tau\rrbracket}\{p(\llbracket\Gamma', y{:}\,\tau \mid \Gamma''\rrbracket) \circ p(\llbracket\Gamma \mid \sigma\rrbracket) \circ p(\llbracket\Gamma, x{:}\,\sigma \mid \Delta\rrbracket)\} \\
= \;& \ldots
\end{aligned}$$

Now we notice that

$$\llbracket\Gamma, x{:}\,\sigma \mid \Delta\rrbracket = \llbracket\Gamma \mid \Delta\rrbracket\{p(\llbracket\Gamma \mid \sigma\rrbracket)\}$$

by induction on $|\Delta|$ and the IH. This in turn implies that

$$p(\llbracket\Gamma \mid \sigma\rrbracket) \circ p(\llbracket\Gamma, x{:}\,\sigma \mid \Delta\rrbracket) = p(\llbracket\Gamma \mid \Delta\rrbracket) \circ q(p(\llbracket\Gamma \mid \sigma\rrbracket), \llbracket\Gamma \mid \Delta\rrbracket)$$

because semantic telescopes form a syntactic category with attributes. We can thus continue the calculation as follows:

$$\begin{aligned}
& \ldots \\
= \;& \upsilon_{\llbracket\Gamma'|\tau\rrbracket}\{p(\llbracket\Gamma', y{:}\,\tau \mid \Gamma''\rrbracket) \circ p(\llbracket\Gamma \mid \Delta\rrbracket) \circ q(p(\llbracket\Gamma \mid \sigma\rrbracket), \llbracket\Gamma \mid \Delta\rrbracket)\} \\
= \;& \upsilon_{\llbracket\Gamma'|\tau\rrbracket}\{p(\llbracket\Gamma, y{:}\,\tau \mid \Gamma'', \Delta\rrbracket) \circ q(p(\llbracket\Gamma \mid \sigma\rrbracket), \llbracket\Gamma \mid \Delta\rrbracket)\} \\
= \;& \llbracket\Gamma', y{:}\,\tau, \Gamma'', \Delta \mid y\rrbracket\{q(p(\llbracket\Gamma \mid \sigma\rrbracket), \llbracket\Gamma \mid \Delta\rrbracket)\}
\end{aligned}$$

On the other hand if $\Delta = \Delta', y{:}\,\tau, \Delta''$ and y does not occur in $\Gamma, \Delta', \Delta''$ then

$$\begin{aligned}
& \llbracket\Gamma, x{:}\,\sigma, \Delta', y{:}\,\tau, \Delta'' \mid y\rrbracket = \\
= \;& \upsilon_{\llbracket\Gamma, x:\sigma, \Delta'|\tau\rrbracket}\{p(\llbracket\Gamma, x{:}\,\sigma, \Delta', y{:}\,\tau \mid \Delta''\rrbracket)\} \\
= \;& \upsilon_{\llbracket\Gamma, \Delta'|\tau\rrbracket\{q(p(\llbracket\Gamma|\sigma\rrbracket), \llbracket\Gamma|\Delta'\rrbracket)\}}\{p(\llbracket\Gamma, x{:}\,\sigma, \Delta', y{:}\,\tau \mid \Delta''\rrbracket)\} && \text{by IH} \\
= \;& \upsilon_{\llbracket\Gamma, \Delta'|\tau\rrbracket}\{q(p(\llbracket\Gamma \mid \sigma\rrbracket), \llbracket\Gamma \mid \Delta', y{:}\,\tau\rrbracket)\}\{p(\llbracket\Gamma, x{:}\,\sigma, \Delta', y{:}\,\tau \mid \Delta''\rrbracket)\} \\
= \;& \ldots
\end{aligned}$$

by Lemma 2.4.1.3 and the definition of $q(-,-)$ on telescopes. Now again by induction on $|\Delta''|$ and the IH we conclude that

$$\llbracket\Gamma, x{:}\,\sigma, \Delta', y{:}\,\tau \mid \Delta''\rrbracket = \llbracket\Gamma, \Delta', y{:}\,\tau \mid \Delta''\rrbracket\{q(p(\llbracket\Gamma \mid \sigma\rrbracket), \llbracket\Gamma \mid \Delta', y{:}\,\tau\rrbracket)\}$$

and thus

$$\begin{aligned}
& q(p(\llbracket\Gamma \mid \sigma\rrbracket), \llbracket\Gamma \mid \Delta', y{:}\,\tau\rrbracket) \circ p(\llbracket\Gamma, x{:}\,\sigma, \Delta', y{:}\,\tau \mid \Delta''\rrbracket) = \\
& p(\llbracket\Gamma, \Delta', y{:}\,\tau \mid \Delta''\rrbracket) \circ q(p(\llbracket\Gamma \mid \sigma\rrbracket), \llbracket\Gamma \mid \Delta', y{:}\,\tau, \Delta''\rrbracket)
\end{aligned}$$

using the commutativity of the square corresponding to the substitution $\llbracket\Gamma, \Delta', y{:}\,\tau \mid \Delta''\rrbracket\{q(p(\llbracket\Gamma \mid \sigma\rrbracket), \llbracket\Gamma \mid \Delta', y{:}\,\tau\rrbracket)\}$ and the observation on nested $q(-,-)$-morphisms on telescopes made above. Now we can finish the calculation as follows

$$\begin{aligned} &\dots \\ = \quad & v_{[\![\Gamma,\Delta'|\tau]\!]}\{p([\![\Gamma,\Delta',y{:}\tau \mid \Delta'']\!])\}\{q(p([\![\Gamma \mid \sigma]\!]),[\![\Gamma \mid \Delta]\!])\} \\ = \quad & [\![\Gamma,\Delta \mid y]\!]\{q(p([\![\Gamma \mid \sigma]\!]),[\![\Gamma \mid \Delta]\!])\} \end{aligned}$$

If the right hand side of the equation for variables is defined then we perform the same case distinction as to whether y occurs in Γ or Δ. Since all the steps in the above calculation are valid if either side is defined this implies that the left hand side is defined, too.

The instances of the form $[\![\Gamma,x{:}\sigma,\Delta \mid M]\!]$ are similar to the instances of the form $[\![\Gamma,x{:}\sigma\Delta \mid \tau]\!]$. If M is a constant then we follow the proof for $\mathbf{N}$, and if the outermost constructor of M is a binder then we argue as in the case of Π.

Lemma 2.5.4 (Substitution). *Let Γ,Δ be pre-contexts, σ,τ pre-types, M,N pre-terms and x a fresh variable. We denote the substitution of M for x in some pre-construction C by $C[M]$. The expression $\overline{[\![\Gamma \mid M]\!]}$ is abbreviated by m. If $[\![\Gamma \mid \sigma]\!]$ and $[\![\Gamma \mid M]\!]$ are both defined and the latter is a section of the former then the following equations hold if either side is defined:*

$$[\![\Gamma,\Delta[M]]\!] = [\![\Gamma]\!]\cdot[\![\Gamma,x{:}\sigma \mid \Delta]\!]\{\overline{[\![\Gamma \mid M]\!]}\}$$

$$[\![\Gamma,\Delta[M] \mid \tau[M]]\!] = [\![\Gamma,x{:}\sigma,\Delta \mid \tau]\!]\{q(\overline{[\![\Gamma \mid M]\!]},[\![\Gamma,x{:}\sigma \mid \Delta]\!])\}$$

$$[\![\Gamma,\Delta[M] \mid N[M]]\!] = [\![\Gamma,x{:}\sigma,\Delta \mid N]\!]\{q(\overline{[\![\Gamma \mid M]\!]},[\![\Gamma,x{:}\sigma \mid \Delta]\!])\}$$

Proof. Again we proceed by simultaneous induction on the weights of the three instances. We abbreviate the morphism $\overline{[\![\Gamma \mid M]\!]}$ by m. It now plays the role of the "weakening morphism" $p([\![\Gamma \mid \sigma]\!])$ in Lemma 2.5.3. There are only two cases the proof of which differs from the one of their companion in Lemma 2.5.3. The first one is the instance $[\![\Gamma,\diamond[M]]\!]$. This equals $[\![\Gamma]\!]\cdot[\![\Gamma,x{:}\sigma \mid \diamond]\!]\{m\}$ iff $[\![\Gamma \mid M]\!]$ is defined and is a section of $[\![\Gamma \mid \sigma]\!]$ because otherwise the substitution would not typecheck and would thus be undefined. This is actually the only place where this assumption is needed; it is propagated to the other cases through the inductive process.

The second case is the instance where N is the variable x, i.e.

$$[\![\Gamma,\Delta[M] \mid M]\!] = [\![\Gamma,x{:}\sigma,\Delta \mid x]\!]\{q(m,[\![\Gamma,x{:}\sigma \mid \Delta]\!])\}$$

Notice here that $x[M]=M$. By the inductive hypothesis and by induction on $|\Delta|$ we may assume that

$$[\![\Gamma \mid \Delta[M]]\!] = [\![\Gamma,x{:}\sigma \mid \Delta]\!]\{m\}$$

and thus

$$m \circ p([\![\Gamma \mid \Delta[M]]\!]) = p([\![\Gamma,x{:}\sigma \mid \Delta]\!]) \circ q(m,[\![\Gamma,x{:}\sigma \mid \Delta]\!])$$

by the commutativity of the square corresponding to the substitution $[\![\Gamma,x{:}\sigma \mid \Delta]\!]\{m\}$. Now we calculate as follows:

$$\begin{array}{lll} & [\![\Gamma, \Delta[M] \mid M]\!] & \\ = & [\![\Gamma \mid M]\!]\{p([\![\Gamma \mid \Delta[M]]\!])\} & \text{by } |\Delta|\text{-fold application of Lemma 2.5.3} \\ = & v_{[\![\Gamma|\sigma]\!]}\{m \circ p([\![\Gamma \mid \Delta[M]]\!])\} & \text{by Lemma 2.4.9} \\ = & v_{[\![\Gamma|\sigma]\!]}\{p([\![\Gamma, x{:}\sigma \mid \Delta]\!]) \circ q(m, [\![\Gamma, x{:}\sigma \mid \Delta]\!])\} & \text{as argued above} \\ = & [\![\Gamma, x{:}\sigma, \Delta \mid x]\!]\{q(m, [\![\Gamma, x{:}\sigma \mid \Delta]\!])\} & \text{by the clause for variables} \end{array}$$

The proofs of all the other cases almost literally follow the proof of Lemma 2.5.3 and are thus left out.

Substitution for arbitrary context morphisms. We can extend the interpretation function to (syntactic) context morphisms (together with their domain and codomain) as follows: If Γ, Δ are pre-contexts, f a tuple of pre-terms and M a pre-term then

- $[\![\Gamma \mid () \mid \diamond]\!] = !_{[\![\Gamma]\!]}$
- $[\![\Gamma \mid (f, M) \mid \Delta, x{:}\sigma]\!] = q([\![\Delta \mid \sigma]\!], [\![\Gamma \mid f\,\Delta]\!]) \circ \overline{[\![\Gamma \mid M]\!]}$

where as before we adopt the convention that expressions which do not type-check are undefined. We now have the following general substitution property.

Lemma 2.5.5 (General substitution). *Let B, Γ, Δ be pre-contexts and f a $|\Gamma|$-tuple of pre-terms, M a pre-term, and σ a pre-type. We denote the simultaneous substitution of the Δ-variables by f in some pre-construction C by $C[f]$. If $[\![B \mid f \mid \Gamma]\!]$ is a morphism from $[\![B]\!]$ to $[\![\Gamma]\!]$ then the following equations hold if either side is defined:*

$$[\![B, \Delta[f]]\!] = [\![B]\!] \cdot [\![\Gamma \mid \Delta]\!]\{[\![B \mid f \mid \Gamma]\!]\}$$

$$[\![B, \Delta[f] \mid \sigma[f]]\!] = [\![\Gamma, \Delta \mid \sigma]\!]\{q([\![B \mid f \mid \Gamma]\!], \Gamma \mid \Delta)\}$$

$$[\![B, \Delta[f] \mid M[f]]\!] = [\![\Gamma, \Delta \mid M]\!]\{q([\![B \mid f \mid \Gamma]\!], \Gamma \mid \Delta)\}$$

Proof. The proof is by induction on the length of f. If $f = ()$ and thus $\Gamma = \diamond$ then the first equation is satisfied by definition of substitution on semantic telescopes. The second equation becomes

$$[\![B, \Delta \mid \sigma]\!] = [\![\Delta \mid \sigma]\!]\{q([\![B \mid () \mid \diamond]\!], [\![\diamond \mid \Delta]\!])\}$$

From a $|B|$-fold application of Lemma 2.5.3 we get

$$[\![B, \Delta \mid \sigma]\!] = [\![\Delta \mid \sigma]\!]\{q(p(\diamond \mid B), [\![\diamond \mid \Delta]\!])\}$$

where we have used compositionality of $q(-,-)$ and the definition of $p(-)$ on telescopes. But now $p(\diamond \mid B)$ as a morphism from $[\![B]\!]$ to $\top$ must equal $!_{[\![\top]\!]}$ by the unicity property of the terminal object. The third equation is analogous.

Now consider the case where $f = (f', N)$ and $\Gamma = \Gamma', y{:}\tau$. If $[\![B \mid f \mid \Gamma]\!]$ is defined and is a morphism from B to Γ then $[\![B \mid N]\!]$ must be defined and be a section of $[\![\Gamma \mid \tau]\!]\{[\![B \mid f' \mid \Gamma]\!]\}$ which by the IH is equal to $[\![B \mid \tau[f']]\!]$. Now for the first equation we calculate as follows:

$$\begin{aligned}
& [\![B, \Delta[(f', N)]]\!] \\
= \; & [\![B, \Delta[f'][y := N]]\!] && \text{by definition of parallel substitution} \\
= \; & [\![B]\!] \cdot [\![B, y\colon \tau[f'] \mid \Delta[f']]\!]\{\overline{[\![B \mid N]\!]}\} && \text{by Lemma 2.5.4} \\
= \; & [\![B]\!] \cdot [\![\Gamma', y\colon \tau \mid \Delta]\!]\{q([\![B \mid f' \mid \Gamma']\!], [\![\Gamma' \mid \tau]\!])\}\{\overline{[\![B \mid N]\!]}\} \\
= \; & [\![B]\!] \cdot [\![\Gamma \mid \Delta]\!]\{[\![B \mid (f', N) \mid \Gamma', y\colon \tau]\!]\}
\end{aligned}$$

Here the second last step uses induction on $|\Delta|$ and the inductive hypothesis, whereas the last step follows by compositionality of $-\{-\}$ and the definition of the interpretation of nonempty context morphisms.

The other cases are similar.

We remark that the inductive argument above does not go through if we restrict ourselves to $\Delta = \diamond$ and thus $q([\![B \mid f \mid \Gamma]\!], [\![\Gamma \mid \Delta]\!]) = [\![B \mid f \mid \Gamma]\!]$.

We have tried to prove the general substitution lemma directly by induction on the weight of the instances. Then all the inductive cases are very much the same as, if not easier than in the weakening case (Lemma 2.5.3), but the case where M is a variable becomes extremely complicated and seems to require weakening and substitution for terms in the first place.

We can now establish a correspondence between syntax and semantics.

Theorem 2.5.6. *The interpretation function enjoys the following soundness properties*

- *If* $\Gamma \vdash$ *then* $[\![\Gamma]\!]$ *is an object of* $\mathbf{C}$.
- *If* $\Gamma \vdash \sigma$ *then* $[\![\Gamma \mid \sigma]\!]$ *is an element of* $Fam([\![\Gamma]\!])$.
- *If* $\Gamma \vdash M : \sigma$ *then* $[\![\Gamma \mid M]\!]$ *is an element of* $Sect([\![\Gamma \mid \sigma]\!])$.
- *If* $\Gamma \vdash$ *and* $\Delta \vdash$ *and* $\Gamma \vdash f : \Delta$ *then* $[\![\Gamma \mid f \mid \Delta]\!] \in \mathbf{C}([\![\Gamma]\!], [\![\Delta]\!])$.
- *If* $\vdash \Gamma = \Delta$ *then* $[\![\Gamma]\!] = [\![\Delta]\!]$.
- *If* $\Gamma \vdash \sigma = \tau$ *then* $[\![\Gamma \mid \sigma]\!] = [\![\Gamma \mid \tau]\!]$.
- *If* $\Gamma \vdash M = N : \sigma$ *then* $[\![\Gamma \mid M]\!] = [\![\Gamma \mid N]\!]$

Proof. The proof is by induction on derivations. Again we only treat selected cases to give the general idea. The cases for context and type formation rules are fairly simple applications of the IH. Assume for example that $\Gamma \vdash \Pi x : \sigma.\tau[x]$ has been derived from $\Gamma \vdash \sigma$ and $\Gamma, x\colon\sigma \vdash \tau[x]$ using Π-Form. Then we may assume that $S := [\![\Gamma \mid \sigma]\!] \in Fam([\![\Gamma]\!])$ and $T := [\![\Gamma, x\colon\sigma \mid \tau]\!] \in Fam([\![\Gamma, x\colon\sigma]\!])$. Now $[\![\Gamma, x\colon\sigma]\!] = [\![\Gamma]\!] \cdot S$ by the semantic clause for context extension. So $\Pi(S, T)$ is defined and equals $[\![\Gamma \mid \Pi x\colon\sigma.\tau]\!]$ by the clause for Π-types.

More interesting are rules involving syntactic substitution. Assume for example that $\Gamma \vdash App_{\sigma,[x:\sigma]\tau}(M, N) : \tau[N]$ has been derived from $\Gamma \vdash M : \Pi x\colon\sigma.\tau[x]$ and $\Gamma \vdash N : \sigma$ using Π-Elim. Then by induction we may assume that $[\![\Gamma \mid M]\!] \in Fam([\![\Gamma \mid \Pi x\colon\sigma.\tau[x]]\!])$ and $[\![\Gamma \mid N]\!] \in Fam([\![\Gamma \mid \sigma]\!])$. By expanding some definitions and using the IH we get from this that $[\![\Gamma \mid App_{\sigma,[x:\sigma]\tau}(M, N)]\!]$ is defined and is a section of the family $[\![\Gamma, x\colon\sigma \mid \tau[x]]\!]\{\overline{[\![\Gamma \mid N]\!]}\}$. We now use Lemma 2.5.4 to conclude that this family equals $[\![\Gamma \mid \tau[N]]\!]$ as required.

We finish our presentation of categorical semantics of dependent type theory with a trivial completeness theorem for syntactic categories with attributes.

Theorem 2.5.7. *Let Γ, Δ be pre-contexts, σ, τ be pre-types, and M, N be pre-terms.*

- *If $[\![\Gamma]\!]$ is defined in all interpretations then $\Gamma \vdash$.*
- *If $[\![\Gamma \mid \sigma]\!]$ is defined in all interpretations then $\Gamma \vdash \sigma$.*
- *If $[\![\Gamma \mid M]\!]$ is defined in all interpretations then $\Gamma \vdash M : \sigma$ for some (unique) σ.*
- *If $[\![\Gamma]\!] = [\![\Delta]\!]$ in all interpretations then $\vdash \Gamma = \Delta$.*
- *If $[\![\Gamma \mid \sigma]\!] = [\![\Gamma \mid \tau]\!]$ in all interpretations then $\Gamma \vdash \sigma = \tau$.*
- *If $[\![\Gamma \mid M]\!] = [\![\Gamma \mid N]\!]$ in all interpretations then $\Gamma \vdash M = N : \sigma$ for some unique σ.*

Proof. Using the term model from Example 2.4.1 and induction on the definition of the interpretation function.

Remark 2.5.8. One can define *morphisms* between syntactic categories with attributes in the obvious way as functions on contexts, morphisms, families, and sections preserving all the structure up to equality. In this way syntactic categories with attributes supporting a given set of type formers form a category in which the respective term model is initial.

2.6 Discussion and related work

Our formulation of the syntax differs from the more modern one in [88] in that it is not based on a logical framework. In the latter approach one first defines a type theory with Π-types (written $(x{:}\sigma)\tau$) and one universe (called *Set*) the way we have done and then one introduces all further constructs as constants of the appropriate type, e.g. for Σ-types one would have a constant $\Sigma : (S{:}\,Set)(T{:}\,(s{:}\,El(S))Set)Set$. In this way many of the rules can be simplified and less meta-notation (like substitution) is needed. The reason why we have not used this notation was that the categorical semantics then becomes more complicated. In retrospect we think that it might have been better to pay this price in exchange for a neater and more up-to-date syntax.

Another basic characteristic of our presentation of the syntax is the handling of substitution as a defined operation on the raw syntax. Recently, there has been an interest in treating substitution as part of the syntax and giving reduction (or equality rules) for it ([23, 24] and unpublished notes by Per Martin-Löf). We have not used this presentation simply because it does not yet seem to be sufficiently settled, see for instance [79].

The categorical semantics of dependent type theory has attracted quite some interest, see [15, 58, 89, 31, 55] for instance, but mostly from an abstract categorical point of view. If we can claim any originality for the material in this chapter then it is for having tried to de-mystify the categorical semantics and to relate it as closely to the syntax as possible. Let us summarise the most important approaches to categorical semantics in the literature. The subject started with Cartmell's pioneering work [15] who invented the notions of *contextual categories* and *categories with attributes*. Categories with attributes have been described in Remark 2.4.7; in contextual categories the "families" are replaced by an additional tree structure on the category of contexts. This latter notion underlies Streicher's work [102, 103] on categorical semantics of the Calculus of Constructions, where the interpretation of the syntax is defined formally and the method of partially interpreting pre-constructions appears for the first time.

In these two notions of model, equality of families and compatibility of the type formers with substitution up to equality is crucial (this is known as the "split case"). In other approaches to categorical semantics this has been weakened to canonical isomorphism (the non-split case), e.g. in locally cartesian closed categories [99], models based on fibrations [31, 58], and display map categories [55, 106]. The interpretation of the syntax in such structures is not obvious at all, however, a fact which has astonishingly been neglected by many authors. An exception is Ehrhard [31] who restricts to the split case and then uses Streicher's method to interpret the syntax of the Calculus of Constructions in his fibrational models. The problem of interpreting the syntax in non-split categorical structures has only recently been taken up by Curien using an intermediate syntax with explicit substitutions [24] and by the author using a categorical construction that turns a non-split structure into a split one [47]. In both approaches problems (in particular the treatment of universes) remain, so that one can conclude that at the current state-of-the-art the non-split models do not properly correspond to the syntax.

Another generalisation offered by models based on fibrations is that morphisms between families are primitive rather than defined through context comprehension as special context morphisms. This has the advantage that context comprehension satisfies a universal property with respect to these morphisms [31] and therefore—if it exists—it is unique up to isomorphism. Morphisms between families also permit a more natural characterisation of Π- and Σ-types as certain adjoint functors. Again, this characterises them up to isomorphism so that the presence of these type formers becomes a property rather than additional structure. However, this only works for models of extensional type theory.

It has already been said that most of the material in this chapter is not original. However, since no homogeneous treatment covering both syntax and semantics of dependent type theory exists in the literature, let alone a textbook (this may be remedied by the forthcoming [92]), we considered it necessary to give a rather detailed account. As novelties, albeit implicit in existing literature, we consider the equational presentation of categories with attributes, the emphasis on stability under substitution rather than universal properties in the formulation of the semantic type formers, e.g. the identity type, the construction achieving equality between the semantics of $Prf(\forall x{:}\,\sigma.S)$ and $\Pi x{:}\,\sigma.Prf(S)$ in Proposition 2.4.32, and the explicit use of telescopes in the correctness proof for the interpretation function.

3. Syntactic properties of propositional equality

The main focus of this Chapter are the proofs of two important properties of extensional type theory, i.e. type theory with propositional and definitional equality identified. The first property is undecidability of this theory; in view of the information loss in the equality reflection rule an obvious thing, which is nevertheless not completely trivial to prove (Section 3.2.2). The second property is the conservativity of extensional type theory over intensional type theory with extensional constructs (Section 3.2.5). These two properties constitute the major justification for the use of intensional type theory with extensional constructs.

On the way we develop some notions and concepts which will be required later such as the formulation of propositional equality using uniqueness of identity and a Leibniz property (Section 3.2.3.1) and the syntactic formulation of extensional constructs relevant in Martin-Löf type theory without a universe of propositions (Sects 3.1.2, 3.1.3, 3.2.6.1). An impatient reader may skip the other parts of this chapter without affecting the understandability of the rest of the thesis.

3.1 Intensional type theory

Intensional type theory is the one described in Section 2.1. Here we present some derived combinators and operations for the inductive identity type Id_σ defined there (Section 3.1.1). Next we look at the extensional constructs of uniqueness of identity (Section 3.1.2) and functional extensionality (Section3.1.3) and establish some syntactic properties of these.

3.1.1 Substitution

From the elimination operator J we can define an operator which allows us to replace propositionally equal objects in a dependent type. Let $x{:}\sigma \vdash \tau[x]$ and $M, M' : \sigma$ and $P : Id_\sigma(M, M')$ and $N : \tau[M]$. We then define

$$\begin{aligned} &\mathit{Subst}_{\sigma,\tau}(M, M', P, N) := \\ &\qquad J_{\sigma,[x,y:\sigma][p:Id_\sigma(x,y)]\tau[x]\to\tau[y]}([x{:}\sigma]\lambda h{:}\tau[x].h\ ,\ M\ ,\ M'\ ,\ P)\ N : \tau[M'] \end{aligned}$$

By the equality rule Id-Comp we have

$$Subst_{\sigma,\tau}(M, M, Refl_\sigma(M), N) = N : \tau[M] \tag{3.1}$$

Intuitively, *Subst* "replaces" M by the propositionally equal M' in the type $\tau[M]$ of N. As opposed to extensional type theory every application of propositional equality must be recorded in the terms, we cannot simply write $N : \tau[M']$, but $Subst_{\sigma,\tau}(M, M', P, N) : \tau[M']$ as indicated.

Since the first two arguments to *Subst* can be inferred from the third one we often suppress these and write $Subst_{\sigma,\tau}(P, N)$. Also, as usual, type annotations may be left out. The same convention applies to other combinators and definitional extensions which we introduce later on.

It is slightly unpleasant that the function type has to be used in order to derive a principle as basic as substitution. This may be avoided by replacing J by another primitive elimination operator J' governed by the typing rule

$$\frac{\begin{array}{l}\Gamma\,,\,x,y:\sigma\,,\,p:Id_\sigma(x,y)\vdash\tau[x,y,p]\\ \Gamma\vdash N_1:\sigma \qquad \Gamma\vdash N_2:\sigma\\ \Gamma\vdash M:\tau[N_1,N_1,Refl_\sigma(N_1)]\\ \Gamma\vdash P:Id_\sigma(N_1,N_2)\end{array}}{\Gamma\vdash J'_{\sigma,\tau}(M,N_1,N_2,P):\tau[N_1,N_2,P]}\ \text{ID-ELIM-J}'$$

and an equality rule analogous to ID-COMP. So instead of having to "prove" $\tau[x, x, Refl_\sigma(x)]$ for all $x : \sigma$ (the second premise to ID-ELIM-J) we only need it for $x := N_1$. This elimination operator is definable from J using function types and gives *Subst* directly without using them. However, since J is more established, we shall stick to it.

It is worth pointing out that identity elimination (J) allows us to "reason" about terms involving instances of *Subst* (an example for such a term occurs in Example 6.6). Intuitively using J one can "replace" the proof P in an instance of *Subst* by an instance of reflexivity and then use the definitional equality Equation 3.1 above to get rid of *Subst*. A more systematic approach to this is provided by the conservativity result in Section 3.2.5 and is exemplified again in Example 6.6.

3.1.1.1 Symmetry and transitivity. Let $M, N : \sigma$ and $P : Id_\sigma(M, N)$. We define

$$Sym_\sigma(M, N, P) := Subst_{\sigma,[x:\sigma]Id_\sigma(x,M)}(M, N, P, Refl_\sigma(M)) : Id_\sigma(N, M)$$

If in addition we have $Q : Id_\sigma(N, O)$ we define

$$Trans_\sigma(M, N, O, P, Q) := Subst_{\sigma,[x:\sigma]Id(M,x)}(N, O, Q, P) : Id_\sigma(M, O)$$

The typing annotations and the inferable arguments may again be elided.

3.1.1.2 Compatibility with function application. Assume $U : \sigma \to \tau$ and $M, N : \sigma$ and $P : Id_\sigma(M, N)$. An element of type $Id_\tau(U\ M, U\ N)$ is constructed as follows:

$$Resp(U, P) := Subst_{\sigma,[x:\sigma]Id_\tau(U\ M,x)}(P, Refl_\tau(U\ M))$$

More generally, if $\vdash \Pi x{:}\sigma.\tau[x]$ and M, N, P are as before then we can find an element of $Id_{\tau[N]}(Subst_{\sigma,\tau}(P, U\ M)\ , U\ N)$.

3.1.2 Uniqueness of identity

The conversion functions from $\tau[M]$ to $\tau[M']$, if $P : Id(M, M')$, obtained from *Subst* depend on the proof supplied. That is, if $\Gamma \vdash P, Q : Id_\sigma(M, N)$ and $\Gamma \vdash L : \tau[M]$ then in general the type

$$\Gamma \vdash Id_{\tau[N]}(Subst_{\sigma,\tau}(P, L)\ ,\ Subst_{\sigma,\tau}(Q, L))$$

need not be inhabited. We shall prove this later in Section 5.2. It is, however, certainly desirable that the above type *is* inhabited[1], or that equivalently the type

$$\Gamma \vdash Id_{Id_\sigma(M,N)}(P, Q)$$

is inhabited. We use the term *uniqueness of identity* to refer to either type being inhabited for every such P, Q, M, N, σ, τ.

Various authors have proposed the addition of axioms to type theory so as to achieve uniqueness of identity [104, 18, 37]. Now as pointed out in [104] it is sufficient to have uniqueness of identity in the case where one of the two proofs is a canonical one by reflexivity. The general case follows using J, see below. So we are led to introduce a family of constants

$$\frac{\Gamma \vdash \sigma \qquad \Gamma \vdash M : \sigma \qquad \Gamma \vdash P : Id_\sigma(M, M)}{\Gamma \vdash IdUni_\sigma(M, P) : Id_{Id_\sigma(M,M)}(P, Refl(M))}\ \textsc{Id-Uni-I}$$

and (in addition to the obvious congruence rules) an equality rule

$$\frac{\Gamma \vdash \sigma \qquad \Gamma \vdash M : \sigma}{\Gamma \vdash IdUni_\sigma(M, Refl(M)) = Refl(Refl(M)) : Id_\sigma(M, M)}\ \textsc{Id-Uni-Comp}$$

In this way an instance of *IdUni* can be eliminated if its arguments are canonical, so that, provided ID-UNI-COMP does not destroy strong normalisation, a type theory extended by ID-UNI-I and ID-UNI-COMP is N-canonical in the sense of Definition 2.1.9. There does not seem to exist a formal proof in the literature that ID-UNI-COMP is indeed strongly normalising, but it seems probable that any of the usual proofs of normalisation carries over.

Where appropriate we omit the type annotation and the first argument to *IdUni*.

For the sake of completeness we give now a proof of the general case of uniqueness of identity using *IdUni*. If $\Gamma \vdash \sigma$ then consider the type

$$\tau[x, y{:}\sigma\ ,\ p{:}\ Id_\sigma(x, y)] := \Pi q{:}\ Id_\sigma(x, y).Id_{Id_\sigma(x,y)}(p, q)$$

Now if $\Gamma \vdash M, N : \sigma$ and $\Gamma \vdash P, Q : Id_\sigma(M, N)$ then we have

$$\Gamma \vdash J_{\sigma,\tau}([x{:}\sigma]\lambda q{:}\ Id_\sigma(x, x).IdUni_\sigma(x, p)\ ,\ M\ ,\ N\ ,\ P)\ Q : Id_{Id_\sigma(M,N)}(P, Q)$$

[1] Inhabited, at any rate, if the identity type is to approximate extensional equality. For an application of a type theory without uniqueness of identity see Section 5.2.4.

3.1.2.1 Uniqueness of identity for definable types. In the presence of a universe one can show that various instances of *IdUni* are definable. In addition to the unit type we need an empty type $\mathbf{0}$ containing no canonical elements and an elimination operator $R^{\mathbf{0}}_\sigma$, where $\Gamma \vdash R^{\mathbf{0}}_\sigma(M) : \sigma$ if $\Gamma \vdash M : \mathbf{0}$. Moreover, we need a universe (U, El) in the sense of 2.3.5 containing codes $\hat{\mathbf{1}}$ and $\hat{\mathbf{0}}$ for the unit type and the empty type. We say that *uniqueness of identity is definable at type* σ if for each $M : \sigma$ and $P : Id_\sigma(M, M)$ the type $Id_{Id_\sigma(M,M)}(P, Refl_\sigma(M))$ is inhabited without using *IdUni*.

Proposition 3.1.1. *In TT extended with an empty type* $\mathbf{0}$ *and a universe* (U, El)*, containing* $\mathbf{0}$ *and* $\mathbf{1}$*, uniqueness of identity is definable at* $\mathbf{0}, \mathbf{1}, \mathbf{N}$*. If* $\Gamma, x{:}\sigma \vdash \tau$ *and uniqueness of identity is definable at* σ *and at* τ*, then it is definable at* $\Sigma x{:}\,\sigma.\tau$*.*

Proof. For the empty type $\mathbf{0}$ the result is trivial, since in the presence of a term $M : \mathbf{0}$ any type is inhabited by $R^{\mathbf{0}}$. For $\mathbf{1}$ we first construct a general proof $x, y{:}\,\mathbf{1} \vdash P_0[x, y] : Id_{\mathbf{1}}(x, y)$ using $R^{\mathbf{1}}$ satisfying $P_0[x, x] = Refl(x)$ and then show using J and $R^{\mathbf{1}}$ that whenever $\Gamma \vdash P : Id_{\mathbf{1}}(M, N)$ then $\Gamma \vdash, Id_{Id_{\mathbf{1}}(M,N)}(P, P_0[M, N])$. Uniqueness of identity at $\mathbf{1}$ is a trivial consequence from this.

For $\mathbf{N}$ we define a term $x, y{:}\,\mathbf{N} \vdash EqNat[x, y] : U$ by induction ($R^{\mathbf{N}}$) in such a way that $\vdash EqNat[0, 0] = \hat{\mathbf{1}} : U$, $x{:}\,\mathbf{N} \vdash EqNat(0, Suc(x)) = EqNat(Suc(x), 0) = \hat{\mathbf{0}} : U$, and $x, y{:}\,\mathbf{N} \vdash EqNat(Suc(x), Suc(y)) = EqNat(x, y)$. Now we construct terms

$$x, y{:}\,\mathbf{N}\ , p{:}\, Id_{\mathbf{N}}(x, y) \vdash \Phi[p] : El(EqNat[x, y])$$

and

$$x, y{:}\,\mathbf{N}\ , p{:}\, El(EqNat[x, y]) \vdash \Psi[p] : Id_{\mathbf{N}}(x, y)$$

The function Φ is defined using identity elimination, whereas Ψ involves induction ($R^{\mathbf{N}}$). Now using identity elimination (J) we can show that

$$x, y{:}\,\mathbf{N}\ , p{:}\, Id_{\mathbf{N}}(x, y) \vdash Id_{Id_{\mathbf{N}}(x,y)}(\Psi(\Phi(p)), p)$$

is inhabited. So uniqueness of identity at $\mathbf{N}$ follows from unicity of proofs of $El(EqNat[x, x])$, which can be established by induction.

For Σ-types consider the type

$$EqSigma[u, v{:}\, \Sigma x{:}\, \sigma.\tau[x]] := \Sigma p{:}\, Id_\sigma(u.1, v.1).Id_{\tau[y]}(Subst_{\sigma,\tau}(p, u.2)\ , v.2)$$

As in the case of $\mathbf{N}$ we can construct an "isomorphism" between $Id_{\Sigma x:\sigma.\tau}(M, N)$ and $EqSigma[M, N]$ and deduce uniqueness of identity from uniqueness of proofs of *EqSigma* which in turn follows from uniqueness of identity at σ and τ.

The methods of the above proof appear to carry over to other inductive types and type formers and under a certain assumption also to Π-types, see below. In order to show that uniqueness of identity is definable at $Id_\sigma(M, N)$ we need a witness for uniqueness of identity at σ which in addition satisfies the equivalent of Equation ID-UNI-COMP. It seems that one can guarantee this stronger requirement in the situation of the above proposition.

Uniqueness of identity is, however, not definable at universes, as emerges from the proof of its independence in Section 5.2.

3.1.2.2 The relationship with pattern-matching. It is worth highlighting the trivial observation that uniqueness of identity is definable at all types using the device of pattern-matching for dependent types introduced by Coquand in [18]. We do not here give the full definition of this mechanism, but only remark that it allows to define a function on an inductive type, such as the identity type, by specifying its value at the canonical elements. In this way we may define a term

$$\Gamma, p{:}\, Id_\sigma(M, M) \vdash IdUni_{\,\sigma}(M, P) : Id_{Id_\sigma(M,M)}(M, Refl_\sigma(M))$$

by the single pattern

$$IdUni_{\,\sigma}(M, Refl_\sigma(M)) = Refl_{Id_\sigma(M,M)}(Refl_\sigma(M))$$

since $Refl_\sigma(M)$ is the only canonical element of $Id_\sigma(M, M)$. Of course, this reduction to canonical elements is also the intention behind the elimination rules $R^{\mathbf{N}}$, J, ..., but there the patterns are given in a parametrised way. This subtle difference is made more explicit in the proof of independence of uniqueness of identity in 5.2.

3.1.3 Functional extensionality

Suppose that $\Gamma \vdash U, V : \Pi x{:}\sigma.\tau$ and $\Gamma, x{:}\sigma \vdash P : Id_\tau(U\ x\ ,\ V\ x)$. It is in general not possible to derive from these hypotheses that $\Gamma \vdash Id_{\Pi x:\sigma.\tau}(U, V)$ is inhabited, that is—in the terminology of the Introduction—intensional type theory does not support functional extensionality. In [104] a formal semantic proof of this is given. Intuitively, one may argue that if functional extensionality was available then in the case $\Gamma = \diamond$ we could deduce $Id_{\Pi x:\sigma.\tau}(U, V)$ *true* (from the existence of P above), but by strong normalisation (see Remark 2.1.6) an identity type in the empty context can only be inhabited by a canonical element $Refl(-)$, so U and V must be definitionally equal, i.e. intensionally equal, which does not follow from the existence of the proof P which may have been obtained using induction. This is an unfortunate problem with the identity type which makes its instances at higher types essentially unusable. To achieve functional extensionality we may add a family of constants $Ext_{\sigma,\tau}(U, V, P)$ obeying the rule

$$\frac{\begin{array}{c}\Gamma \vdash U, V : \Pi x{:}\sigma.\tau \\ \Gamma, x{:}\sigma \vdash P : Id_\tau(U\ x, V\ x)\end{array}}{\Gamma \vdash Ext_{\sigma,\tau}(U, V, P) : Id_{\Pi x:\sigma.\tau}(U, V)} \quad \text{EXT-FORM}$$

The type annotations and the first two arguments to Ext may be omitted.

The introduction of these constants is essentially the solution proposed by Turner in [111]. Clearly, the addition of these constants Ext is consistent as may be seen from the set-theoretic or the ω-set model (Examples 2.4.3 and 2.4.4). Its serious drawback, immediately pointed out by Martin-Löf in a subsequent discussion also in [111], is that this introduces non-canonical elements in each type, since we have not specified how the eliminator J should behave when applied to a proof having *Ext* as outermost constructor. For example, consider the (constant) family $f : \mathbf{N} \to \mathbf{N} \vdash \mathbf{N}$. Now if $x{:}\,\mathbf{N} \vdash P[x] : \mathit{Id}_\tau(U\ x, V\ x)$ for two functions $U, V : \mathbf{N} \to \mathbf{N}$ then $\mathit{Subst}_{\mathbf{N}\to\mathbf{N},[x:\mathbf{N}\to\mathbf{N}]\mathbf{N}}(\mathit{Ext}(P), 0)$ is an element of $\mathbf{N}$ in the empty context which does not reduce to canonical form. We address this problem later in Chapter 5 where we give syntactic models in which functional extensionality holds and which induce a decidable definitional equality on the syntax under which terms like the above are indeed definitionally equal to an element in canonical form. One may try to achieve the same thing by adding reduction rules for *Ext* under which the above term would for example reduce to 0. Yet no satisfactory set of such rules has been found to date. Notice, however, that we can show that the term in question is *propositionally equal* to 0 because using J we can "replace" $\mathit{Ext}(P)$ by an instance of *Refl*.

In *loc. cit.* Turner proposes to add the (definitional) equation

$$\Gamma \vdash \mathit{Ext}_{\sigma,\tau}([x{:}\,\sigma]\mathit{Refl}_\tau(U)) = \mathit{Refl}_{\Pi x:\sigma.\tau}(\lambda x{:}\,\sigma.U) : \mathit{Id}_{\Pi x:\sigma.\tau}(\lambda x{:}\,\sigma.U, \lambda x{:}\,\sigma.U) \tag{3.2}$$

i.e. if we use *Ext* only to establish propositional equality of definitionally equal terms then we may use *Refl* straightaway. The equation is 'incomplete' (i.e. does not solve the problem with non-canonical elements) because for example even in its presence the above term is not equal to 0 or any other canonical natural number. We remark that the propositional version of Turner's equation is an instance of *IdUni*, but does not seem to follow from J alone. Interestingly, using Turner's equation we can define uniqueness of identity at Π-types as well.

Proposition 3.1.2. *Consider an extension of TT containing a family of term formers* $\mathit{Ext}_{\sigma,\tau}$ *satisfying Turner's equation (3.2) above. If uniqueness of identity is definable at* $\Gamma, x{:}\,\sigma \vdash \tau$ *then it is definable at* $\Gamma \vdash \Pi x{:}\,\sigma.\tau$.

Proof. Consider the type $\mathit{EqPi}[u, v : \Pi x{:}\,\sigma.\tau] := \Pi x{:}\,\sigma.\mathit{Id}_\tau(u\ x, v\ x)$. We define

$$\Phi[u, v{:}\,\Pi x{:}\,\sigma.\tau\ ,\ p{:}\,\mathit{Id}_{\Pi x:\sigma.\tau}(u, v)] := \lambda x{:}\,\sigma.\mathit{Resp}\,([u : \Pi x{:}\,\sigma.\tau]u\ x\ ,\ p) : \mathit{EqPi}[u, v]$$

and

$$\Psi[u, v{:}\,\Pi x{:}\,\sigma.\tau\ ,\ p{:}\,\mathit{EqPi}[u, v]] := \mathit{Ext}_{\sigma,\tau}([x{:}\,\sigma]p\ x) : \mathit{Id}_{\Pi x:\sigma.\tau}(u, v)$$

From uniqueness of identity at τ and *Ext* it follows that any two elements of *EqPi* are propositionally equal. On the other hand an element $p : \mathit{Id}_{\Pi x:\sigma.\tau}(u, v)$ is propositionally equal to $\Psi[u, v, \Phi[u, v, p]]$ using first J and then Turner's equation. (Notice that $\mathit{Resp}\,(f, \mathit{Refl}(x)) = f\ x$ by ID-COMP.) We conclude using *Resp* for Ψ.

Note that this proof still goes through if Equation 3.2 only holds propositionally, e.g. by *IdUni*.

In the remainder of this chapter we shall leave aside the issue of non-canonical elements and study the logical implications of functional extensionality together with uniqueness of identity. We shall see that in a certain sense these two allow to recover the strength of extensional type theory in an intensional setting.

3.2 Extensional type theory

As described in the Introduction, in *extensional type theory* the definitional equality is identified with propositional equality and thereby becomes extensional. This is achieved by adding the following two rules [88, p. 65]:

$$\frac{\Gamma \vdash \sigma \qquad \Gamma \vdash M : \sigma \qquad \Gamma \vdash N : \sigma \qquad \Gamma \vdash P : Id_\sigma(M,N)}{\Gamma \vdash M = N : \sigma} \quad \text{ID-DEFEQ}$$

$$\frac{\Gamma \vdash \sigma \qquad \Gamma \vdash M : \sigma \qquad \Gamma \vdash N : \sigma \qquad \Gamma \vdash P : Id_\sigma(M,N)}{\Gamma \vdash P = \mathit{Refl}_\sigma(M) : Id_\sigma(M,N)} \quad \text{ID-UNI}$$

The first rule (ID-DEFEQ) is the equality reflection rule discussed in the Introduction. Notice that this rule is needed to make the other one (ID-UNI) "typecheck". For technical reasons we also assume an η-rule for Π-types:

$$\frac{\Gamma \vdash M : \Pi x{:}\sigma.\tau}{\Gamma \vdash M = \lambda x{:}\sigma.M\ x : \Pi x{:}\sigma.\tau} \quad \Pi\text{-ETA}$$

In [88] it is pointed out that in the presence of rules ID-DEFEQ and ID-UNI, identity elimination becomes definable by putting

$$J_{\sigma,\tau}(M, N_1, N_2, P) := M[x := N_1]$$

The operators *IdUni* and *Ext* are also definable, see Proposition 3.2.2; in fact they are just instances of reflexivity! Therefore, extensional type theory does not need explicit elimination operators for identity. For these reasons, extensional type theory is intuitively quite appealing. It also is very close to set theory and extensional constructs like quotient types are easily added (see [16] and Section 3.2.6.1). Moreover, it avoids the unusual coexistence of two different notions of equality. The rest of this Chapter is devoted to refute these apparent advantages of extensional type theory and to argue in favour of intensional type theory with extensional constructs. The two main arguments we put forward are the undecidability of extensional type theory (Section 3.2.2) and the conservativity of extensional type theory over intensional type theory (Section 3.2.5) which shows that nothing is lost by using intensional type theory.

3.2.1 Comparison with Troelstra's presentation

In order to prevent possible confusion let us remark that in Troelstra and van Dalen's book [109, Section 4–5], the predicates "intensional" and "extensional" are used in a different way. Their intensional theory $\mathbf{ML}_0^i$ is a version of our *extensional type theory*, that is it has rule ID-DEFEQ, but with type equality confined to basically syntactic identity. So no definitional conversion is permitted inside a type and thus the introduction rule for the identity type must be extended to permit one to conclude $\Gamma \vdash \mathit{Refl}(M) : \mathit{Id}_\sigma(M, M')$ from $\Gamma \vdash M = M' : \sigma$. Moreover, $\mathbf{ML}_0^i$ neither contains a congruence rule for abstraction ("ξ-rule") nor an η-rule for Π-types. On the other hand, their extensional type theory $\mathbf{ML}_0$ agrees mostly with extensional type theory in our sense (and in the sense of [88]). Intensional type theory in our sense (and in the sense of [88]) is only mentioned briefly as a note [*loc. cit.*, p. 633].

3.2.2 Undecidability of extensional type theory

The equality reflection principle ID-DEFEQ is problematic since upon its application the proof P for the propositional equality is lost. Therefore, a syntax-directed decision procedure for definitional equality in extensional type theory would have to "guess" this proof P. Our aim in this section is to prove that this is not possible by showing that definitional equality in extensional type theory is undecidable. Before doing so we describe how this entails undecidability of all other kinds of judgements. Assume $\Gamma \vdash M : \sigma$ and $\Gamma \vdash N : \sigma$. Now $\Gamma \vdash \mathit{Refl}_\sigma(M) : \mathit{Id}_\sigma(M, N)$ iff $\Gamma \vdash M = N : \sigma$ so typechecking is equivalent to deciding definitional equality. Furthermore, we have $\Gamma \vdash \mathit{Id}_\sigma(M, M)$ iff $\Gamma \vdash M : \sigma$ so typehood is equivalent to typechecking. The same goes for the remaining three kinds of judgements: type equality, context equality, and well-formedness of contexts.

Let us now turn to the undecidability of definitional equality. Although the rule ID-DEFEQ represents an information loss, undecidability is not entirely obvious because it is not clear how many propositional equalities are actually provable. In particular, there does not seem to be an obvious reduction to the halting problem, but one must use the tool of recursively inseparable sets [25]. In order to avoid the introduction of too much machinery we shall, however, avoid explicit mention of this notion. We also avoid explicit formalisation of proofs and algorithms in Martin-Löf type theory and appeal to a variant of Church's thesis and the mathematical intuition of the reader.

If $n \in \omega$ is a natural number let $\hat{n}$ stand for the numeral $\mathit{Suc}^n(0)$ in type theory. We have $\vdash \hat{n} : \mathbf{N}$. A k-ary function f on the natural numbers is called *numeral-wise representable* if there exists a term $\vdash \hat{f} : \mathbf{N} \to \mathbf{N} \to \ldots \to \mathbf{N}$ such that for every k-tuple of natural numbers $(x_1, \ldots, x_k)$ we have $\vdash \hat{f}\ \hat{x}_1 \ldots \hat{x}_k = \widehat{f(x_1, \ldots, x_k)} : \mathbf{N}$. It is well-known and actually obvious from the form of the recursor $R^{\mathbf{N}}$ that every primitive recursive function is numeral-wise representable. Compare this to the interpretation of Heyting arithmetic in type theory [109, Theorem 4.9].

Fix some Gödel numbering of Turing machines (TMs) in order to allow for self-application of Turing machines and consider the following function

$$f(e,t) := \begin{cases} 1, \text{ if Case 1} \\ 0, \text{ otherwise} \end{cases}$$

where Case 1 applies if TM e applied to itself halts after less than t steps with result 0. Since the number of steps required to evaluate $f(e,t)$ is linear in e and t, this function is primitive recursive. Let $\vdash \hat{f} : \mathbf{N} \to \mathbf{N} \to \mathbf{N}$ be its numeral-wise representation.

Now suppose e is such that e applied to itself halts after some number of steps with a result other than 0. Then $f(e, -)$ is constantly 0 and this fact is actually provable in type theory, i.e. in this case the following type is inhabited by induction ($R^{\mathbf{N}}$) and finite case distinction (also $R^{\mathbf{N}}$):

$$t : \mathbf{N} \vdash Id_{\mathbf{N}}(\hat{f}\ \hat{e}\ t\ ,\ 0)$$

Using Id-DefEq this implies that

$$t : \mathbf{N} \vdash \hat{f}\ \hat{e}\ t = 0 : \mathbf{N}$$

holds.

On the other hand, if TM e applied to itself halts with result 0 then

$$t : \mathbf{N} \vdash \hat{f}\ \hat{e}\ t = 0 : \mathbf{N}$$

does not hold because this would contradict the definition of f.

Notice that in case e applied to itself does not halt, the above equation may or may not hold according to whether nontermination is provable in type theory.

Now if definitional equality in extensional type theory were decidable we could construct a Turing machine which when applied to some number e returns 0 if

$$t : \mathbf{N} \vdash \hat{f}\ \hat{e}\ t = 0 : \mathbf{N}$$

and returns 1 otherwise. Let e_0 be the Gödel number of this machine and consider the application of e_0 to itself. By definition of e_0 this computation terminates. We arrive at a contradiction whatever the result is.

Summing up, we obtain:

Theorem 3.2.1. *In extensional type theory definitional equality, typechecking and typehood are equivalent and undecidable.*

The same argument shows that inhabitation of identity types in nonempty contexts is undecidable in intensional type theory.

We remark that the undecidability of extensional type theory is part of the "folklore" in the field, but apparently has never been explicitly proved in the literature.

3.2.3 Interpreting extensional type theory in intensional type theory

In view of the above undecidability results, implementations of extensional type theory like e.g. Nuprl [16] deal with *derivations* rather than terms and judgements. The question arises whether derivations in extensional type theory can be translated into terms in intensional type theory with functional extensionality and uniqueness of identity. For example, we may want to know whether if $x{:}\mathbf{N} \vdash U, V : \mathbf{N}$ in intensional type theory and $x{:}\mathbf{N} \vdash U = V : \mathbf{N}$ in extensional type theory then the type $x{:}\mathbf{N} \vdash \mathit{Id}_{\mathbf{N}}(U,V)$ is inhabited in intensional type theory. This is indeed the case and follows from a more general correspondence between intensional and extensional type theory which we prove below.

3.2.3.1 The type theories $\mathbf{TT}_I$ and $\mathbf{TT}_E$. Streicher has observed in [104] that in the presence of uniqueness of identity the elimination operator J can be defined in terms of its particular instance *Subst* defined above (Section 3.1.1). Indeed, if $\Gamma, x{:}\sigma \vdash H : \tau[x, x, \mathit{Refl}(x)]$ and $\Gamma \vdash P : \mathit{Id}_\sigma(M, N)$ then we form the type

$$\rho[x{:}\sigma] := \Pi p{:}\, \mathit{Id}_\sigma(M, x).\tau[M, x, p]$$

Now we observe that $\Gamma \vdash \rho[M]$ is inhabited by

$$L[\gamma] := \lambda p{:}\, \mathit{Id}(M, M).\mathit{Subst}\,(\mathit{Sym}\,(\mathit{IdUni}\,(P)))\,,\, H[M])$$

Therefore, $\mathit{Subst}_{\sigma,\rho}(P, L)\ P$ is an inhabitant of $\Gamma \vdash \tau[M, N, P]$ as required.

This suggests to take *Subst* instead of J as a primitive if one has uniqueness of identity. For the remainder of this chapter, by "intensional type theory" or TT_I we mean the theory TT from above together with the constants *IdUni* and *Ext*, and with the elimination operator J with its associated rules replaced by a newly introduced term former *Subst* governed by the following two rules:

$$\frac{\begin{array}{l}\Gamma, x{:}\sigma \vdash \tau[x] \\ \Gamma \vdash M_1, M_2 : \sigma \\ \Gamma \vdash P : \mathit{Id}_\sigma(M_1, M_2) \\ \Gamma \vdash N : \tau[M_1]\end{array}}{\Gamma \vdash \mathit{Subst}_{\sigma,\tau}(M_1, M_2, P, N) : \tau[M_2]}\ \textsc{Leibniz}$$

$$\frac{}{\Gamma \vdash \mathit{Subst}_{\sigma,\tau}(M, M, \mathit{Refl}_\sigma(M), N) = N : \tau[M]}\ \textsc{Leibniz-Comp}$$

Thus, the term formers associated to propositional equality in TT_I are: *Refl*, *Subst* , *IdUni* , and *Ext*. The type annotations and the first two arguments to *Subst* may be omitted. If so stated in the running text, TT_I may be augmented by further type formers and rules.

By "extensional type theory" or TT_E we mean the core type theory TT with the identity elimination rules (the rules ID-ELIM-J and ID-COMP J) replaced by ID-DEFEQ and ID-UNI, and with Π-ETA added. In order to distinguish the intensional from the extensional type theory we write $\vdash_I$ and $\vdash_E$ for the two respective judgement relations. Recall from Section 2.1.3 that (for $\vdash \in \{\vdash_I, \vdash_E\}$)

we write $\Gamma \vdash \sigma$ *true* if there exists M with $\Gamma \vdash M : \sigma$ and that $\equiv$ denotes syntactic identity modulo renaming of variables.

It was mentioned before that the operators of TT_I are definable in TT_E. More formally, we can define a "stripping map" $| - |$ by

- $|\mathit{Subst}_{\sigma,\tau}(M_1, M_2, P, N)| := |N|$
- $|\mathit{IdUni}_{\sigma}(M, P)| := \mathit{Refl}_{\mathit{Id}_{|\sigma|}(|M|,|M|)}(\mathit{Refl}_{|\sigma|}(|M|))$
- $|\mathit{Ext}_{\sigma,\tau}(U, V, P)| := \mathit{Refl}_{\Pi x:|\sigma|.|\tau|}(|U|)$
- homomorphically extended to all other terms, types, contexts. and judgements

This mapping enjoys the following trivial soundness property.

Proposition 3.2.2. *If $\Gamma \vdash_I J$ then $|\Gamma| \vdash_E |J|$ for all contexts Γ and judgements J.*

Proof. By induction on derivations in TT_I or equivalently by showing that the term model of TT_E forms a model of TT_I with the settings prescribed by the definition of $| - |$. The only interesting case is the interpretation of *Ext*. If $\Gamma, x{:}\sigma \vdash_E P{:}\, \mathit{Id}_\tau(U\ x, V\ x)$ then by ID-DEFEQ we have $\Gamma, x{:}\sigma \vdash_E U\ x = V\ x : \Pi x{:}\sigma.\tau$. So by the congruence rule for abstraction we get $\Gamma \vdash_E \lambda x{:}\sigma.U\ x = \lambda x{:}\sigma.V\ x : \Pi x{:}\sigma.\tau$ and hence using Π-ETA and transitivity $\Gamma \vdash_E U = V : \Pi x{:}\sigma.\tau$ and finally $\Gamma \vdash_E \mathit{Refl}(U) : \mathit{Id}(U, V)$ as required. Without Π-ETA we could obtain functional extensionality only for abstractions but not for arbitrary terms of Π-type.

In order to compare the strengths of extensional and intensional type theory we are now interested in possible converses of this proposition. In particular:

INV	$\Gamma \vdash_E J$ implies $\Gamma' \vdash_I J'$ for some Γ', J' with $\lvert\Gamma'\rvert \equiv \Gamma$ and $\lvert J'\rvert \equiv J$.
CONS-TY	$\lvert\Gamma\rvert \vdash_E \lvert\sigma\rvert$ *true* implies $\Gamma \vdash_I \sigma$ *true* .
CONS-ID	$\lvert\Gamma\rvert \vdash_E \lvert M\rvert = \lvert N\rvert : \lvert\sigma\rvert$ implies $\Gamma \vdash_I \mathit{Id}_\sigma(M, N)$ *true*.

Clearly, CONS-TY implies CONS-ID. The first property INV which in contrast to the other two speaks about arbitrary judgements in TT_E, is not true. A counterexample is the judgement

$$x{:}\mathbf{N} \vdash_E x = \mathit{Suc}^x(0) : \mathbf{N}$$

where $\mathit{Suc}^x(0) := R^{\mathbf{N}}_{\mathbf{N}}(0, [x, y{:}\mathbf{N}]\mathit{Suc}(y), x)$ is the x-th iteration of the successor function on start value 0. If we decorate both sides of this equation with instances of *Subst* we get more complicated terms, but certainly no true equation in TT_I. The reason is that (apart from congruence) the only way that a *Subst*-expression can be definitionally equal to something else is by use of LEIBNIZ-COMP, but then the outermost *Subst* disappears and no equation has been added which wasn't there before. However, we certainly have

$$x{:}\mathbf{N} \vdash_I \mathit{Id}_{\mathbf{N}}(x, \mathit{Suc}^x(0))\ \mathit{true}$$

since the required inhabitant of the identity type may be constructed by induction. Together with ID-DEFEQ this gives a proof of the above judgement in TT_E and corroborates CONS-ID.

Our aim is to prove CONS-TY, and thus CONS-ID and to define a mild extension of TT_I under which INV becomes true.

3.2.4 An extension of TT_I for which the interpretation in TT_E is surjective

The above counterexample would no longer go through if there were a term whose stripping equals x and which nevertheless is definitionally equal to $Suc^x(0)$ in TT_I. To cure the problem with INV we may simply add such a construct. More precisely, we introduce a new term former $\mathfrak{R}_\sigma(M, N, P)$ described by the following two rules:

$$\frac{\Gamma \vdash_I M, N : \sigma \qquad \Gamma \vdash_I P : Id_\sigma(M, N)}{\Gamma \vdash_I \mathfrak{R}_\sigma(M, N, P) : \sigma}\ \mathfrak{R}\text{-FORM}$$

$$\frac{\Gamma \vdash_I \mathfrak{R}_\sigma(M, N, P) : \sigma}{\Gamma \vdash_I \mathfrak{R}_\sigma(M, N, P) = N : \sigma}\ \mathfrak{R}\text{-EQ}$$

Let us write $TT_\mathfrak{R}$ for TT_I extended with $\mathfrak{R}$ and $\vdash_\mathfrak{R}$ for the corresponding judgement relation. In view of $\mathfrak{R}$-EQ, $\mathfrak{R}$ is a simple definitional extension comparable with the addition of an explicit function symbol for multiplication which is definitionally equal to its primitive recursive coding using $R^{\mathbf{N}}$. The power of $\mathfrak{R}$ stems from its interpretation in TT_E. Namely, we decree that:

$$|\mathfrak{R}_\sigma(M, N, P)| := |M|$$

Now Proposition 3.2.2 continues to hold:

Proposition 3.2.3. *If $\Gamma \vdash_\mathfrak{R} J$ then $|\Gamma| \vdash_E |J|$ for all contexts Γ and judgements J.*

Proof. As before by induction on derivations; the only interesting case is the rule $\mathfrak{R}$-EQ. If $\Gamma \vdash_\mathfrak{R} \mathfrak{R}_\sigma(M, N, P) : \sigma$ then we must have $\Gamma \vdash_\mathfrak{R} M, N : \sigma$ and $\Gamma \vdash_\mathfrak{R} P : Id_\sigma(M, N)$. So we may inductively assume that $|\Gamma| \vdash_E |P| : Id_{|\sigma|}(|M|, |N|)$ and thus $|\Gamma| \vdash_E |M| = |N| : |\sigma|$ by ID-DEFEQ. But this is the stripping of the conclusion of rule $\mathfrak{R}$-EQ.

The interesting aspect of $TT_\mathfrak{R}$ is that the interpretation $|-|$ of it in TT_E is actually surjective, which means that INV holds.

Theorem 3.2.4. *Whenever $\Gamma \vdash_E J$ then there exist Γ' and J' in the language of $TT_\mathfrak{R}$ such that $\Gamma' \vdash_\mathfrak{R} J'$ and $|\Gamma'| \equiv \Gamma$ and $|J'| \equiv J$.*

Proof. By induction on derivations in TT_E. The interesting cases are the rules ID-DEFEQ, ID-UNI, and Π-ETA. In the other cases we just follow the derivation in TT_E. For ID-DEFEQ assume $\Gamma \vdash_E P : Id_\sigma(M, N)$ and (inductively) $\Gamma' \vdash_\mathfrak{R} P' : Id_{\sigma'}(M', N')$ where $|C'| \equiv C$ for $C \in \{\Gamma, \sigma, P, M, N\}$. Now the conclusion of ID-DEFEQ is $\Gamma \vdash_E M = N : \sigma$. Putting $M'' := \Re_{\sigma'}(M', N', P')$ we have $\Gamma' \vdash_\Re M'' = N' : \sigma'$ but on the other hand $|M''| \equiv |M'| \equiv M$.

The rule ID-UNI has the same premises so we keep the variables introduced so far. The conclusion of the rule is $\Gamma \vdash_E P = \mathit{Refl}_\sigma(M) : \mathit{Id}_\sigma(M, N)$. Therefore, we must find a term P'' with stripping equal to P and which itself is definitionally equal (in TT_I) to $\mathit{Refl}(M')$. Using $\Re$ we can relax this to propositional equality. This suggests to define

$$P'' := \mathit{Subst}_{\sigma,[x:\sigma']\mathit{Id}(M',x)}(N', M', \mathit{Sym}(P'), P')$$

We have $\Gamma' \vdash_\Re P'' : \mathit{Id}_\sigma(M, M)$ and $\Gamma' \vdash_\Re \mathit{IdUni}(P'') : \mathit{Id}(P'', \mathit{Refl}(M))$. Moreover, $|P''| \equiv |P'| \equiv P$. Therefore with

$$P''' := \Re_{\mathit{Id}_{\sigma'}(M',M')}(P'', \mathit{Refl}(M'), \mathit{IdUni}(P''))$$

we have $\Gamma' \vdash_\Re P''' = \mathit{Refl}_\sigma(M') : \mathit{Id}_{\sigma'}(M', M')$ and the stripping of this judgement is the conclusion of rule ID-UNI.

Finally, for Π-ETA we use a particular instance of *Ext*.

The particular instance of *IdUni* used in this proof can be defined using J as well. So if we had used J instead of *Subst* we could have avoided the use of uniqueness of identity. Then, however, $|P''|$ would not be identical to P but only β-reduce to P. One could circumvent this using J' (defined above in Section 3.1.1) as primitive. The only use of functional extensionality was in order to mimic the rule Π-ETA. If one were not interested in this rule, one could get surjectivity of $|-|$ with $\Re$ alone and no extensional constructs at all. In particular, in $\mathrm{TT}_\Re$ we can derive a propositional counterpart of the λ-congruence rule (ξ-rule) as follows: Suppose that $\Gamma, x{:}\sigma \vdash_\Re U, V : \tau$ and $\Gamma, x{:}\sigma \vdash_\Re P : \mathit{Id}_\tau(U, V)$. Then we have

$$\Gamma \vdash_\Re \mathit{Refl}_{\Pi x:\sigma.\tau}(\lambda x{:}\sigma.V) : \mathit{Id}_{\Pi x:\sigma.\tau}(\lambda x{:}\Re_\tau(U, V, P)\,,\, \lambda x{:}\sigma.V) \tag{3.3}$$

This shows quite well how $\Re$ works. In a certain sense, the above judgement is a triviality because it is an instance of reflexivity. But if one does not identify definitionally equal terms and instead reads an expression $\Re_\sigma(M, N, P)$ as something like "means M but is written N for reasons of intensionality and this is justified by virtue of P" then the above judgement really corresponds to the conclusion of a propositional ξ-rule. One can do even more and—using extensional constructs and instances of *Resp*—move all the instances of $\Re$ to the root of a term, with the stripping unchanged, which may be used to obtain a variant of property CONS-ID as follows: If $\Gamma \vdash_E M = N : \sigma$ we can find Γ', σ' and $\Re$-free terms M_1, M_2, N_1, N_2, and proofs P_1, P_2 such that

$$\Gamma' \vdash_\Re \Re_{\sigma'}(M_1, M_2, P_1) = \Re_{\sigma'}(N_1, N_2, P_2) : \sigma'$$

where $|\Gamma'| \equiv \Gamma$, $|\sigma'| \equiv \sigma$, $|M_1| \equiv M$, and $|N_1| \equiv N$. From the definition of $\Re$ we then get

$$\Gamma' \vdash_\Re \mathit{Trans}(P_1, \mathit{Sym}(P_2)) : \mathit{Id}_{\sigma'}(M_1, M_2)$$

which we can see as the conclusion of CONS-ID in case Γ' and σ' are sufficiently simple so that they equal their stripping. Also the type annotations of M_1, M_2 must not contain any instances of $\mathfrak{R}$ or *Subst* which may be difficult to guarantee. We will not pursue this line of thought any further and prove CONS-TY and CONS-ID by other methods below. However, we want to advocate the use of $\mathfrak{R}$ as a possible addition to intensional type theory even in the absence of extensional constructs. This requires a slightly unusual approach to theorem proving in type theory because in the course of trying to prove a goal it may become necessary to alter it by interspersing instances of $\mathfrak{R}$. For instance, we cannot prove that $\lambda x{:}\sigma.U$ and $\lambda x{:}\sigma.V$ from above are definitionally equal or (without functional extensionality) that they are propositionally equal, but by altering the goal using $\mathfrak{R}$ this becomes possible as we have just seen in the derivation of Judgement 3.3.

3.2.5 Conservativity of TT_E over TT_I

Our aim is now to prove property CONS-TY of which CONS-ID is a consequence. In fact we shall establish the following mild generalisation of CONS-TY.

Theorem 3.2.5. *If $\Gamma \vdash_I \sigma$ and $|\Gamma| \vdash_E P : |\sigma|$ for some P then there exists P' such that $\Gamma \vdash_I P' : \sigma$. Moreover, $|\Gamma| \vdash_E |P'| = P : |\sigma|$.*

Before embarking on the proof we remark that it does not follow from INV, i.e. for example in the presence of $\mathfrak{R}$, because if $\Gamma \vdash_{\mathfrak{R}} \sigma$ and $|\Gamma| \vdash_E |\sigma|$ *true* then we have $\Gamma' \vdash_{\mathfrak{R}} \sigma'$ *true* for some Γ' and σ' with $|\Gamma'| \equiv |\Gamma|$ and $|\sigma'| \equiv |\sigma|$, but there is no reason why $\Gamma \vdash_{\mathfrak{R}} \sigma$ *true* should hold, too. It would, however, be true if types (and contexts) with identical stripping could be shown to be isomorphic. On the other hand, this fact (at least for types) is a consequence of CONS-TY because if $|\sigma| \equiv |\sigma'|$ then the stripping of

$$\Sigma f{:}\sigma \to \sigma'.\Sigma f^{-1}{:}\sigma' \to \sigma.$$
$$Id(\lambda x{:}\sigma.f^{-1}(f\ x), \lambda x{:}\sigma.x) \times Id(\lambda x{:}\sigma'.f(f^{-1}\ x), \lambda x{:}\sigma'.x)$$

is inhabited in TT_E by taking the identity on σ for f and f^{-1}, and CONS-TY then gives an isomorphism between σ and σ' in TT_I. These isomorphisms are indeed the key to proving the conservativity theorem, but it turns out that the use of $\mathfrak{R}$ can be avoided.

The idea for the proof is to show that the term model of TT_I factored by propositional equality forms a model of TT_E. Factoring a model of type theory is, however, not an easy thing to do because as soon as one identifies two morphisms $f, g : \Gamma \to \Delta$ one also has to identify families over Γ which come from the same family over Δ through substitution along f and g. This identification of families also forces identification of certain contexts and therefore of new morphisms which were incomparable before. We do not know whether there exists a general theory of quotients of syntactic categories with attributes and more generally of fibrations and generalised algebraic theories, but it seems that such a theory would have to be fairly complicated. Fortunately, in the particular case at hand we have a good definition for when two types or contexts are to be identified, so that we can get away without looking at the most general case.

Our strategy consists of first extending propositional equality to contexts so that we can say when two context morphisms are to be identified. Next, by structural induction we construct possibly undefined isomorphisms (up to propositional equality) between types and between contexts in such a way that the isomorphism between two types (and between two contexts) is defined iff their strippings are equal in TT_E. The key idea of the proof is that this semantic property of types and contexts can be defined by structural induction,

We identify two types or contexts if the corresponding isomorphism is defined, and show that this gives a model of TT_E. We then show that stripping lifts to an operation on the induced equivalence classes and defines a structure-preserving map between this model and the term model of TT_E. Its composition with the interpretation of TT_E in the quotient model must therefore be the identity. This, together with the fact that equality in the quotient model is isomorphism, gives the desired result.

3.2.5.1 Propositional equality of context morphisms. Let Γ be a context in TT_I. By induction on the length of Γ we define a type $\vdash_I \overline{\Gamma}$ in the empty context and substitutions $x{:}\overline{\Gamma} \vdash_I \mathit{out}_\Gamma \Rightarrow \Gamma$ and $\Gamma \vdash_I \mathit{in}_\Gamma \Rightarrow x{:}\overline{\Gamma}$ as follows:

$$\begin{array}{lcl} \overline{\diamond} & := & \mathbf{1} \\ \mathit{in}_\diamond & := & \star \\ \mathit{out}_\diamond & := & () \\ & & \\ \overline{\Gamma, x{:}\sigma} & := & \Sigma g{:}\overline{\Gamma}.\sigma[\mathit{out}_\Gamma[g]] \\ \mathit{in}_{\Gamma,x:\sigma}[\gamma{:}\Gamma, x{:}\sigma] & := & \mathit{pair}_{\overline{\Gamma},\sigma[\mathit{out}_\Gamma]}(\mathit{in}_\Gamma[\gamma], x) \\ \mathit{out}_{\Gamma,x:\sigma}[z{:}\overline{\Gamma, x{:}\sigma}] & := & (\mathit{out}_\Gamma[z.1]\,,\, z.2) \end{array}$$

From the equation Σ-COMP we get the definitional equality

$$\gamma{:}\Gamma \vdash_I \mathit{out}_\Gamma[\mathit{in}_\Gamma[\gamma]] = \gamma : \Gamma \tag{3.4}$$

This equation is needed for the definitions above to typecheck, so strictly speaking we have to establish it along with the inductive definition above.

The converse only holds propositionally: If $\vdash_I M : \overline{\Gamma}$ then by induction on the length of Γ and using R^Σ we can construct a term $\mathit{srj}_\Gamma(M)$ such that

$$\vdash_I \mathit{srj}_\Gamma(M) : \mathit{Id}_{\overline{\Gamma}}(\mathit{in}_\Gamma[\mathit{out}_\Gamma[M]]\,,\, M)$$

This reflection of contexts into types allows us to compare substitutions up to propositional equality. If $\Gamma \vdash_I f, g \Rightarrow \Delta$ then we can form

$$\Gamma \vdash_I \mathit{Id}_{\overline{\Delta}}(\mathit{in}_\Delta \circ f, \mathit{in}_\Delta \circ g) \qquad (\star)$$

and if $\Delta \vdash_I \sigma$, $\Gamma \vdash_I P : \mathit{Id}_{\overline{\Delta}}(\mathit{in}_\Delta \circ f, \mathit{in}_\Delta \circ g)$, and $\Gamma \vdash_I M{:}\sigma[f]$ then since $\Gamma \vdash_I \sigma[f] = \sigma[\mathit{out}_\Delta][\mathit{in}_\Delta \circ f]$ we have

$$\Gamma \vdash_I \mathit{Subst}_{\overline{\Delta},\sigma[\mathit{out}_\Delta]}(\mathit{in}_\Delta \circ f, \mathit{in}_\Delta \circ g, P, M) : \sigma[g]$$

So the above type ($\star$) behaves like an identity type at contexts. We thus introduce the abbreviations

$$Id_{\Delta}(f, g) := Id_{\overline{\Delta}}(in_{\Delta} \circ f, in_{\Delta} \circ g)$$

and

$$Subst_{\Delta,\sigma}(f, g, P, M) := Subst_{\overline{\Delta},\sigma[out_{\Delta}]}(in_{\Delta} \circ f, in_{\Delta} \circ g, P, M)$$

for these types and terms. We also write

$$Refl_{\Delta}(f) := Refl_{\overline{\Delta}}(in_{\Delta} \circ f)$$

and

$$IdUni_{\Delta}(f, P) := IdUni_{\overline{\Delta}}(in_{\Delta} \circ f, P)$$

As usual, we allow the omission of arguments which can be inferred or are clear from the context.

From LEIBNIZ-COMP and IDUNI-COMP we obtain the definitional equalities

$$\Gamma \vdash_I Subst_{\Delta,\sigma}(Refl(f), M) = M : \sigma[f]$$

$$\Gamma \vdash_I IdUni_{\Delta}(f, Refl(f)) = Refl(Refl(f)) : Id_{\Delta}(f, f)$$

Remark 3.2.6. This definition of propositional equality of context morphisms can still be carried out when Σ-COMP is a replaced by a propositional equality (as is the case in the type theory $\mathbf{S}_1$ to be introduced in Section 5.3), but then one must construct a witness for the propositional counterpart of Equation 3.4 along with the definition of in_{Γ} and out_{Γ}. This is possible, but messy to write down.

3.2.5.2 Propositional isomorphisms. A pair of syntactic context morphisms $\Gamma \vdash_I f \Rightarrow \Delta$ and $\Delta \vdash_I f' \Rightarrow \Gamma$ is called a *propositional isomorphism* between Γ and Δ if $\gamma{:}\,\Gamma \vdash_I Id_{\Gamma}(f^{-1}[f[\gamma]]\ ,\ \gamma)$ *true* and $\delta{:}\,\Delta \vdash_I Id_{\Delta}(f[f^{-1}[\delta]]\ ,\ \delta)$ *true*. In this situation we write $(f, f^{-1}) : \Gamma \cong \Delta$. If $\Gamma \vdash_I \sigma$ and $\Gamma \vdash_I \tau$ then a pair of terms $\Gamma, x{:}\sigma \vdash_I f[x] : \tau$ and $\Gamma, x{:}\tau \vdash_I f^{-1}[x] : \sigma$ is called a *propositional isomorphism* between σ and τ if $\Gamma, x{:}\sigma \vdash_I Id_{\sigma}(f^{-1}[f[x]]\ ,\ x)$ *true* and $\Gamma\ ,\ y{:}\tau \vdash_I Id_{\tau}(f[f^{-1}[y]]\ ,\ y)$ *true*. We write $\Gamma \vdash_I (f, f^{-1}) : \sigma \cong \tau$ to indicate this.

We shall now partially define symbols $co_{\Gamma,\Delta}$ and $ty_{\Gamma,\Delta,\sigma,\tau}$ by structural induction on pre-terms and pre-contexts in such a way that if these symbols are defined they constitute propositional isomorphisms in the following way

$$(co_{\Gamma,\Delta}, co^{-1}_{\Gamma,\Delta}) : \Gamma \cong \Delta$$

and

$$\Gamma \vdash_I (ty_{\Gamma,\Delta,\sigma,\tau}, ty^{-1}_{\Gamma,\Delta,\sigma,\tau}) : \sigma \cong \tau[co_{\Gamma,\Delta}]$$

The inductive clauses for these symbols are as follows:

$$\begin{array}{lcl} co_{\diamond,\diamond} & = & co^{-1}_{\diamond,\diamond} = () \\ co_{\Gamma,x:\sigma,\Delta,y:\tau}[\gamma{:}\,\Gamma, x{:}\,\sigma] & = & (co_{\Gamma,\Delta}[\gamma], ty_{\Gamma,\Delta,\sigma,\tau}[\gamma, x]) \\ co^{-1}_{\Gamma,x:\sigma,\Delta,y:\tau}[\delta{:}\,\Delta, y{:}\,\tau] & = & (co^{-1}_{\Gamma,\Delta}[\delta], ty^{-1}_{\Gamma,\Delta,\sigma,\tau}[co^{-1}_{\Gamma,\Delta}, Subst_{\Delta,\tau}(P, y)]) \end{array}$$

where $\delta\colon \Delta \vdash_I P : Id(\delta, co_{\Gamma,\Delta}[co^{-1}_{\Gamma,\Delta}[\delta]])$ is obtained from the assumption that $(co_{\Gamma,\Delta},\ co^{-1}_{\Gamma,\Delta}) : \Gamma \cong \Delta$. In the cases not covered by these clauses the context morphisms co and co^{-1} are undefined. In particular, $co_{\Gamma,\Delta}$ is undefined if Γ and Δ have different length.

The propositional isomorphisms $ty_{\Gamma,\Delta,\sigma,\tau}$ can only be defined if σ and τ share the same outermost type former. We have a clause for each of these.

$$ty_{\Gamma,\Delta,\mathbf{N},\mathbf{N}}[\gamma, x] = \begin{cases} x, \text{ if } co_{\Gamma,\Delta} \text{ is defined} \\ \text{undefined otherwise} \end{cases}$$

$$ty^{-1}_{\Gamma,\Delta,\mathbf{N},\mathbf{N}} = ty_{\Gamma,\Delta,\mathbf{N},\mathbf{N}}$$

For the definition of $ty_{\Gamma,\Delta,\Pi x:\sigma_1.\sigma_2,\Pi y:\tau_1.\tau_2}$ assume $\gamma\colon \Gamma$, $f\colon \Pi x\colon \sigma_1.\sigma_2$ and $y\colon \tau_1[co_{\Gamma,\Delta}[\gamma]]$. We put $U := ty^{-1}_{\Gamma,\Delta,\sigma_1,\tau_1}[\gamma, y] : \sigma_1[\gamma]$ and $V := f\ v : \sigma_2[\gamma, U]$. Now

$$W := ty_{(\Gamma,x:\sigma_1),(\Delta,y:\tau_1),\sigma_2,\tau_2}[(\gamma, U), V] : \tau_2[co_{\Gamma,\Delta}[\gamma]\ ,\ ty_{\Gamma,\Delta,\sigma_1,\tau_1}[\gamma, U]]$$

by expanding the definition of $co_{(\Gamma,x:\sigma_1),(\Delta,y:\tau_1)}$. Finally,

$$Subst\,(P, W) : \tau_2[co_{\Gamma,\Delta}[\gamma]\ ,\ y]$$

where

$$\begin{aligned}\gamma\colon \Gamma\ ,\ y\colon \tau_1[co_{\Gamma,\Delta}] \vdash_I P : Id_{\Gamma,y:\tau_1[co_{\Gamma,\Delta}]}(&\\ (co_{\Gamma,\Delta}[\gamma]\ ,\ ty_{\Gamma,\Delta,\sigma_1,\tau_1}[\gamma, ty^{-1}_{\Gamma,\Delta,\sigma_1,\tau_1}[\gamma, y]])\ ,&\\ (co_{\Gamma,\Delta}[\gamma]\ ,\ y))&\end{aligned}$$

is obtained from reflexivity and the assumption $\Gamma \vdash_I (ty_{\Gamma,\Delta,\sigma_1,\tau_1}, ty^{-1}_{\Gamma,\Delta,\sigma_1,\tau_1}) : \sigma_1 \cong \tau_1[co_{\Gamma,\Delta}]$ and thus

$$\begin{aligned}&ty_{\Gamma,\Delta,\Pi x:\sigma_1.\sigma_2,\Pi y:\tau_1.\tau_2}[\gamma, f] :=\\ &\quad \lambda y\colon \tau_1[co_{\Gamma,\Delta}[\gamma]].Subst\,(P, W) : (\Pi y\colon \tau_1.\tau_2)[co_{\Gamma,\Delta}[\gamma]]\end{aligned}$$

The inverse $ty^{-1}_{\Gamma,\Delta,\Pi x:\sigma_1.\sigma_2,\Pi y:\tau_1.\tau_2}$ is defined in the same way, interchanging the roles of ty and ty^{-1}. The proofs that a propositional isomorphism has been constructed are obtained using *Ext*.

For $ty_{\Gamma,\Delta,\Sigma x:\sigma_1.\sigma_2,\Sigma y:\tau_1.\tau_2}$ we assume $\gamma\colon \Gamma$ and $f\colon \Sigma x\colon \sigma_1.\sigma_2$. Now

$$U := ty_{\Gamma,\Delta,\sigma_1,\tau_1}[\gamma, f.1] : \tau_1[co_{\Gamma,\Delta}[\gamma]]$$

and

$$V := ty_{(\Gamma,x:\sigma_1),(\Delta,y:\tau_1),\sigma_2,\tau_2}[(\gamma, f.1), f.2] : \tau_2[co_{\Gamma,\Delta}[\gamma], U]$$

We thus define

$$ty_{\Gamma,\Delta,\Sigma x:\sigma_1.\sigma_2,\Sigma y:\tau_1.\tau_2}[\gamma, f] := (U, V) : (\Sigma y\colon \tau_1.\tau_2)[co_{\Gamma,\Delta}[\gamma]]$$

The inverse is again obtained by interchanging ty and ty^{-1}. For the proof of the isomorphism property one uses that surjective pairing holds propositionally (see Section 2.3.2).

For $ty_{\Gamma,\Delta,Id_\sigma(M_1,M_2),Id_\tau(N_1,N_2)}$ we check whether there exist P_1, P_2 with

$$\gamma{:}\,\Gamma \vdash_I P_1 : Id_{\tau[co_{\Gamma,\Delta}[\gamma]]}(ty_{\Gamma,\Delta,\sigma,\tau}[\gamma, M_1], N_1[co_{\Gamma,\Delta}[\gamma]])$$
$$\gamma{:}\,\Gamma \vdash_I P_2 : Id_{\tau[co_{\Gamma,\Delta}[\gamma]]}(ty_{\Gamma,\Delta,\sigma,\tau}[\gamma, M_2], N_2[co_{\Gamma,\Delta}[\gamma]])$$

If not, then $ty_{\Gamma,\Delta,Id_\sigma(M_1,M_2),Id_\tau(N_1,N_2)}$ is undefined. If yes, then we define

$$\begin{aligned} &ty_{\Gamma,\Delta,Id_\sigma(M_1,M_2),Id_\tau(N_1,N_2)}[\gamma, p{:}\, Id_\sigma(M_1, M_2)] := \\ &\qquad Trans\,(Sym\,(P_1),\, Trans\,(Resp\,(\lambda x{:}\,\sigma.ty_{\Gamma,\Delta,\sigma,\tau}[\gamma, x]\ , p)\ , P_2)) \\ &\qquad\quad : (Id_\tau(N_1\ , N_2)[co_{\Gamma,\Delta}[\gamma]] \end{aligned}$$

The inverse is defined analogously; the proof of the isomorphism property is an instance of *IdUni* .

Notice that this definition depends in various places on choices of particular proofs of propositional equalities. Some of the choices, e.g. in the definition of $co^{-1}_{\Gamma,x:\sigma,\Delta,y:\tau}$, could be avoided by carrying through explicit witnesses for the various isomorphism properties. Others, like the one in the clause for the identity type, are more difficult to eliminate.

These propositional isomorphisms enjoy various properties all of which follow by straightforward structural induction. First (for convenience) we restate the isomorphism properties.

Lemma 3.2.7. *Let $\Gamma \vdash_I \sigma$ and $\Delta \vdash_I \tau$.*

i. If $co_{\Gamma,\Delta}$ is defined so is $co^{-1}_{\Gamma,\Delta}$ and the two constitute a propositional isomorphism between Γ and Δ.
ii. If $ty_{\Gamma,\Delta,\sigma,\tau}$ is defined so are $co_{\Gamma,\Delta}$ and $ty^{-1}_{\Gamma,\Delta,\sigma,\tau}$ and

$$\Gamma \vdash_I (ty_{\Gamma,\Delta,\sigma,\tau}, ty^{-1}_{\Gamma,\Delta,\sigma,\tau}) : \sigma \cong \tau[co_{\Gamma,\Delta}]$$

Next we state that definedness of the propositional isomorphisms induces an equivalence relation on contexts and on types.

Lemma 3.2.8. *i. If $\vdash_I \Gamma$ then $co_{\Gamma,\Gamma}$ is defined and*

$$\gamma{:}\,\Gamma \vdash_I Id_\Gamma(\gamma, co_{\Gamma,\Gamma}[\gamma])\ true$$

ii. If $\Gamma \vdash_I \sigma$ then $ty_{\Gamma,\Gamma,\sigma,\sigma}$ is defined and

$$\Gamma, x{:}\,\sigma \vdash_I Id_\sigma(x, ty_{\Gamma,\Gamma,\sigma,\sigma}[x])\ true$$

iii. If $co_{\Gamma,\Delta}$ is defined then so is $co_{\Delta,\Gamma}$.
iv. If $ty_{\Gamma,\Delta,\sigma,\tau}$ is defined then so is $ty_{\Delta,\Gamma,\tau,\sigma}$.
v. If $co_{\Gamma,\Delta}$ and $co_{\Delta,\Theta}$ are both defined then so is $co_{\Gamma,\Theta}$ and

$$\gamma{:}\,\Gamma \vdash_I Id_\Theta(co_{\Delta,\Theta}[co_{\Gamma,\Delta}[\gamma]]\ , co_{\Gamma,\Theta}[\gamma])\ true$$

vi. If $ty_{\Gamma,\Delta,\rho,\sigma}$ and $ty_{\Delta,\Theta,\sigma,\tau}$ are both defined then so is $ty_{\Gamma,\Theta,\rho,\tau}$ and

$$\begin{aligned} \gamma{:}\,\Gamma, x{:}\,\rho \vdash_I\ & Id_{\tau[co_{\Gamma,\Theta}]}(\\ & \quad Subst_{\,\Theta,\tau}(P, ty_{\Delta,\Theta,\sigma,\tau}[co_{\Gamma,\Delta}[\gamma], ty_{\Gamma,\Delta,\rho,\sigma}[\gamma, x]])\ , \\ & \quad ty_{\Gamma,\Theta,\rho,\tau}[\gamma, x])\ true \end{aligned}$$

for every

$$\gamma{:}\,\Gamma \vdash_I P : Id_\Theta(co_{\Delta,\Theta}[co_{\Gamma,\Delta}[\gamma]]\ , co_{\Gamma,\Theta}[\gamma])\ true$$

We also need that the propositional isomorphisms are stable under definitional equality and under syntactic substitution.

Lemma 3.2.9. *i. If* $\vdash_I \Gamma = \Gamma'$ *and* $\vdash_I \Delta = \Delta'$ *then if* $co_{\Gamma,\Delta}$ *is defined so is* $co_{\Gamma',\Delta'}$ *and*

$$\gamma{:}\,\Gamma \vdash_I Id_\Delta(co_{\Gamma,\Delta}[\gamma], co_{\Gamma',\Delta'})\ true$$

ii. If $\vdash_I \Gamma = \Gamma'$ *and* $\vdash_I \Delta = \Delta'$ *and* $\Gamma \vdash_I \sigma = \sigma'$ *and* $\Delta \vdash_I \tau = \tau'$ *then if* $ty_{\Gamma,\Delta,\sigma,\tau}$ *is defined so is* $ty_{\Gamma',\Delta',\sigma',\tau'}$ *and*

$$\gamma{:}\,\Gamma, x{:}\,\sigma \vdash_I Id_\tau(Subst_{\Delta,\tau}(P, ty_{\Gamma,\Delta,\sigma,\tau}[\gamma,x])\ ,\ ty_{\Gamma',\Delta',\sigma',\tau'}[\gamma,x])\ true$$

for every $\gamma{:}\,\Gamma \vdash_I P : Id_\Delta(co_{\Gamma,\Delta}, co_{\Gamma',\Delta'})$.

Lemma 3.2.10. *Suppose that* $ty_{\Gamma,\Delta,\sigma,\tau}$ *is defined and that* $\Gamma' \vdash_I f \Rightarrow \Gamma$ *and* $\Delta' \vdash_I g \Rightarrow \Delta$ *are syntactic context morphisms satisfying*

$$\Gamma' \vdash_I Id_\Delta(co_{\Gamma,\Delta} \circ f\ ,\ g \circ co_{\Gamma',\Delta'})\ true$$

so in particular $co_{\Gamma',\Delta'}$ *defined, then* $ty_{\Gamma',\Delta',\sigma[f],\tau[g]}$ *is defined and*

$$\gamma'{:}\,\Gamma', x{:}\,\sigma[f] \vdash_I Id_{\tau[g \circ co_{\Gamma',\Delta'}]}(ty_{\Gamma',\Delta',\sigma[f],\tau[g]}[\gamma', x]\ ,\ ty_{\Gamma,\Delta,\sigma,\tau}[f[\gamma'], x])\ true$$

The relationship to extensional type theory is described by the following lemma.

Lemma 3.2.11. *i. If* $co_{\Gamma,\Delta}$ *is defined then* $\vdash_E |\Gamma| = |\Delta|$ *and* $\gamma{:}\,|\Gamma| \vdash_E \gamma = |co_{\Gamma,\Delta}|[\gamma] : |\Gamma|$.

ii. If $ty_{\Gamma,\Delta,\sigma,\tau}$ *is defined then* $|\Gamma| \vdash_E |\sigma| = |\tau|$ *and*

$$\gamma{:}\,|\Gamma|, x{:}\,|\sigma| \vdash_E x = |ty_{\Gamma,\Delta,\sigma,\tau}|[\gamma, x] : |\sigma|$$

3.2.5.3 A model for TT_E. We come to the construction of a model for TT_E by factoring the term model of TT_I. We write $\sim$ for an equivalence relation determined by the (linguistic) context and $[-]$ for equivalence classes.

Definition 3.2.12. The category **C** has as objects equivalence classes of contexts $\vdash_I \Gamma$ where Γ and Δ are considered equivalent if $co_{\Gamma,\Delta}$ is defined (this forms an equivalence relation by Lemma 3.2.8). A morphism between $[\Gamma]$ and $[\Delta]$ is an equivalence class of triples (A, B, f) where $A \in [\Gamma]$, $B \in [\Delta]$, and $A \vdash_I f \Rightarrow B$. Two such triples (A, B, f) and (A', B', f') are equivalent if $co_{A,A'}$ and $co_{B,B'}$ are defined and

$$\alpha{:}\,A \vdash_I Id_{B'}(co_{B,B'}[f[\alpha]]\ ,\ f'[co_{A,A'}[\alpha]])\ true$$

Again, it follows from Lemma 3.2.8 together with properties of *Id* that this defines an equivalence relation. The identity morphism on $[\Gamma]$ is given by $[(\Gamma, \Gamma, \gamma{:}\,\Gamma \vdash_I \gamma \Rightarrow \Gamma)]$. The composition of $[(B', \Xi, g)] : [\Delta] \to [\Theta]$ and $[(A, B, f)] : [\Gamma] \to [\Delta]$ is given by $[(A, \Xi, \alpha{:}\,A \vdash_I g[co_{B,B'}[f[\alpha]]] : \Xi)]$. Observe that $co_{B,B'}$ is defined by Lemma 3.2.8.

Proposition 3.2.13. **C** *is a category with terminal object* $[\diamond]$.

Proof. First we check that homsets, identities, and composition are well-defined, i.e. independent of the choice of representatives. For the homsets this is obvious, the case of identities is an instance of reflexivity. For composition we consider the diagram

$$\begin{array}{ccccccc}
A & \xrightarrow{f} & B & \xrightarrow{co} & B' & \xrightarrow{g} & \Xi \\
\downarrow{\scriptstyle co} & & \downarrow{\scriptstyle co} & & \downarrow{\scriptstyle co} & & \downarrow{\scriptstyle co} \\
A_1 & \xrightarrow{f_1} & B_1 & \xrightarrow{co} & B_1' & \xrightarrow{g_1} & \Xi_1
\end{array}$$

If the two outer squares commute up to propositional equality, i.e. if $(A, B, f) \sim (A_1, B_1, f_1)$ and $(B', \Xi, g) \sim (B_1', \Xi_1, g_1)$, then the whole rectangle commutes up to propositional equality because the inner square does by Lemma 3.2.8. It is obvious that identities are neutral. For associativity of composition assume $A \vdash_I f \Rightarrow B$, $B' \vdash_I g \Rightarrow \Xi$, $\Xi' \vdash_I h \Rightarrow \Theta$. A common representative for both ways of composing the triple is given by

$$\alpha{:}\, A \vdash_I h[co_{\Xi,\Xi'}[g[co_{B,B'}[f[\alpha]]]]] : \Theta$$

Finally, the obviously unique morphism from $[\Gamma]$ to $[\diamond]$ is given by $[(\Gamma, \diamond, ())]$.

Definition 3.2.14 (The model Q). A syntactic category with attributes **Q** over the category **C** is defined as follows: Let $[\Gamma] \in Ob(\mathbf{C})$. A family over $[\Gamma]$ is an equivalence class of pairs (Δ, σ) with $\Delta \vdash_I \sigma$, where two such pairs (Δ, σ) and (Δ', σ') are considered equivalent if $ty_{\Delta,\Delta',\sigma,\sigma'}$ is defined. The set of families is denoted $Fam([\Gamma])$.

If $[(\Delta, \sigma)] \in Fam([\Gamma])$ then the morphism $p([(\Delta, \sigma)]) : [\Gamma] \cdot [\Delta, \sigma] \to [\Gamma]$ is represented by

$$[((\Delta, x{:}\,\sigma)\ ,\ \Gamma\ ,\ \delta{:}\, \Delta, x{:}\,\sigma \vdash_I co_{\Delta,\Gamma}[\delta : \Gamma])]$$

If $[(A, B, f)] : [\Theta] \to [\Gamma]$ then the substitution $[(\Delta, \sigma)]\{[(A, B, f)]\}$ is represented by $(A\ ,\ (\alpha{:}\, A \vdash_I \sigma[co_{B,\Delta}[f[\alpha]]]))$. The morphism $q([(A, B, f)], [(\Delta, \sigma)])$ is represented by

$$\begin{array}{l}((A, x{:}\,\sigma[co_{B,\Delta} \circ f])\ ,\ (\Delta, x{:}\,\sigma)\ , \\ \qquad \alpha{:}\, A, x{:}\,\sigma[co_{B,\Delta} \circ f] \vdash_I (co_{B,\Delta} \circ f, x) : (\Delta, x{:}\,\sigma))\end{array}$$

A section of $[(\Delta, \sigma)]$ is an equivalence class of triples (Θ, τ, M) with $(\Theta, \tau) \sim (\Delta, \sigma)$ and $\Theta \vdash_I M : \tau$, where two such triples (Θ, τ, M) and (Θ', τ', M') are equivalent if

$$\vartheta{:}\, \Theta \vdash_I Id_{\tau'}(ty_{\Theta,\Theta',\tau,\tau'}[\vartheta, M]\ ,\ M'[co_{\Theta,\Theta'}[\vartheta]])\ true$$

If $[(\Theta, \tau, M)]$ is a section of $Sect([(\Delta, \sigma)])$ then a morphism $\overline{[(\Theta, \tau, M)]} : [\Gamma] \to [\Gamma] \cdot [(\Delta, \sigma)]$ is represented by

$$(\Theta\ ,(\Theta,x{:}\tau)\ ,$$
$$\vartheta{:}\Theta \vdash_I (\vartheta,M):(\Theta,x{:}\tau))$$

Conversely, if $[(A,B,f)]:[\Gamma]\to[\Gamma]\cdot[(\Delta,\sigma)]$ then we must have $B\equiv B',x{:}\rho$ since equivalent contexts have equal lengths. Thus $f\equiv(f',M)$ for $B\vdash_I M:\rho[f']$. We put $Hd([(A,B,f)]):=[(B,\sigma[f'],M)]$.

Proposition 3.2.15. **Q** *is a syntactic category with attributes.*

Proof. Routine verification. As an example we check that substitution in families is well-defined and commutes with composition. Suppose that $\mathbf{g}=[(B',\Xi,g)]:[\Delta]\to[\Theta]$ and $\boldsymbol{\sigma}=[(\Omega,\sigma)]\in Fam([\Theta])$. We have

$$\boldsymbol{\sigma}\{\mathbf{g}\}=[(B',\sigma[co_{\Xi,\Omega}\circ g])]$$

Now let $(\Omega',\sigma')\sim(\Omega,\sigma)$ be another representative for $\boldsymbol{\sigma}$. We must show that

$$(B',\sigma[co_{\Xi,\Omega}\circ g])\sim(B',\sigma'[co_{\Xi,\Omega'}\circ g])$$

By definition this is equivalent to $ty_{B',B',\sigma[co_{\Xi,\Omega}\circ f],\sigma'[co_{\Xi,\Omega'}\circ f]}$ being defined. By Lemma 3.2.10 this holds if $ty_{\Omega,\Omega',\sigma,\sigma'}$ is defined and

$$\beta'{:}B'\vdash_I Id_{\Omega'}(co_{\Omega,\Omega'}\circ co_{\Xi,\Omega}\circ g\ ,\ co_{\Omega',\Xi}\circ g\circ co_{B',B'})\ \mathit{true}$$

The former follows from the assumption $(\Omega',\sigma')\sim(\Omega,\sigma)$; the latter follows by equality reasoning from Lemma 3.2.8.

Now suppose that $(B'',\Xi',g')\sim(B',\Xi,g)$ is another representative for $\mathbf{g}$. We must show that $ty_{B',B'',\sigma[co_{\Xi,\Omega}\circ g],\sigma[co_{\Xi',\Omega}\circ g']}$ is defined. Using Lemmas 3.2.8 and 3.2.10 this boils down to

$$B'\vdash_I Id_\Omega(co_{\Omega,\Omega}\circ co_{\Xi,\Omega}\circ g\ ,\ co_{\Xi',\Omega}\circ g'\circ coB',B'')\ \mathit{true}$$

Using the fact that Id respects composition and using Lemma 3.2.8 this follows from

$$B'\vdash_I Id_{\Xi'}(co_{\Xi,\Xi'}\circ g\ ,\ g'\circ coB',B'')\ \mathit{true}$$

which is the definition of $(B'',\Xi',g')\sim(B',\Xi,g)$.

For compatibility with composition assume in addition that $\mathbf{f}=[(A,B,f)]:[\Gamma]\to[\Delta]$ and $\boldsymbol{\sigma}=[(\Omega,\sigma)]\in Fam([\Theta])$. We have $\mathbf{g}\circ\mathbf{f}=[(A,\Xi,g\circ co_{B,B'}\circ f)]$ and thus

$$\boldsymbol{\sigma}\{\mathbf{g}\circ\mathbf{f}\}=[(A,\sigma[co_{\Xi,\Omega}\circ g\circ co_{B,B'}\circ f])]$$

which equals $\boldsymbol{\sigma}\{\mathbf{g}\}\{\mathbf{f}\}$ as required.

Proposition 3.2.16. *The stripping map* $|-|$ *from* TT_I *to* TT_E *lifts to a structure preserving map from* **Q** *to the term model of* TT_E *(constructed according to Example 2.4.1) defined by* $|[\Gamma]|=|\Gamma|$, $|[(\Gamma,\sigma)]|=|\sigma|$, *and* $|[(\Gamma,\sigma,M)]|=|M|$

Proof. Immediate from Lemma 3.2.11 (for contexts and types) and rule DefEq and the definition of stripping (for context morphisms and terms).

Proposition 3.2.17. *The model* $\mathbf{Q}$ *supports* Π*-types,* Σ*-types, natural numbers, and extensional identity types, and these are preserved by* $|-|$.

Proof. Let $\boldsymbol{\sigma} = [(\Gamma, \sigma)] \in Fam(\Gamma)$ and $\boldsymbol{\tau} = [(\Gamma', x{:}\sigma')\ ,\ \tau] \in Fam(\Gamma \cdot \boldsymbol{\sigma})$. We represent $\Pi(\boldsymbol{\sigma}, \boldsymbol{\tau}) \in Fam(\Gamma)$ by

$$\gamma{:}\,\Gamma \vdash_I \Pi x{:}\,\sigma.\tau[co_{\Gamma,\Gamma'}[\gamma], ty_{\Gamma,\Gamma',\sigma,\sigma'}[\gamma, x]]$$

It follows from the definition of ty for Π-types that this is well-defined.

If $\mathbf{M} = [((\Gamma'', x{:}\sigma'')\ ,\ \tau'\ ,\ M)] \in Sect(\boldsymbol{\tau})$ we represent its abstraction $\lambda_{\boldsymbol{\sigma},\boldsymbol{\tau}}(\mathbf{M}) \in Sect(\Pi(\boldsymbol{\sigma}, \boldsymbol{\tau}))$ by

$$\gamma''{:}\,\Gamma'' \vdash_I \lambda x{:}\,\sigma''.M : \Pi x{:}\,\sigma''.\tau'$$

This is indeed a section of $\Pi(\boldsymbol{\sigma}, \boldsymbol{\tau})$ because $ty_{\Gamma,\Gamma'',\Pi x:\sigma.\tau[co_{\Gamma,\Gamma'}],ty_{\Gamma,\Gamma',\sigma,\sigma'},\Pi x:\sigma''.\tau'}$ is defined by assumption and the definition of ty for Π-types. For well-definedness we use again the definition of ty for Π-types.

Now let $\mathbf{M} = [(\Gamma'', \Pi x{:}\,\sigma''.\tau', M)] \in Sect(\Pi(\boldsymbol{\sigma}, \boldsymbol{\tau}))$ (observe that the type of M must be a dependent product) and $\mathbf{N} = [(\Gamma''', \sigma''', N)] \in Sect(\boldsymbol{\sigma})$. The application $App_{\boldsymbol{\sigma},\boldsymbol{\tau}}(\mathbf{M}, \mathbf{N}) \in Sect(\boldsymbol{\tau}\{\overline{\mathbf{N}}\})$ is represented by

$$\gamma'''{:}\,\Gamma''' \vdash_I M\ ty_{\Gamma''',\Gamma'',\sigma''',\sigma''}[\gamma''', N] : \tau'[ty_{\Gamma''',\Gamma'',\sigma''',\sigma''}[\gamma''', N]]$$

Now the required equations are readily checked, and since the Π-type and its associated term formers are modelled by their syntactic companions interspersed with canonical isomorphisms, it follows that stripping preserves the dependent product structure because the canonical isomorphisms are mapped to identities by Lemma 3.2.11.

For the identity type assume $\boldsymbol{\sigma} = [(\Gamma, \sigma)] \in Fam(\Gamma)$ and $\mathbf{M} = [(\Gamma', \sigma', M)] \in Sect(\boldsymbol{\sigma})$ and $\mathbf{N} = [(\Gamma'', \sigma'', M)] \in Sect(\boldsymbol{\sigma})$. We represent the identity type $Id_{\boldsymbol{\sigma}}(\mathbf{M}, \mathbf{N}) \in Fam(\Gamma)$ by

$$\gamma'{:}\,\Gamma' \vdash_I Id_{\sigma''}(ty_{\Gamma',\Gamma'',\sigma',\sigma''}[\gamma', M]\ ,\ N[co_{\Gamma',\Gamma''}])$$

Now if this family has a section then $\mathbf{M} = \mathbf{N}$ by definition of equality on sections and any two sections of this family are equal by *IdUni* and the definition of equality on sections.

We follow the same pattern for Σ-types and natural numbers.

Proof of Theorem 3.2.5. Consider the following informal diagram of interpretations

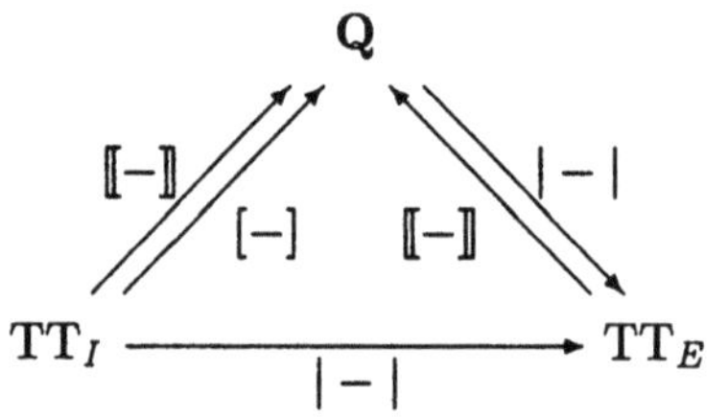

Here TT_I and TT_E denote the term models associated to the type theories TT_I and TT_E. The informal map $[\![-]\!] : \mathrm{TT}_I \to \mathbf{Q}$ denotes the interpretation

of TT_I in $\mathbf{Q}$ which being a model for TT_E also is a model for TT_I. The mapping $|-| : \mathbf{Q} \to \mathrm{TT}_E$ is the lifting of $|-| : \mathrm{TT}_I \to \mathrm{TT}_E$ given by Proposition 3.2.16. Finally, $[-] : \mathrm{TT}_I \to \mathbf{Q}$ is the canonical mapping associating equivalence classes, which is structure preserving by definition of $\mathbf{Q}$.

These maps preserve all the type and term formers up to semantic equality; thus, by induction on derivations or more elegantly by an initiality argument[2] we find that any two paths in this diagram with common source and target and starting from either TT_I or TT_E are equal. In more elementary terms this gives in particular the following identities:

i. If $\Gamma \vdash_I$ then $[\Gamma] = [\![|\Gamma|]\!]$ in $\mathbf{Q}$.
ii. If $\Gamma \vdash_I \sigma$ then $[(\Gamma, \sigma)] = [\![|\Gamma| \mid |\sigma|]\!]$ in $\mathbf{Q}$.
iii. If $\Gamma \vdash_I M : \sigma$ then $[(\Gamma, \sigma, M)] = [\![|\Gamma| \mid |M|]\!]$ in $\mathbf{Q}$.
iv. If $\Gamma \vdash_E$ then $\vdash_E |[\![\Gamma]\!]| = \Gamma$.
v. If $\Gamma \vdash_E \sigma$ then $\Gamma \vdash_E |[\![\Gamma \mid \sigma]\!]| = \sigma$.
vi. If $\Gamma \vdash_E M : \sigma$ then $\Gamma \vdash_E |[\![\Gamma \mid M]\!]| = M : \sigma$.

Now suppose that $\Gamma \vdash_I \sigma$ and $|\Gamma| \vdash_E M : |\sigma|$. The interpretation yields

$$[\![|\Gamma| \mid M]\!] \in \mathit{Sect}([\![|\Gamma| \mid |\sigma|]\!])$$

Thus $\mathit{Sect}([(\Gamma, \sigma)]) \neq \emptyset$ by Equation ii. By definition of $\mathbf{Q}$ this means that there exist Γ', σ', M' with $\Gamma' \vdash_I P' : \sigma'$ and $co_{\Gamma,\Gamma'}$ and $ty_{\Gamma,\Gamma',\sigma,\sigma'}$ are defined. Therefore we have

$$\gamma{:}\, \Gamma \vdash_I ty^{-1}_{\Gamma,\Gamma',\sigma,\sigma'}[\gamma, M'[co_{\Gamma,\Gamma'}[\gamma]]] : \sigma$$

and therefore $\Gamma \vdash_I \sigma$ *true*. Call this term M'', i.e. $\Gamma \vdash M'' : \sigma$. By definition of $\mathbf{Q}$ we have $(\Gamma, \sigma, M'') \in [\![|\Gamma| \mid M]\!]$ and thus $|\Gamma| \vdash_E |M''| = M : |\sigma|$ by Equation vi.

3.2.6 Discussion and extensions

The relative complexity and clumsiness of the present proof is regrettable and it is our hope that in the future a shorter and more elegant proof will be found. Notice, however, that the described method is fairly robust with respect to extensions of the type theory. For it to be extensible to a new type former it is enough that this type former admits an action on propositional isomorphisms. To demonstrate this we consider the addition of quotient types and of a universe closed under Π-types and natural numbers below.

A possible point of criticism is the non-constructive nature of the proof. Not only has the axiom of choice been used in the definition of the canonical isomorphisms *co* and *ty*; more seriously the interpretation of TT_E in $\mathbf{Q}$ associates equivalence classes to contexts, types, and terms. In order to get an inhabitant of type σ in the proof above we must arbitrarily choose a representative of the corresponding class. So the present proof does not directly give

[2] We cannot properly formulate initiality of $\mathrm{TT_I}$ and $\mathrm{TT_E}$ since we have not defined morphisms of models. See however Remark 2.5.8.

rise to an algorithm which effectively computes an inhabitant of σ in TT_I from a derivation of $|\sigma|$ *true* in TT_E. Such algorithm trivially exists by Markov's principle: we simply try out all possible terms and derivations and from the non-constructive proof of existence we know that this search always succeeds. But of course one would like a more efficient algorithm which makes use of the derivation in TT_E. It is, however, not clear whether the described argument gives rise to such an algorithm. An idea would be to carry out the construction of **Q** in a setoid model similar to the ones described in Chapter 5, where quotients come with a canonical choice of representatives. In the present framework of a set-theoretic presentation of syntactic categories with attributes this is, however, not possible.

Definitional equality has played a minor role in the present proof and it appears that the whole development goes through if TT_I was replaced by a type theory without definitional equality at all and rules like β replaced by corresponding constants of identity types in the style of IDUNI. The construction of the model **Q** would remain unchanged since propositionally equal objects are identified in **Q**.

3.2.6.1 Quotient types. The Nuprl version [16] of TT_E contains a quotient type former governed by the following rules.

$$\frac{\Gamma \vdash \sigma \quad \Gamma\,,\, x, x'\!:\sigma \vdash \rho[x,x']}{\Gamma \vdash \sigma/\rho} \ \text{Q-E-FORM}$$

$$\frac{\Gamma \vdash M\!:\sigma \quad \Gamma \vdash \sigma/\rho}{\Gamma \vdash [M]_\rho\!:\sigma/\rho} \ \text{Q-E-INTRO}$$

$$\frac{\Gamma \vdash M, N : \sigma \qquad \Gamma \vdash H : \rho[M,N]}{\Gamma \vdash [M]_\rho = [N]_\rho : \sigma/\rho} \ \text{Q-E-EQ}$$

$$\frac{\begin{array}{l}\Gamma, x\!:\sigma/\rho \vdash \tau[x] \qquad \Gamma, x\!:\sigma \vdash M[x] : \tau[[x]_\rho] \\ \Gamma\,,\, x, x'\!:\sigma\,,\, p\!:\rho[x,x'] \vdash M[x] = M[x'] : \tau[[x']_\rho] \\ \Gamma \vdash N : \sigma/\rho\end{array}}{\Gamma \vdash \mathit{plug}_\rho\, N\ \mathit{in}\ M : \tau[N]} \ \text{Q-E-ELIM}$$

$$\frac{}{\Gamma \vdash \mathit{plug}_\rho\, [N]_\rho\ \mathit{in}\ M = M[N] : \tau[[N]_\rho]} \ \text{Q-E-COMP}$$

Here ρ is viewed as a binary relation on σ and the idea is that σ/ρ is the type of equivalence classes of the least equivalence relation on σ containing ρ. The operator $[\,-\,]_\rho$ associates an equivalence class to an element of σ. The rule Q-E-EQ states that equivalence classes of ρ-related elements are equal in σ/ρ. Notice that this rule introduces an information loss in a way similar to ID-DEFEQ—upon its application the proof H is lost. The lifting operator "plug_ρ – *in* –" permits one to define functions on the quotient type as usual in mathematical practice by defining it on representatives and proving independence of the particular representative chosen. Notice that Q-E-EQ is required for Q-E-ELIM to make sense. Finally, the rule Q-E-COMP states intuitively that the underlying algorithm of a lifted function is the function itself.

A version of quotient types for TT_I has to replace the conclusion of Q-E-EQ by a propositional equality and include instances of *Subst* in order to make rule Q-E-ELIM typecheck. Formally, we extend TT_I by the following rules for intensional quotient types.

$$\frac{\Gamma \vdash \sigma \qquad \Gamma\ , x, x'\!:\sigma \vdash \rho[x, x']}{\Gamma \vdash \sigma/\rho} \quad \text{Q-I-FORM}$$

$$\frac{\Gamma \vdash M\!:\sigma \qquad \Gamma \vdash \sigma/\rho}{\Gamma \vdash [M]_\rho\!:\sigma/\rho} \quad \text{Q-I-INTRO}$$

$$\frac{\Gamma \vdash M, N : \sigma \qquad \Gamma \vdash H : \rho[M, N]}{\Gamma \vdash \mathit{Qax}_\rho(H) : \mathit{Id}_{\sigma/\rho}([M]_\rho, [N]_\rho)} \quad \text{Q-I-AX}$$

$$\frac{\begin{array}{l}\Gamma, x\!:\sigma/\rho \vdash \tau[x] \qquad \Gamma, x\!:\sigma \vdash M[x] : \tau[[x]_\rho] \\ \Gamma\ , x, x'\!:\sigma\ , p\!:\rho[x, x'] \vdash \\ \qquad H : \mathit{Id}_{\tau[[x']_\rho]}(\mathit{Subst}_{\sigma/\rho,\tau}(\mathit{Qax}_\rho(p), M[x])\ , M[x']) \\ \Gamma \vdash N : \sigma/\rho\end{array}}{\Gamma \vdash \mathit{plug}_\rho\ N\ \mathit{in}\ M\ \mathit{using}\ H : \tau[N]} \quad \text{Q-I-ELIM}$$

$$\frac{}{\Gamma \vdash \mathit{plug}_\rho\ [N]_\rho\ \mathit{in}\ M\ \mathit{using}\ H = M[N] : \tau[[N]_\rho]} \quad \text{Q-I-COMP}$$

Let TT_{EQ}, TT_{IQ} refer to the respective extensions of TT_E and TT_I with quotient types and let $\vdash_{EQ}, \vdash_{IQ}$ denote the corresponding judgement relations.

We notice that, like *Ext*, the term former *Qax* introduces non-canonical elements in the identity type and thus in all types. In Chapter 5 we study models[3] which induce further equations for *Qax* so that these non-canonical elements disappear, but for now TT_{IQ} is any extension of TT_I supporting the rules for quotient types. So in particular this may be one of the type theories to be studied in Chapter 5 or it may be a simple addition of the rules for quotient types to TT_I where we ignore the problem with non-canonical elements.

Let us now study the relationship between TT_{IQ} and TT_{EQ}. The interpretation $|-|$ of TT_I in TT_E extends to quotient types by setting $|\mathit{Qax}_\rho(H)| := \mathit{Refl}_{|\sigma|/|\rho|}(|[M]_R|)$ where $\Gamma \vdash_I H : \rho[M, N]$, and homomorphically extending to the other type and term formers. For example, $|\sigma/\rho| := |\sigma|/|\rho|$. Now again Proposition 3.2.2 continues to hold, i.e. if $\Gamma \vdash_{IQ} J$ then $|\Gamma| \vdash_{EQ} |J|$. Moreover, if $TT_{\Re Q}$ and $\vdash_{\Re Q}$ denote the extension of TT_{IQ} by $\Re$ as in Section 3.2.4, we obtain an analogue of Theorem 3.2.4, i.e. if $\Gamma \vdash_{EQ} J$ then $\Gamma' \vdash_{\Re Q} J'$ for some Γ', J' with $|\Gamma'| \equiv \Gamma$ and $|J'| \equiv J$. Here the only new nontrivial case arises from rule Q-E-EQ. In order to mimic this in $TT_{\Re Q}$ we have to use Q-I-AX and an instance of $\Re$.

More interestingly, TT_{EQ} is again conservative over TT_{IQ} with respect to type inhabitation.

[3] In the first of these models (Section 5.1) Q-I-FORM is restricted to relations ρ of the form $\mathit{Prf}(R)$ for some $\Gamma\ , x, x'\!:\sigma \vdash R : \mathit{Prop}$, and Q-I-ELIM is split into a non-dependent elimination rule and an induction axiom.

Theorem 3.2.18. *If $\Gamma \vdash_{IQ} \sigma$ and $|\Gamma| \vdash_{EQ} P : |\sigma|$ for some P then there exists P' such that $\Gamma \vdash_{IQ} P' : \sigma$.*

Proof. The proof follows the same pattern as the proof of Theorem 3.2.5. First, we extend the definition of the *ty*-isomorphisms as follows: Assume $\Gamma \vdash_{IQ} \sigma/\rho$ and $\Delta \vdash_{IQ} \tau/\varphi$. In this context we put $U[x{:}\sigma] := ty_{\Gamma,\Delta,\sigma,\tau}[\gamma, x] : \tau[co_{\Gamma,\Delta}[\gamma]]$. Now, in the context $\gamma{:}\, \Gamma\ ,\ x, x'{:}\, \sigma, p{:}\, \rho[x, x']$ we put

$$V := ty_{\Gamma,x,x':\sigma,\Delta,y,y':\tau,\rho,\varphi}[\gamma, x, x'\ , p] : \\ \varphi[co_{\Gamma,\Delta}[\gamma]\ ,\ ty_{\Gamma,\Delta,\sigma,\tau}[\gamma, x]\ ,\ ty_{\Gamma,\Delta,\sigma,\tau}[\gamma, x']]$$

and therefore we have

$$\gamma{:}\, \Gamma, x, x'{:}\, \sigma\ ,\ p{:}\, \rho[x, x'] \vdash_{IQ} V : \varphi[co_{\Gamma,\Delta}[\gamma]\ ,\ U[x]\ ,\ U[x']]$$

and hence

$$\gamma{:}\, \Gamma, x, x'{:}\, \sigma\ ,\ p{:}\, \rho[x, x'] \vdash_{IQ} Qax_{\varphi}(V)[co_{\Gamma,\Delta}[\gamma]] : Id([U[x]]_{\varphi}\ ,\ [U[x']]_{\varphi})$$

Therefore we may use rule Q-I-ELIM and conclude

$$\gamma{:}\, \Gamma, q{:}\, \sigma/\rho \vdash_{IQ} plug_{\rho}\, q\ in\ [U]_{\varphi}\ using\ V : (\tau/\varphi)[co_{\Gamma,\Delta}]$$

If all the participating isomorphisms were defined we make this term the value of $ty_{\Gamma,\Delta,\sigma/\rho,\tau/\varphi}$. The inverse is defined analogously.

Next we must show that Lemmas 3.2.7–3.2.11 carry over to this extension. The proofs are difficult to write down, but are essentially straightforward. As an example we show Part ii of Lemma 3.2.8. In the situation $\Gamma \vdash_{IQ} \sigma/\rho$ we must establish

$$\gamma{:}\, \Gamma, q{:}\, \sigma/\rho \vdash_{IQ} Id(q, ty_{\Gamma,\Gamma,\sigma/\rho,\sigma/\rho}[\gamma, q])\ true$$

After expanding the definitions and simplifying by using the Lemma inductively for σ and ρ, this boils down to the following property which may be compared to the (η)-rule for quotients in simple type theory [57].

Lemma 3.2.19. *If $\Gamma \vdash_{IQ} \sigma/\rho$ then*

$$\Gamma, q{:}\, \sigma/\rho \vdash_{IQ} \\ Id_{\sigma/\rho}(plug_{\rho}\, q\ in\ [x{:}\,\sigma][x]_{\rho}\ using\ [x, x'{:}\, \sigma, p{:}\, \rho[x, x']]\, Qax_{\rho}(p)\ ,\ q)\ true$$

holds. In other words the lifting of the projection $[-]_{\varphi}$ equals the identity.

Proof of Lemma. Let

$$\tau[\gamma{:}\,\Gamma, q{:}\,\sigma/\rho] := \\ Id_{\sigma/\rho}(plug_\rho\, q\; in\, [x{:}\,\sigma][x]_\rho\; using\, [x, x'{:}\,\sigma, p{:}\,\rho[x, x']]\, Qax_\rho(p)\;,\, q)$$

By rule Q-I-COMP we have

$$\gamma{:}\,\Gamma\;,\, x{:}\,\sigma \vdash_{IQ} Refl_{\sigma/\rho}([x]_\rho) : \tau[[x]_\rho]$$

Moreover, since τ is an identity type, we have

$$\gamma{:}\,\Gamma\;,\, x, x'{:}\,\sigma\;,\, p{:}\,\rho[x, x'] \vdash_{IQ} \\ H : Id_{\tau[x']}(Subst_{\,\sigma/\rho,\tau}(Qax_\rho(p), Refl_{\sigma/\rho}([x]_\rho))\;,\, Refl_{\sigma/\rho}([x']_\rho))$$

for $H := IdUni_{\,\tau[x']}(Subst_{\,\sigma/\rho,\tau}(Qax_\rho(p), Refl_{\sigma/\rho}([x]_\rho)))$. (In fact, this identity may also be proved using J alone.) Therefore, we may conclude

$$\gamma{:}\,\Gamma, q{:}\,\sigma/\rho \vdash_{IQ} plug_\rho\, q\; in\, [x{:}\,\sigma] Refl_{\sigma/\rho}([x]_\rho)\; using\, H : \tau[q]$$

as required. Having defined the canonical isomorphisms *co* and *ty* for TT_{IQ} we construct a model $\mathbf{Q}$ from TT_{IQ} in exactly the same way as in the proof of Theorem 3.2.5. In order to get an interpretation of TT_{EQ} in this model we must show that it supports the rules for quotient types from TT_{EQ}. Suppose that $\boldsymbol{\sigma} \in Fam(\Gamma)$ and $\boldsymbol{\rho} = [(\Gamma, x{:}\,\sigma, x'{:}\,\sigma'\;,\, \rho)] \in Fam(\Gamma \cdot \boldsymbol{\sigma} \cdot \boldsymbol{\sigma}^+)$. Recall that this means $\Gamma, x{:}\,\sigma, x'{:}\,\sigma' \vdash_{IQ} \rho$ and that $ty_{\Gamma,\Gamma,\sigma,\sigma'}$ is defined and that both (Γ, σ) and (Γ, σ') are representatives for $\boldsymbol{\sigma}$. We define the interpretation of the quotient type as the following family over Γ:

$$\boldsymbol{\sigma}/\boldsymbol{\rho} := [(\Gamma\;,\, \sigma/\rho')]$$

where

$$\rho' = [\gamma{:}\,\Gamma, x, x'{:}\,\sigma]\rho[x, ty_{\Gamma,\Gamma,\sigma,\sigma'}[\gamma, x']]$$

If $\mathbf{M} = [(\Gamma', \sigma'', M)] \in Sect(\boldsymbol{\sigma})$ then we put

$$[\mathbf{M}]_{\boldsymbol{\rho}} := [(\Gamma\;,\, \sigma/\rho'\;,\, [\gamma{:}\,\Gamma][ty^{-1}_{\Gamma,\Gamma',\sigma,\sigma''}[\gamma, M[co_{\Gamma,\Gamma'}[\gamma]]]]_\rho)]$$

which is a section of $\boldsymbol{\sigma}/\boldsymbol{\rho}$. If $\mathbf{M}, \mathbf{N} \in Sect(\boldsymbol{\sigma})$ and $\mathbf{H} \in Sect(\rho\{\mathbf{M}^+\}\{\mathbf{N}\})$ then by suitably composing with the *ty*-isomorphisms and using Q-I-AX and the definition of equality for sections we obtain that $[\mathbf{M}]_{\boldsymbol{\rho}} = [\mathbf{N}]_{\boldsymbol{\rho}}$, i.e. rule Q-E-EQ is validated. We continue similarly for the remaining term formers and rules and then proceed as in the proof of Theorem 3.2.5.

3.2.6.2 Universes. We consider an extension of both TT_I and TT_E by the rules for a universe closed under Π-types and natural numbers from Section 2.3.5. It is known that extending TT_E by universes allows non-normalising terms to be typed. For example, in the context $\Gamma \equiv d{:}\ U\ ,\ p{:}\ \mathit{Id}_U(d, \hat{\Pi}(d, [x{:}d]d)$ we can derive the type equality

$$\Gamma \vdash_E D = D \to D$$

for $D := \mathit{El}(d)$ using rules ID-DEFEQ and U-EQ-Π and therefore we have the judgement

$$\Gamma \vdash_E (\lambda x{:}\ D.x\ x)\ (\lambda x{:}\ D.x\ x) : D$$

In TT_I we can only construct a propositional isomorphism between D and $D \to D$ using instances of $\mathit{Subst}_{U,[x:U]\mathit{El}(x)}$. This allows one to construct a term M such that $|M|$ is non-normalising, but where M itself does not contain any redexes at all because the instances of *Subst* contain the variable p which prevents reduction.

Our aim is to show that the development in the previous sections goes through for this extension. Certainly, Theorem 3.2.4 characterising the extension of TT_I by the $\mathfrak{R}$-operator continues to hold because we make the same extension to TT_I and TT_E. More importantly, Theorem 3.2.5 establishing the conservativity of TT_E over TT_I as far as inhabitation of types is concerned, carries over as well.

Theorem 3.2.20. *The extension of TT_E by a universe closed under Π and natural numbers is conservative over TT_I with the same extension in the sense of Theorem 3.2.5.*

Proof. As in the case of quotient types we have to extend the definition of the *ty*-isomorphisms so as to account for the newly introduced types U and $\mathit{El}(M)$. A certain complication arises from the fact that it is no longer the case that if $\Gamma \vdash_I \sigma = \tau$ then σ and τ share the same outermost type former because of the "non-logical" type equalities U-EQ-Π and U-EQ-N. We may introduce an additional clause for *ty* to account for this equality, but then in order to retain transitivity (Lemma 3.2.8) we must close up under composition of *ty* from the left and from the right. Then, however, *ty* is no longer uniquely defined and one would have to establish coherence up to propositional equality. To account for arbitrary non-logical type equalities this seems indeed to be the only possible way; in the particular case of a universe we can get away by treating types of the form $\mathit{El}(M)$ separately.

We say that $\Gamma \vdash_I \sigma$ is a *small type* if $\Gamma \vdash_I \sigma = \mathit{El}(M)$ for some $\Gamma \vdash_I M : U$. Notice that in this case we must have either $\Gamma \vdash_I \sigma = \mathbf{N}$ or $\sigma \equiv \Pi x_1{:}\tau_1.\cdots \Pi x_k{:}\tau_k.\mathit{El}(N)$ for some $k \geq 0$ and $\Gamma, x_1{:}\tau_1, \cdots x_k{:}\tau_k \vdash_I N : U$ and small $\tau_1, \ldots, \tau_k$. Now we introduce the following two clauses to deal with U and small types:

$$ty_{\Gamma,\Delta,U,U}[\gamma{:}\ \Gamma, x{:}\ U] = \begin{cases} x, \text{ if } co_{\Gamma,\Delta} \text{ is defined} \\ \text{undefined otherwise} \end{cases}$$

and

$$ty_{\Gamma,\Delta,\sigma,\tau}[\gamma{:}\,\Gamma, x{:}\,\sigma] = Subst_{[x:U]El(x)}(P, x)$$

if there exist M, N, P with $\Gamma \vdash_I \sigma = El(M)$ and $\Delta \vdash_I \tau = El(N)$, i.e. σ and τ are small, and $\Gamma \vdash_I P : Id_U(M, N[co_{\Gamma,\Delta}])$, and undefined otherwise. In the former case M, N, P are chosen arbitrarily. The inverses are defined analogously.

The clause for $\mathbf{N}$ in Section 3.2.5.2 is removed and the clause for Π-types in Section 3.2.5.2 is restricted to those cases which are not yet dealt with by the above clause for small types, i.e. $ty_{\Gamma,\Delta,\Pi x:\sigma_1.\sigma_2,\Pi y:\tau_1.\tau_2}$ is defined by the clause given there only if $\Pi x{:}\,\sigma_1.\sigma_2$ and $\Pi x{:}\,\tau_1.\tau_2$ are not small.

Now it is easy to show that the auxiliary properties of these canonical isomorphisms continue to hold. Notice that if $ty_{\Gamma,\Delta,\sigma,\tau}$ is defined and σ is small so is τ.

It remains to show that the model $\mathbf{Q}$ constructed from the extended syntax as in Definition 3.2.14 still supports Π-types, natural numbers, and a universe.

The type of natural numbers is given by $\mathbf{N} := [(\Gamma, \mathrm{N})]$. We have $[(\Gamma, \mathrm{N}, 0)] \in Sect(\mathbf{N})$. If $\mathbf{M} = [(\Delta, \sigma, M)] \in Sect(\mathbf{N})$ then we must have $\Delta \vdash_I \sigma = El(S)$ and $\Delta \vdash_I P : Id(S, \hat{\mathrm{N}})$ for some P. Now we define

$$Suc(\mathbf{M}) := [(\Delta, \mathrm{N}, Suc(Subst\,(P, M)))]$$

In a similar way we define $R^{\mathbf{N}}$ using pre-composition with instances of $Subst_{U,El}$ where necessary.

For the Π-type we only need to consider the case $\Pi(\boldsymbol{\sigma}, \boldsymbol{\tau})$ where both $\boldsymbol{\sigma}$ and $\boldsymbol{\tau}$ are small since in all other cases we may proceed as in the proof of Proposition 3.2.17 above. So assume $\boldsymbol{\sigma} = [\Gamma, El(S)]$ and w.l.o.g. $\boldsymbol{\tau} = [\Gamma, x{:}\,El(S)\ ,\ El(T[x])]$. We define $\Pi(\boldsymbol{\sigma}, \boldsymbol{\tau})$ as $[(\Gamma, El(\hat{\Pi}(S, T)))]$. That this is well-defined and also the definition of the associated combinators follow from the following Lemma which states that the ty-isomorphisms for Π-types remain essentially unchanged by the alteration for small types.

Lemma 3.2.21. *Suppose that* $\Gamma \vdash_I S : U$ *and* $\Delta \vdash_I T : U$ *and* $\Gamma, x{:}\,El(S) \vdash_I M : U$ *and* $\Delta, x{:}\,El(T) \vdash_I N : U$ *and* $\Gamma \vdash_I P : Id(S, T[co_{\Gamma,\Delta}])$ *and* $\Gamma, x{:}\,El(S) \vdash_I Q : Id(M, N[Subst\,(P, x)])$, *i.e.* $ty_{\Gamma,x:El(S),\Delta,x:El(T),El(M),El(N)}$ *is defined. Then*

$$\begin{aligned}&\gamma{:}\,\Gamma\ ,\ u{:}\,\Pi x{:}\,El(S).El(M) \vdash_I Id(\\ &\quad ty'_{\Gamma,\Delta,\Pi x:El(S).El(M),\Pi x:El(T).El(N)}[\gamma, x]\ ,\\ &\lambda x{:}\,\tau[co_{\Gamma,\Delta}].Subst_{U,El}(Q, u\ ty^{-1}_{\Gamma,\Delta,El(S),El(T)}[\gamma, x]))\ true\end{aligned}$$

where ty' is the propositional isomorphism given by the (now overridden) clause of Section 3.2.5.2.

Proof of Lemma. Using *Ext* and elementary equality reasoning. □

We may now replace instances of *ty* at small Π-types by *ty'* and proceed as in the proof of Proposition 3.2.17.

The universe and its externalisation are defined as the equivalence classes of U and El, respectively. The code $\hat{\mathbf{N}}$ in context $[\Gamma]$ is given by $[\Gamma, U, \hat{\mathbf{N}}]$. Similarly, we define the code of Π as the lifting to equivalence classes of the syntactic term former $\hat{\Pi}$. That this is well-defined is readily checked using RESP and *Ext*.

We conclude as in the Proof of Theorem 3.2.5.

Remark 3.2.22. We believe that this result extends to an *impredicative* universe as defined in Section 2.3.4, but we have not been able to prove this because it is not clear how the $\forall$-operator lifts to equivalence classes. That is, if $ty_{\Gamma,\Delta,\sigma,\tau}$ is defined and $\Gamma, x{:}\tau \vdash_I M : Prop$ then it is not obvious why one should have

$$\Gamma \vdash_I Id_{Prop}(\forall x{:}\sigma.M[ty_{\Gamma,\Delta,\sigma,\tau}]\,,\, \forall x{:}\tau.M)\ true$$

although this can probably be shown by induction on the structure of σ.

We also remark that the existence of a term M in TT_I such that $|M|$ is non-normalising, is now an immediate consequence of the conservativity theorem 3.2.5 extended to universes.

In Chapters 4 and 5 we also consider *Prop*-valued Leibniz equality as an implementation of propositional equality. In general, Leibniz equality is much weaker than the identity type in TT_I because substitution (*Subst*) is possible only for "propositional" families of the form $x{:}\sigma \vdash Prf(P)$. Thus, the results reported in this chapter do not apply to Leibniz equality. However, the type theory $\mathbf{S}_0$ we consider in Chapter 5 has the unusual property that its Leibniz equality *behaves* like the identity type so that the results do carry over.

3.2.7 Conservativity of quotient types and functional extensionality

We have now studied in much detail the question of conservativity of TT_E over TT_I. Another important question is the conservativity of TT_I over pure type theory without extensional constructs added. Let TT_p denote the type theory without *Ext* but with *IdUni*, and let $\vdash_p$ denote the corresponding judgement relation. Certainly, we cannot in general have that $\Gamma \vdash_p \sigma$ and $\Gamma \vdash_I \sigma\ true$ implies $\Gamma \vdash \sigma\ true$ because the type of *Ext* is valid in TT_p. However, we believe that the following holds.

Conjecture 3.2.23. If $\Gamma \vdash_p \sigma$ and neither Γ nor σ contain instances of the Π-type then $\Gamma \vdash_I \sigma\ true$ implies $\Gamma \vdash_p \sigma\ true$.

A possible proof of this would use one of the models to be presented in Chapter 5 and show that a Π-free type is (propositionally) isomorphic to its denotation in the model. We shall try to answer this question in future work. In [32, Section 4.4.2] such a conservativity result is proved for first-order predicate logic over simple type theory using a similar method, but the situation is much simpler there because one only has to consider bi-implication instead of isomorphism.

As shown in [104], uniqueness of identity is not conservative over type theory without it in any reasonable sense, because for example we have

$$x{:}\,\sigma\ ,\ u, v{:}\,\tau[x] \vdash_I Id_{\Sigma x:\sigma.\tau[x]}(pair(x,u), pair(x,v)) \rightarrow Id_{\tau[x]}(u,v)\ true$$

but this does not hold in the pure type theory with J alone. Probably one has conservativity for types not containing the identity type, but this is logically uninteresting.

Surprisingly, quotient types are not conservative over TT_p because in their presence a weak form of functional extensionality is derivable. Suppose that $\vdash_p F, G : \Pi x{:}\,\sigma.\tau$ and $\vdash_p H : \Pi x{:}\,\sigma.Id_\tau(F\ x, G\ x)$. We claim that

$$\vdash_{IQ} Id_{\Pi x:\sigma.\tau}(\lambda x{:}\,\sigma.F\ x, \lambda x{:}\,\sigma.G\ x)\ true$$

using quotient types, but of course without using functional extensionality. To get the conclusion of functional extensionality from this (in the sense of Section 3.1.3) an η-rule for Π-types would be required.

Define

$$ExtEq[u, v{:}\,\Pi x{:}\,\sigma.\tau] := \Pi x{:}\,\sigma.Id_\tau(u\ x, v\ x)$$

and consider the quotient type $(\Pi x{:}\,\sigma.\tau)/ExtEq$. Now we have

$$x{:}\,\sigma, u{:}\,\Pi x{:}\,\sigma.\tau \vdash (u\ x) : \tau[x]$$

and

$$x{:}\,\sigma\ ,\ u, v{:}\,\Pi x{:}\,\sigma.\tau\ ,\ p{:}\,ExtEq[u,v] \vdash p\ x : Id_{\tau[x]}(u\ x, v\ x)$$

Thus we can "lift" application to the quotient type and get

$$\begin{array}{l} x{:}\,\sigma, q{:}\,(\Pi x{:}\,\sigma.\tau)/ExtEq \vdash_{IQ} \\ \quad plug_{ExtEq}\ q\ in\ [u{:}\,\Pi x{:}\,\sigma.\tau]u\ x\ using\ [u, v{:}\,\Pi x{:}\,\sigma.\tau\ ,\ p{:}\,ExtEq[u,v]]p\ x : \\ \qquad \tau[x] \end{array}$$

and by abstracting from $x{:}\,\sigma$ we obtain

$$\begin{array}{l} q{:}\,(\Pi x{:}\,\sigma.\tau)/ExtEq \vdash_{IQ} \\ \quad \lambda x{:}\,\sigma.plug_{ExtEq}\ q\ in\ [u{:}\,\Pi x{:}\,\sigma.\tau]u\ x \\ \qquad\qquad using\ [u, v{:}\,\Pi x{:}\,\sigma.\tau\ ,\ p{:}\,ExtEq[u,v]]p\ x \\ \quad : \Pi x{:}\,\sigma.\tau \end{array}$$

Call this function *Extract*, i.e.

$$q{:}\,(\Pi x{:}\,\sigma.\tau)/ExtEq \vdash_{IQ} Extract[q] : \Pi x{:}\,\sigma.\tau$$

Notice that by rule Q-I-COMP we have

$$f{:}\,\Pi x{:}\,\sigma.\tau \vdash_{IQ} Extract[[f]_{ExtEq}] = \lambda x{:}\,\sigma.f\ x : \Pi x{:}\,\sigma.\tau$$

We have

$$Qax_{ExtEq}(H) : Id_{(\Pi x:\sigma.\tau)/ExtEq}([F]_{ExtEq}, [G]_{ExtEq})$$

and therefore $Id_{\Pi x:\sigma.\tau}(\lambda x{:}\,\sigma.F\ x, \lambda x{:}\,\sigma.G\ x)$ *true* by RESP.

We do not know whether quotient types are conservative over extensional type theory (TT_E).

3.3 Related work

Astonishingly, the meta-theory of propositional equality, in particular in the context of intensional type theory, has attracted very little attention in the literature. In the standard reference [88] the intensional (TT_I) and the extensional (TT_E) versions of type theory are introduced and compared by way of example.

Streicher [104] was the first to see the need for uniqueness of identity as an additional principle for propositional equality in intensional type theory. He also realised that the complicated elimination rule J may be replaced by the Leibniz principle *Subst* and uniqueness of identity, if one is interested in the latter. In *loc. cit.* an η-rule for the identity type is introduced and it is shown that it entails equality reflection, i.e. the rule ID-DEFEQ. The η-rule is as follows: If $x, y{:}\,\sigma\ ,\ p{:}\ Id_\sigma(x,y) \vdash M[x,y,p] : \tau[x,y,p]$ then

$$x, y{:}\,\sigma\ ,\ p{:}\ Id_\sigma(x,y) \vdash J([x{:}\,\sigma]M[x,x,\mathit{Refl}(x)]\ ,\ x, y\ ,\ p) = M : \tau[x,y,p]$$

Streicher's work has been taken up in [37] and syntactic consequences of uniqueness of identity in the context of Pure Type Systems are studied.

Luo [68, 70] studies a few meta-properties of propositional equality in the context of the Extended Calculus of Constructions. He discusses the so-called *equality reflection principle* which states that propositional and definitional equality agree *in the empty context*. This principle (not to be confused with the equality reflection *rule* ID-DEFEQ) is (in the absence of extensional constructs) an immediate consequence of normalisation. Luo also considers formulations of the Calculus of Constructions which have an identity type inside *Prop* and nevertheless allow for elimination (*Subst*, J) over arbitrary families, not only propositional ones. He observes that these identity types and the definable Leibniz equality imply each other. However, as shown by Streicher in [104] this bi-implication only forms a retraction and not a propositional isomorphism between the two incarnations of propositional equality and so in particular it is not possible to define a J-like elimination rule for Leibniz equality in this case.

Recently, Gabbay and de Queiroz [34] have looked at propositional equality in the context of *labelled deductive systems*. In their setting not only propositional equality but also definitional equality is witnessed by a proof term ("label"), written as a subscript to =, which in this case records instances of reflexivity, symmetry, transitivity, and of the congruence rules. They have the following introduction rule for identity types (given here in our notation):

$$\frac{\Gamma \vdash M =_s N : \sigma}{\Gamma \vdash \mathit{Refl}_\sigma(M,N,s) : Id_\sigma(M,N)}$$

and the elimination rule

$$\frac{\Gamma \vdash \tau \qquad \Gamma\ ,\ M =_t N{:}\,\sigma \vdash L : \tau \qquad \Gamma \vdash P : Id_\sigma(M,N)}{\Gamma \vdash \mathit{TEST}(P,L) : \tau}$$

where t is a variable that becomes bound in the *TEST*-expression. They claim that this elimination rule for propositional equality is simpler than ID-ELIM-J, but they do not compare to the formulation with *Subst* and *IdUni*. It is difficult to give a satisfactory comparison between their system and TT_I, say, because the framework is very different; for example the hypothetical use of definitional equalities in the elimination rule is not possible in TT_I and in fact we consider it the *raison d'être* of propositional equality that it may be assumed hypothetically. We conjecture that one may give a translation between Gabbay and de Queiroz' calculus and TT_I by mapping both annotated definitional equality and propositional equality to the identity type, but since their calculus is not given in a fully formal way it is hard to give more detail. As an application they give a proof of a constructive variant of Leisenring's formula $\exists x.\forall y.P(x) \Rightarrow P(y)$, that is they show that the type (in our notation)

$$\sigma \to \Sigma x{:}\, \sigma.\Pi y{:}\, \sigma.Id_\sigma(x,y) \to \tau[x] \to \tau[y]$$

is inhabited. Obviously, an inhabitant may be constructed using *Subst* as well.

Extensional quotient types like the ones considered in Section 3.2.6.1 have been theoretically investigated by Mendler [82] who finds that they correspond to categorical coequalisers. See also the section on related work in Chapter 5.

The problem with the failure of strong normalisation in the presence of extensional identity types and universes (Section 3.2.6.2) has also been noticed in [41] where it is concluded that for this reason identity types cannot be used to describe sharing constraints in Standard ML functor's [83] and that sharing by parametrisation [12, 11] ought to be used instead. It seems that the intensional identity type provides a reasonable alternative to sharing by parametrisation. We leave this to further research.

4. Proof irrelevance and subset types

Our aim in this chapter is to present a syntactic model for the extensional constructs of proof irrelevance and subset types. The model we give not only provides these concepts, but also relates two approaches to program development in Martin-Löf type theory [88] or the Extended Calculus of Constructions (ECC) [68]. Under the first one specifications and types are freely mixed using Σ-types. This methodology underpins a project for development of correct software at the University of Ulm [112]. It is also the most natural approach and is probably employed by most users of type systems like the ones under consideration.

In the second approach types and specifications are kept completely distinct. This way of using type theory has been studied systematically in McKinna's thesis [78, 13]. We consider both approaches in detail. For definiteness we work in the Calculus of Constructions extended with Σ-types and natural numbers which enables us to use the defined logical connectives and Leibniz equality ($\stackrel{L}{=}$). Much of the development to follow can also be carried out in Martin-Löf type theory replacing Leibniz equality by the identity type and the logical connectives by product, sum, etc..

4.1 The refinement approach

Let us start with an example. A type of even numbers can be formed as

$$Even := \Sigma n{:}\,\mathbf{N}.\exists m{:}\,\mathbf{N}.n \stackrel{L}{=} 2m$$

An element of type *Even* is thus a natural number together with a proof that it is even. More generally, a function with result type *Even* can be viewed as an $\mathbf{N}$-valued function together with a proof that it only takes on even values. Such a function can thus be viewed as a "verified program". For lack of a better name we call this methodology the *refinement approach*[1]. There are basically two problems with this approach. First, the projection from a refined type such as *Even* to its underlying type of elements (here $\mathbf{N}$) is in general not injective[2],

[1] In the terminology of Hayashi [43] *Even* is a *refinement* of $\mathbf{N}$.

[2] If we formulate this example using the identity type instead of Leibniz equality and Σ instead of $\exists$, then we could prove using Proposition 3.1.1 and R^{Σ} that the projection *is* injective. However, for a predicate like $[n{:}\,\mathbf{N}]\exists m, m'{:}\,\mathbf{N}.m > 1 \wedge m' > 1 \wedge n \stackrel{L}{=} m \times m'$ stating that n is composite this is not possible.

for two elements of type *Even* may have the same underlying number, but two different proofs of evenness. More seriously, a function out of a refined type even into a base type like $\mathbf{N}$ may depend on the proof component. For example if $f : Even \rightarrow Even$ then f decomposes into

$$f_1 : \Pi n{:}\,\mathbf{N}.Prf(\exists m{:}\,\mathbf{N}.n \stackrel{L}{=} 2m) \rightarrow \mathbf{N}$$

and

$$f_2 : Prf(\forall n{:}\,\mathbf{N}.\forall p{:}\,\exists m{:}\,\mathbf{N}.n \stackrel{L}{=} 2m.\exists m'{:}\,\mathbf{N}.(f_1\ n\ p) \stackrel{L}{=} 2m')$$

So even in order to compute the underlying natural number of $(f\ x)$ for some $x : Even$, both components of x are required. Even if by strong normalisation we know that $(f\ x).1$ eventually converts to a natural number, this computation may in principle require evaluation of the proof part of the argument, $x.2$. Hence one cannot honestly claim that f is a verified program; it is at most a computation which bears some extensional resemblance to an intended algorithm.

The situation is not quite as serious as it might appear, since one can show that computations of type $\mathbf{N}$ (and other basic inductive types) cannot really depend on proofs.

Proposition 4.1.1. *Let $\diamond \vdash P : Prop$ be a proposition in the empty context and $p : Prf(P) \vdash M : \mathbf{N}$ be a term of type $\mathbf{N}$ containing a variable of type $Prf(P)$. Then M is definitionally equal to a numeral, i.e. $p : Prf(P) \vdash M = Suc^n(0)$ for some $n \in \omega$.*

Proof. We give the proof for the Calculus of Constructions with natural numbers, but without Σ-types, i.e. the only type formers are $\mathbf{N}$, *Prop*, *Prf*, Π. The extension with Σ-types and other inductive types is similar. We reason by induction on the length of the normal form of M. If $M = 0$ or $M = Suc(M')$ the result follows directly or from the induction hypothesis. If $M = R^{\mathbf{N}}_{[x:\mathbf{N}]\mathbf{N}}(M_z, M_s, N)$ then by induction N is a numeral, thus M was not in normal form for it admits a reduction using NAT-COMP-ZERO or NAT-COMP-SUC. Now assume that the normal form of M is of the form $(M_1\ M_2\ \ldots M_n)$ where M_1 is not an application. If M_1 is a variable then by assumption it would have to be $p : Prf(P)$. But then for M to typecheck we would have to have

$$Prf(P) = \Pi x_2{:}\,\sigma_2.\ldots.\Pi x_n{:}\,\sigma_n.\mathbf{N}$$

for suitable types $\sigma_2 \ldots \sigma_n$. But this is possible only if $\mathbf{N} = Prf(\hat{\mathbf{N}})$ for some $\hat{\mathbf{N}} : Prop$ which is not the case.

On the other hand if

$$M_1 = R^{\mathbf{N}}_{[x:\mathbf{N}]\Pi x_2:\sigma_2.\ldots.\Pi x_n:\sigma_n.\mathbf{N}}(M_z, M_s, N)$$

then again by induction we can assume that N is a numeral whereby M was not in normal form. This exhausts all possible cases for a normal form of M.

We observe that the proof depended upon the fact that $\mathbf{N}$ is not a proposition and that M does not contain any free variables other than than $p : Prf(P)$. Indeed, it is not necessarily the case that an *open* term of type $\mathbf{N}$ containing p is definitionally equal to a term not containing p. For example

$$M[p{:}\, Prf(P), x{:}\,\mathbf{N}] :=$$
$$R^{\mathbf{N}}_{[y:\mathbf{N}]Prf(P)\to\mathbf{N}}(\lambda f{:}\, Prf(P).0, \lambda y{:}\,\mathbf{N}.\lambda f{:}\, Prf(P) \to \mathbf{N}.f\ , x)\ p$$

is in normal form. Now clearly every instance $M[p, N]$ for some numeral N converts to 0, but M itself does not convert to 0. The reason is that the application to p cannot be moved inside the recursion operator.

Of course, using induction ($R^{\mathbf{N}}$) we find $p{:}\, Prf(P), x{:}\,\mathbf{N} \vdash Id_{\mathbf{N}}(M[p, x]\ ,\ 0)$ *true*.

4.2 The deliverables approach

In [78] and [13] another approach to program development is advocated under which a *specification* is a type σ together with a predicate $P : \sigma \to Prop$ and a function between two specifications (σ, P) and (σ', P') is a function $f_1 : \sigma \to \sigma'$ together with a "proof" $f_2 : \Pi a{:}\,\sigma.(P\ a) \to P'\ (f_1\ a)$. Such a pair (f_1, f_2) is called a *deliverable* for it corresponds to a program together with its verification. Thus a deliverable contains an actual algorithm (f_1) which does not depend on proof components; by forgetting the second component of a deliverable one gets rid of all computationally irrelevant information. In [78] various examples of program development in this framework have successfully been carried out.

What we consider a serious disadvantage of the deliverables approach is that the user has to do a lot of bookkeeping. He or she has to keep track of which proof corresponds to which program, probably by means of a suitable choice of identifiers. Also a lot of work has to be done twice. For example if two deliverables are composed one has to compose the function and proof parts separately. This problem was realised in *loc. cit.* too, and as a remedy it is shown there that specifications and deliverables form a *semi-cartesian closed category* whereby one may use categorical combinators like composition, abstraction, and application to obtain new deliverables from already constructed ones.

Our aim in this chapter is to extend this result substantially by showing how deliverables can be organised into a full-blown model of the Calculus of Constructions where specifications interpret types and deliverables interpret terms. Moreover, we shall see that in this model the underlying type of the specification interpreting a Σ-type $\Sigma x{:}\,\sigma.P\ x$ for some *Prop*-valued predicate P is the same as the underlying type of the specification interpreting σ. Thus type refinement using Σ-types over *Prop*-valued predicates only changes the predicate component, but leaves the underlying type unchanged. This means that we can basically use the refinement approach as an *internal language* for deliverables thereby unifying the advantages of the two approaches described above. Furthermore, we shall see that in the deliverables model the following type is inhabited

$$\Gamma \vdash \Pi p, q : Prf(P).p \stackrel{L}{=} q$$

for any proposition $\Gamma \vdash P : Prop$. Thus in the model the projection from a refined type to its underlying "algorithmic" type is provably injective.

4.3 The deliverables model

In this section we construct a syntactic category with attributes whose contexts and families are (modulo some syntactic clutter) pairs of types and Prop-valued predicates, whereas morphisms and thus terms are pairs consisting of an ordinary term of the underlying type and a proof that the predicates are respected. In this model we shall identify a family **Prop** which has as underlying type the type *Prop* and the trivial predicate $[x{:}\,Prop]tt$ (recall that tt stands for the true proposition $\forall P{:}\,Prop.Prf(P) \rightarrow Prf(P))$. If $P : Prop$ and P' is some proof that this trivial predicate is satisfied for P, i.e. $P' : Prf(tt)$, then we define a specification $\mathbf{Prf}(P, P')$ with underlying type $\mathbf{1}_E$ (extensional unit type) and predicate $\lambda x{:}\,\mathbf{1}_E.P$. We shall see that with this choice we obtain a (loose) model of the Calculus of Constructions which meets the requirements set out in the previous section. Let us now describe the model in detail. We use the generic notations $\mathbf{C}$, *Fam*, etc. to denote the entities in the deliverables model.

4.3.1 Contexts

Let Γ be a (syntactic) context. A *propositional telescope* over Γ is a telescope Δ with respect to Γ, i.e. $\Gamma \vdash \Delta$, such that all the types in Δ are of the form $Prf(M)$ for some term M. In particular, the empty telescope is propositional.

A *context of specifications* is a pair (Γ, Δ) such that Γ is a valid (syntactic) context and Δ is a propositional telescope over Γ. For example the pair $(\diamond, \diamond)$ forms a context of specifications which is denoted $\top$. Another example is $(x{:}\,\mathbf{N}, y{:}\,\mathbf{N}\ ,\ p : Prf(x + y \stackrel{L}{=} 5), q : Prf(x - y \stackrel{L}{=} 1))$.

If (Γ, Δ) and (Γ', Δ') are contexts of specifications then a *morphism* (a deliverable) from the former to the latter is a pair (f, g) where f is a (syntactic) context morphism from Γ to Γ', i.e. $\gamma{:}\,\Gamma \vdash f[\gamma] : \Gamma'$, and g is an element of the telescope $\Delta'[f]$ over Γ, Δ, i.e.

$$\gamma{:}\,\Gamma, \delta{:}\,\Delta \vdash g[\gamma, \delta] : \Delta'[f[\gamma]]$$

For example the pair $(3, 2)$ forms the first component of a morphism from the empty context of specifications $\top$ to the example with natural numbers considered above. The second component would contain two instances of reflexivity.

It is obvious that the morphisms of contexts of specifications contain identities and are closed under component-wise composition so that they form a category. It is also clear that this category $\mathbf{C}$ has a terminal object, viz. $\top = (\diamond, \diamond)$.

Notation. If $\Gamma \in \mathbf{C}$ is a context of specifications we denote by Γ_{set} its underlying type and by Γ_{pred} its second component, the propositional telescope over Γ_{set}. If $f : \Gamma \to \Delta$ is a morphism of contexts of specifications we denote by f_{fun} its first component, the syntactic context morphism from Γ_{set} to Δ_{set}; we denote by f_{resp} the second component, the element $\Gamma_{set}, \Gamma_{pred} \vdash f_{resp} : \Delta_{pred}[f_{fun}]$.

Equality of contexts of specifications and their morphisms is component-wise definitional equality.

4.3.2 Families of specifications

Let Γ be a context of specifications. A *family of specifications* over Γ is a pair $(\sigma_{set}, \sigma_{pred})$ where

$$\Gamma_{set} \vdash \sigma_{set}$$

and

$$\Gamma_{set}, x : \sigma_{set} \vdash \sigma_{pred} : Prop$$

for instance, if $\Gamma_{set} = \mathbf{N}$ and $\Gamma_{pred}[n{:}\,\mathbf{N}] = Even[n]$ then $\sigma_{set}[m{:}\,\Gamma_{set}] = \mathbf{N}$ and $\sigma_{pred}[m{:}\,\Gamma_{set}, n{:}\,\mathbf{N}] = (m \stackrel{L}{=} n)$ is a family of specifications.

The set of these families is denoted $Fam(\Gamma)$. Equality of families of specifications is again point-wise definitional equality. If $\sigma \in Fam(\Gamma)$ then we define the comprehension $\Gamma \cdot \sigma$ by

$$(\Gamma \cdot \sigma)_{set} := \Gamma_{set}, x{:}\,\sigma_{set}$$

$$(\Gamma \cdot \sigma)_{pred}[\gamma{:}\,\Gamma_{set}, x{:}\,\sigma_{set}] := \Gamma_{resp}[\gamma], p{:}\,Prf(\sigma_{pred}[\gamma, x])$$

The projection $p(\sigma)$ is given by

$$p(\sigma)_{fun}[\gamma{:}\,\Gamma_{set}, x{:}\,\sigma_{set}] := \gamma$$

$$p(\sigma)_{resp}[\gamma{:}\,\Gamma_{set}, x{:}\,\sigma_{set}\ , p{:}\,\Gamma_{pred}[\gamma], q{:}\,Prf(\sigma_{pred}[\gamma, x])] := p$$

If $f : B \to \Gamma$ and $\sigma \in Fam(\Gamma)$ we define the substitution of f in σ by

$$\sigma\{f\}_{set}[\beta{:}\,B_{set}] := \sigma_{set}[f_{fun}[\beta]]$$

$$\sigma\{f\}_{pred}[\beta{:}\,B_{set}, x{:}\,\sigma_{set}[f_{fun}[\beta]]] := \sigma_{pred}[f_{fun}[\beta], x]$$

We also define the morphism $q(f, \sigma)$ by

$$q(f, \sigma)_{fun}[\beta{:}\,B_{set}, x{:}\,\sigma_{set}[f_{fun}[\beta]]] := (f[\beta], x)$$

$$q(f, \sigma)_{resp}[\beta{:}\,B_{set}, x{:}\,\sigma_{set}[f_{fun}[\beta]], p{:}\,B_{pred}[\beta], q{:}\,Prf(\sigma_{pred}[f_{fun}[\beta], x])] := (f_{resp}[p], q)$$

So the first components of all the definitions made follow exactly the term model (Example 2.4.1), whereas in the second component we carry through the proofs that the predicates are respected.

4.3.3 Sections of specifications (deliverables)

Let $\sigma \in Fam(\Gamma)$. A *section* of σ is a pair (M_{fun}, M_{resp}) where

$$\gamma{:}\, \Gamma_{set} \vdash M_{fun}[\gamma] : \sigma_{set}[\gamma]$$

$$\gamma{:}\, \Gamma_{set}, p{:}\, \Gamma_{pred}[\gamma] \vdash M_{resp}[p] : Prf(\sigma_{pred}[M_{fun}[\gamma]])$$

A section M gives rise to a context morphism $\overline{M}$ from Γ to $\Gamma \cdot \sigma$ by decreeing

$$\overline{M}_{fun}[\gamma{:}\, \Gamma_{set}] := (\gamma, M_{fun}[\gamma])$$

$$\overline{M}_{resp}[\gamma{:}\, \Gamma_{set}, p{:}\, \Gamma_{pred}[\gamma]] := (p, M_{resp}[\gamma, p])$$

As in the term model, composition with the projection $p(\sigma)$ yields the identity.

If $((f,M)\,,\,(p,Q)) : \Gamma \to \Delta \cdot \sigma$ is a morphism of contexts of specifications then from the definition of such morphisms we obtain that

$$(f,p) : \Gamma \to \Delta \qquad\qquad \gamma{:}\, \Gamma_{set} \vdash M : \sigma_{set}[f[\gamma]]$$
$$\gamma{:}\, \Gamma_{set}, r{:}\, \Gamma_{pred}[\gamma] \vdash Q : Prf(\sigma_{pred}[f[\gamma], M[\gamma]])$$

and thus $Hd((f,M)\,,\,(p,Q)) := (M,Q)$ is a section of $\sigma\{(f,p)\}$. By straightforward equality reasoning we now obtain:

Proposition 4.3.1. *Contexts of specifications together with families of specifications and their sections form a syntactic category with attributes.*

We call this model the *deliverables model*, or **D** for short. Proving properties about syntactic models like the above proposition is often elementary, but requires some bookkeeping effort due to the relatively complex syntax. It is therefore ideally suited for machine-supported reasoning. In the following we describe how the Lego system can be used for mechanised proof of such properties.

4.4 Model checking with Lego

Our definition of syntactic categories with attributes contains no conditional equations. This makes it particularly easy to verify the correctness of an instance of this definition by means of an implementation of type theory, e.g. Lego [71]. The idea is to code the structure using higher universes as a metalanguage. One problem is that Lego (or any other implementation of type theory) does not support contexts and telescopes, which means that we have to encode these using Σ-types. These encodings, however, are slightly weaker than actual telescopes for they do not have surjective pairing up to conversion. Therefore in order to check certain equations we have to perform η-expansions, i.e. replace x by $(x.1, x.2)$. Since surjective pairing holds on the level of contexts, any equation which holds in Lego modulo such expansions does hold in the actual model. Also in Lego we encode meta-level pairs, like the ones occurring in the definition of contexts of specifications, as internal Σ-types.

Again, η expansions may become necessary. These issues will become clear by way of example. In order to reduce redundancy we adopt a record notation for Lego which unfortunately is not (yet) implemented.[3] For the syntax and the pragmatics of Lego we refer the reader to [71].

4.4.1 Records in Lego

The declaration

$[R = \texttt{<<}l_1 : S_1, \ldots, l_n : S_n\texttt{>>}]$

expands to the following sequence of declarations.

$[R = \texttt{<}l_1 : S_1\texttt{><}l_2 : S_2\texttt{>} \ldots \texttt{<}l_{n-1} : S_{n-1}\texttt{>}S_n]$ — Σ-type

$[\texttt{r}:R]$ $[l_1 = r.1 : S_1]$ — 1st projection

$[l_2 = r.2.1 : S_2]$

...

$[l_{n-1} = r\underbrace{.2.2.\ldots.2}_{n-2 \text{ times}}.1]$

$[l_n = r\underbrace{.2.2.\ldots.2}_{n-1 \text{ times}}]$ — nth projection

Discharge r;

$[\texttt{Mk}R[l_1 : S_1] \ldots [l_n : S_n] = (l_1, \ldots, l_n : R)]$ — tupling

So if $[\texttt{r}:R]$ we can access its components by applying the defined projection functions, e.g. $(l_2\ \texttt{r})$ is the second component. Using Lego's postfix notation for function application this expression may also be written as $\texttt{r}.l_2$ which is more suggestive of the intended record semantics. The function $\texttt{Mk}R$ allows one to collect n items of the appropriate types to form an object of type R. The β-rule for the Σ-type (Σ-COMP) gives the equations

$(\texttt{Mk}R\ x_1 \ldots x_n).l_i == x_i$

We now come to the actual implementation of the deliverables model in Lego.

4.4.2 Deliverables in Lego

The contexts of specifications are defined as elements of the following record type.

```
[CON = <<set:Type(0), pred:set->Prop>>]
```

So in Lego a context of specifications is just a type and a `Prop`-valued predicate on it. As we see below we then use the internal Σ-type instead of context extension for comprehension. The morphisms are defined analogously:

```
[MOR[G,D:CON] = <<fun:G.set->D.set,
                  resp:{g|G.set}(G.pred g)->(D.pred (fun g))>>]
```

[3] In the meantime this situation has changed and records, albeit with a different syntax, are available in Lego.

The empty context must be defined using a unit type and a true proposition, i.e. we have

```
[Emp = MkCON unit ([x:unit]tt)]
```

where `unit:Type(0)` is an inductive type with sole constructor `star:unit` and `tt` is the proposition `{X:Prop}X->X`— the higher-order encoding of truth. We have a (not necessarily unique) morphism into `Emp` from any other context.

```
[bang[G:CON] = MkMOR G Emp ([_:G.set]star)
                  ([g:G.set][_:G.pred g][X:Prop][x:X]x)]
```

Thus in order to prove that the category of contexts of specifications has a terminal object the Lego encoding is of no use; we must look at its definition in terms of contexts and telescopes. We can, however, prove in Lego that morphisms admit an associative composition and identities.

```
[Id[G:CON] = MkMOR ([g:G.set]g) ([g:G.set][p:G.pred g]p)]
[Comp[G,D,T|CON][f:MOR D T][g:MOR G D] =
              MkMOR ([x:G.set]f.fun(g.fun x))
                    ([x:G.set][p:G.pred x]f.resp (g.resp p))]

Goal {G,D,T,X:CON}{f:MOR G D}{g:MOR D T}{h:MOR T X}
     Q (Comp (Comp h g) f) (Comp h (Comp g f));
Intros G D T X f g h;
Refine Q_refl;
(*** QED ***)
```

We have formulated associativity in terms of propositional equality (`Q`). However, since we were able to prove it using reflexivity alone, we know that the two terms in question have identical normal forms and thus are definitionally equal as required. Alternatively, we can use Lego to calculate these normal forms and compare them by hand, but the use of propositional equality appears more economical. We see that Lego is used here only as a proof assistant. The full proof that deliverables form a syntactic category with attributes is done by hand.

The remaining definitions are as follows:

```
[FAM[G:CON] = <<set:G.set->Type(0), pred:{g|G.set}(set g)->Prop>>]
[SECT[G|CON][S:FAM G] =
     <<fun:{g:G.set}S.set g,
       resp:{g|G.set}(G.pred g) -> (S.pred (fun g))>>]
```

For simplicity we use the same name for the first components of contexts and of families. In the actual Lego implementation we have to choose a different name each time.

Context comprehension is defined using Σ-types and conjunction.

```
[Dot[G:CON][S:FAM G] =
    MkCON (<g:G.set>S.set g)
          ([x:<g:G.set>S.set g](G.pred x.1) /\ (S.pred x.2)]
[p[G|CON][S:FAM G] = MkMOR (Dot G S) G
                           ([x:(Dot G S).set]x.1)
```

```
                          ([x|(Dot G S).set][p:(Dot G S).pred x]fst p)
[Subst[G,D|CON][S:FAM D][f:MOR G D] =
              MkFAM G ([g:G.set]S.set
                     (f.fun g))
                     ([g|G.set][s:S.set (f.fun g)]S.pred s)]
[q[G,D|CON][f:MOR G D][S:FAM D] =
          MkMOR (Dot G (Subst S f)) (Dot D S)
                ([x:(Dot G (Subst S f)).set](f.fun x.1,x.2)
                ([x|(Dot G (Subst S f)).set]
                          [p:(Dot G (Subst f)).pred x]
                          (pair (f.resp (fst p)) (snd p)))]
```

We use conjunction in Lego (/\) to concatenate parts of a propositional telescope. Now we can use Lego to check for example coherence of substitution where in the case of the identity law we must make an η-expansion.

```
Goal {G|CON}{S:FAM G}Q (Subst S (Id G)) (MkFAM S.set S.pred);
Intros G S;Refine Q_refl;
(*** QED ***)

Goal {B,G,D|CON}{f:MOR B G}{g:MOR G D}{S:FAM D}
     Q (Subst S (Comp g f)) (Subst (Subst S g) f);
intros;Refine Q_refl;
(*** QED ***)
```

Since the η-expansion we made is an identity in the true deliverables model we can conclude that the equations we are able to prove in Lego do indeed hold in the actual model. But for example due to the deficiencies of the unit type as a surrogate for the empty context we cannot expect completeness.

4.5 Type formers in the model D

We have now set out the structure of the model, it remains to define various type constructors. In fact it seems that the deliverables model can be endowed with whatever features the original type theory has; we shall, however, limit ourselves to Π- and Σ-types, the natural numbers, a universe, and of course the proof-irrelevant type of propositions.

4.5.1 Dependent products

Let Γ be a context of specifications and $\sigma \in Fam(\Gamma)$ and $\tau \in Fam(\Gamma\cdot\sigma)$. Their dependent product $\Pi(\sigma,\tau) \in Fam(\Gamma)$ is defined by

$$\Pi(\sigma,\tau)_{set}[\gamma : \Gamma_{set}] = \Pi s{:}\, \sigma_{set}[\gamma].\tau_{set}[\gamma, s]$$

$$\Pi(\sigma,\tau)_{pred}[\gamma : \Gamma_{set}, f : \Pi(\sigma,\tau)_{set}] = \forall s : \sigma_{set}[\gamma].\sigma_{pred}[\gamma, s] \Rightarrow \tau_{pred}[\gamma, s, (f\ s)]$$

So the set-component of the dependent product is just the syntactic product over the set-components of σ and τ, and the associated predicate expresses that a function in this Π-type sends arguments meeting σ_{pred} to values meeting τ_{pred}.

The associated combinators are almost forced. If $M \in Sect(\tau)$ then $\lambda_{\sigma,\tau}(M)$ in $Sect(\Pi(\sigma,\tau))$ is defined by

$$\lambda_{\sigma,\tau}(M)_{fun}[\gamma : \Gamma_{set}] = \lambda s{:}\, \sigma_{set}[\gamma].M_{fun}[\gamma, s]$$

$$\lambda_{\sigma,\tau}(M)_{resp}[\gamma : \Gamma_{set}, p{:}\, \Gamma_{pred}[\gamma]] = \lambda s{:}\, \sigma_{set}[\gamma].\lambda q{:}\, Prf(\sigma_{pred}[\gamma, s])M_{resp}[\gamma, s, p, q]$$

Similarly we define application for $M \in Sect(\Pi(\sigma,\tau))$ and $N \in Sect(\sigma)$.

$$App_{\sigma,\tau}(M, N)_{fun}[\gamma{:}\, \Gamma_{set}] = M_{fun}[\gamma]\ N_{fun}[\gamma]$$

$$App_{\sigma,\tau}(M, N)_{resp}[\gamma{:}\, \Gamma_{set}, p{:}\, \Gamma_{pred}[\gamma]] = M_{resp}[\gamma, p]\ (N_{fun}[\gamma])\ (N_{resp}[\gamma, p])$$

Proposition 4.5.1. *The above assignments determine dependent products in the deliverables model.*

Proof. All the equations follow by simply rewriting the definitions. For compatibility with substitution one uses the fact that syntactic substitution commutes by definition with all type and term formers. The equations can also be checked automatically using Lego.

4.5.2 Dependent sums

Let Γ be a context of specifications and $\sigma \in Fam(\Gamma)$ and $\tau \in Fam(\Gamma \cdot \sigma)$. Their dependent sum $\Sigma(\sigma,\tau) \in Fam(\Gamma)$ is defined by

$$\Sigma(\sigma,\tau)_{set}[\gamma : \Gamma_{set}] = \Sigma s{:}\, \sigma_{set}[\gamma].\tau_{set}[\gamma, s]$$

$$\begin{aligned}\Sigma(\sigma,\tau)_{pred}[\gamma : \Gamma_{set}, f : \Sigma(\sigma,\tau)_{set}] = \\ R^{\Sigma}_{\sigma_{set},\tau_{set},[f:\Sigma x:\sigma_{set}.\tau_{set}]Prop} \\ ([x : \sigma_{set}, y : \tau_{set}[x]]\sigma_{pred}[x] \wedge \tau_{pred}[x, y]\ ,\ f)\end{aligned}$$

That is, $\Sigma(\sigma,\tau)_{pred}$ holds for a canonical element (x, y) if σ_{pred} holds for x and τ_{pred} holds for y and it is extended to arbitrary elements of $\Sigma(\sigma,\tau)_{set}$ using the Σ-elimination operator R^{Σ}. Alternatively, one could define $\Sigma(\sigma,\tau)_{pred}$ using a higher-order encoding or the defined projections.

Next we define the pairing morphism $pair : \Gamma \cdot \sigma \cdot \tau \to \Gamma \cdot \Sigma(\sigma,\tau)$. Its first component is defined by

$$pair_{fun}[\gamma, s, t] = (\gamma, \langle s, t\rangle)$$

The second component is a bit more difficult. We start from the context

$$\gamma{:}\, \Gamma_{set}, s{:}\, \sigma_{set}, t{:}\, \tau_{set}, p{:}\, Prf(\Gamma_{pred}[\gamma]), q{:}\, Prf(\sigma_{pred}[s]), r{:}\, Prf(\tau_{pred}[s, t])$$

In this context we must find an element M of $S := Prf(\Sigma(\sigma,\tau)_{pred})[\langle s, t\rangle]$. We then put

$$pair[\gamma, s, t, p, q, r] = (p, M)$$

For the construction of M we observe that by virtue of the computation rule Σ-COMP, S equals $Prf(\sigma_{pred}[s] \wedge \tau_{pred}[s,t])$. We thus put $M := \wedge\text{-Intro}(s,t)$ where $\wedge$-Intro is the pairing combinator associated to $\wedge$.

For Σ-elimination assume $\rho \in Fam(\Gamma \cdot \Sigma(\sigma,\tau))$ and $M \in Sect(\rho\{pair\})$. We must construct a section $R^{\Sigma}(M)$ of ρ. The first component is just as in the term model:

$$R^{\Sigma}(M)_{fun}[\gamma{:}\,\Gamma_{set}, x{:}\,\Sigma(\sigma,\tau)_{set}[\gamma]] = R^{\Sigma}_{\sigma_{set},\tau_{set},\rho_{set}}(M_{fun}, x)$$

For the second component we use Σ-elimination as well. We must find an inhabitant of the type

$$Prf(\rho_{pred}[R^{\Sigma}_{\sigma_{set},\tau_{set},\rho_{set}}(M_{fun}, x)])$$

in the context

$$\gamma{:}\,\Gamma_{set}, x{:}\,\Sigma(\sigma,\tau)_{set}, p{:}\,\Gamma_{pred}[\gamma], q{:}\,Prf(\Sigma(\sigma,\tau)_{pred}[x])$$

Using "cut" this may be reduced to finding an inhabitant of

$$\rho'[x] := Prf(\Sigma(\sigma,\tau)_{pred}[x] \Rightarrow \rho_{pred}[R^{\Sigma}_{\sigma_{set},\tau_{set},\rho_{set}}(M_{fun}, x)])$$

in context

$$\gamma{:}\,\Gamma_{set}, x{:}\,\Sigma(\sigma,\tau)_{set}, p{:}\,\Gamma_{pred}[\gamma]$$

Using R^{Σ} we can reduce this task to finding an inhabitant of $\rho'[\langle s,t\rangle]$ in context

$$\gamma{:}\,\Gamma_{set}, x{:}\,\Sigma(\sigma,\tau)_{set}, p{:}\,\Gamma_{pred}[\gamma], s{:}\,\sigma_{set}[\gamma], t : \tau_{set}[s]$$

But using Σ-COMP, $\rho'[\langle s,t\rangle]$ is equal to

$$Prf(\sigma_{pred}[s] \wedge \sigma_{pred}[t] \Rightarrow \rho_{pred}[M_{fun}[s,t]])$$

This latter type is inhabited by

$$K := \lambda h : Prf(\sigma_{pred}[s] \wedge \sigma_{pred}[t]).M_{resp}[s,t,fst(h),snd(h)]$$

where *fst* and *snd* are the projections associated with $\wedge$. So putting things together we get

$$R^{\Sigma}(M)_{resp}[\gamma,x,p,q] = (R^{\Sigma}_{\sigma_{set},\tau_{set},\rho'}(K,x))\ p$$

Proposition 4.5.2. *The above data endow* **D** *with Σ-types.*

Proof. The pairing morphism satisfies

$$p(\sigma) \circ p(\tau) = p(\Sigma(\sigma,\tau)) \circ pair$$

since the Γ-components are simply copied in both parts of *pair*. Stability of all components under substitutions follows again directly from the split property of syntactic substitution. It remains to check equation Σ-COMP. In the $-_{fun}$ component it is identical to its syntactic companion, for the $-_{resp}$-component assume

$$\gamma{:}\,\Gamma_{set}, s{:}\,\sigma_{set}, t{:}\,\tau_{set}, p{:}\,\Gamma_{pred}[\gamma], q{:}\,Prf(\sigma_{pred}[s]), r{:}\,Prf(\tau_{pred}[t])$$

Then we calculate as follows

$$\begin{aligned} & (R^{\Sigma}(M)\{pair\})_{resp}[\gamma, s, t, p, q, r] & \\ = \ & K\ \wedge\text{-Intro}(q, r) & \text{by } \Sigma\text{-COMP} \\ = \ & M_{resp}[s, t, fst(\wedge\text{-Intro}(q, r)), snd(\wedge\text{-Intro}(q, r))] & \text{by } \Pi\text{-BETA} \\ = \ & M_{resp}[s, t, q, r] & \end{aligned}$$

Using this technique we can show that the deliverables model admits other inductive type constructors. We content ourselves by describing natural numbers in $\mathbf{D}$.

4.5.3 Natural numbers

If $\Gamma \in \mathbf{C}$ we define $(\mathbf{N}_\Gamma)_{set}[\gamma] = \mathbf{N}$ and $(\mathbf{N}_\Gamma)_{pred}[\gamma, n] = tt$. The morphisms 0 and *Suc* are readily constructed from their syntactic companions. We have $(0_\Gamma)_{fun}[\gamma] = 0$ and $(Suc_\Gamma)_{fun}[\gamma, n] = Suc(n)$. The $-_{resp}$-components are constants returning the canonical proof of tt. For $\mathbf{N}$-elimination let $\rho \in Fam(\Gamma \cdot \mathbf{N}_\Gamma)$ and $M_z \in Sect(\sigma\{\overline{0}\})$, and $M_s \in Sect(\sigma\{Suc_\Gamma \circ p(\sigma)\})$. We must construct a section $R^{\mathbf{N}_\Gamma}(M_z, M_s)$ of ρ. The $-_{fun}$ component is again the same as in the term model, $R^{\mathbf{N}_\Gamma}(M_z, M_s)_{fun}[\gamma, n] = R^{\mathbf{N}}((M_z)_{fun}, (M_s)_{fun}, n)$. For the $-_{resp}$-part we must "show" that ρ_{pred} "holds" for this term. By analogy to the case of Σ-types we use $R^{\mathbf{N}}$. This time there is no need for strengthening the inductive hypothesis since the predicate on $\mathbf{N}$ is trivial.

Proposition 4.5.3 (N-canonicity). *If $M \in Sect(\mathbf{N}_\top)$ then M is canonical, i.e. $M = Suc^n(0)$ for some (set-theoretic) natural number n.*

Proof. We have $\vdash M_{fun} : \mathbf{N}$ so by Remark 2.1.6 we have $\vdash M_{fun} = Suc^n(0 : \mathbf{N})$ for some $n \in \omega$. Moreover, we have $\vdash M_{resp} : Prf(tt)$, but by strong normalisation for the Calculus of Constructions there is only one element of $Prf(tt)$ in the empty context, so $M = Suc^n(0)$ as required.

4.5.4 The type of propositions

Now we come to the main feature of the deliverables model — the proof-irrelevant type of propositions. We show that the model is a loose model of the Calculus of Constructions in the sense of Definition 2.4.31. For the moment we assume that the underlying type theory supports an extensional unit type obeying the rules

$$\frac{\Gamma \vdash}{\Gamma \vdash \mathbf{1}_E}\ \text{UNIT-FORM} \qquad \frac{\Gamma \vdash}{\Gamma \vdash \star : \mathbf{1}_E}\ \text{UNIT-INTRO}$$

$$\frac{\Gamma \vdash M : \mathbf{1}_E}{\Gamma \vdash M = \star : \mathbf{1}_E}\ \text{UNIT-EQ}$$

We shall see below how this assumption can be eliminated.

Now we define the family **Prop** $\in Fam(\top)$ by

$$\mathbf{Prop}_{set} = Prop$$
$$\mathbf{Prop}_{pred}[P\colon Prop] = tt$$

The generic family of proof types **Prf** $\in Fam(\top \cdot \mathbf{Prop})$ is defined by

$$\mathbf{Prf}_{set}[P\colon Prop] = \mathbf{1}_E$$
$$\mathbf{Prf}_{pred}[P\colon Prop, x\colon \mathbf{1}_E] = P$$

So the proof type associated to a proposition P is always the unit type, but the predicate associated to P is P itself.

Therefore, a section of **Prop** is already determined by its $-_{fun}$ component; a section of $\mathbf{Prf}\{s\}$ is already determined by its $-_{resp}$-component.

We come to the definition of universal quantification. Let $\sigma \in Fam(\Gamma)$ and $s : \Gamma \cdot \sigma \to \mathbf{Prop}$. Recall that **Prop** here denotes $\top \cdot \mathbf{Prop}$. We define the morphism $\forall_\sigma(s) : \Gamma \to \mathbf{Prop}$ by

$$\forall_\sigma(s)_{fun}[\gamma\colon \Gamma_{set}] = \forall x\colon \sigma_{set}.\sigma_{pred}[x] \Rightarrow s[\gamma, x]$$

For the $-_{resp}$-part we take some constant function returning a proof of tt.

If $M \in Sect(\mathbf{Prf}\{s\})$, i.e.

$$\gamma\colon \Gamma_{set}, x\colon \sigma_{set} \vdash M_{fun} : \mathbf{1}_E$$
$$\gamma\colon \Gamma_{set}, x\colon \sigma_{set}, p\colon \Gamma_{pred}[\gamma], q\colon Prf(\sigma_{pred}[x]) \vdash M_{resp}[p, q] : Prf(s[\gamma, x])$$

then we define $\lambda_{\sigma,s}(M) \in Sect(\mathbf{Prf}\{\forall_\sigma(s)\})$ by

$$\lambda_{\sigma,s}(M)_{fun}[\gamma\colon \Gamma_{set}] = \star : \mathbf{1}_E$$
$$\lambda_{\sigma,s}(M)_{resp}[\gamma\colon \Gamma_{set}, p\colon \Gamma_{pred}[\gamma]] = \lambda x\colon \sigma_{set}.\lambda q\colon \sigma_{pred}[x].M_{resp}[p, q] : Prf(\forall x\colon \sigma_{set}.\sigma_{pred}[x] \Rightarrow s[\gamma, x])$$

The evaluation morphism $ev_{\sigma,s} : \Gamma \cdot \sigma \cdot \mathbf{Prf}\{\forall_\sigma(s) \circ p(\sigma)\} \to \Gamma \cdot \sigma \cdot \mathbf{Prf}\{s\}$ is defined by

$$(ev_{\sigma,s})_{fun}[\gamma, x, p] = (\gamma\ , \star)$$
$$(ev_{\sigma,s})_{resp}[\ \gamma\colon \Gamma_{set}, x\colon \sigma_{set}, f\colon \mathbf{1}_E\ \ p\colon \Gamma_{pred}[\gamma], r\colon Prf(\sigma_{pred}[x]), q\colon Prf(\forall y\colon \sigma_{set}.\sigma_{pred}[y] \Rightarrow s[\gamma, y])] = (p\ ,\quad q\ x\ r)$$

Proposition 4.5.4. *The above data endow* **D** *with the structure of a loose model for the Calculus of Constructions.*

Proof. For the β-equation let $M \in Sect(\mathbf{Prf}\{s\})$. We want to show that

$$ev_{\sigma,s} \circ \overline{\lambda_{\sigma,s}(M)\{p(\sigma)\}} = \overline{M}$$

The $-_{fun}$-part is an equation of type $\mathbf{1}_E$ and therefore holds by UNIT-EQ. The $-_{resp}$-part follows immediately from the definitions and Π-BETA and PROP-EQ. The other laws including the substitutivity laws are also immediate consequences of the definitions.

We can now establish logical consistency of the deliverables model in the sense that there exists a morphism $ff : \top \to \mathbf{Prop}$ such that $\mathbf{Prf}\{ff\}$ has no sections.

Proposition 4.5.5 (Consistency). *The family* $\mathbf{Prf}\{ff\}$ *with*

$$ff := \forall_{\mathbf{Prop}}(id_{\mathbf{Prop}}) : \top \to \mathbf{Prop}$$

has no sections.

Proof. We have $\mathbf{Prf}\{ff\}_{set} = \mathbf{1}_E$ and $\mathbf{Prf}\{ff\}_{pred}[x] = \forall p\colon Prop.tt \Rightarrow p$. A section of $\mathbf{Prf}\{ff\}$ thus consists of an element of $\mathbf{1}_E$ (necessarily $\star$) and a term of type

$$\diamond \vdash Prf(\forall p\colon Prop.tt \Rightarrow p)$$

but no such term exists by consistency of the syntax.

4.5.4.1 Eliminating extensional unit types. One may object against the use of extensional unit types as they are not part of the original syntax and because their behaviour in terms of rewriting and in particular normalisation is dubious. For instance they lead to non-confluence in the presence of η-reductions for functions and require a typed notion of reduction (for otherwise every term could be reduced to $\star$). Although we believe that their addition does not cause any substantial problems (and indeed this has been done in the formulation of the Calculus of Constructions used in [31]) we prefer to show two ways in which their use can be eliminated.

First, we observe that the only use of the extensional unit type was to ascribe a $-_{set}$-component to the families of the form $\mathbf{Prf}\{s\}$. So a possible solution consists of not giving these families a $-_{set}$-component at all. We then have to refine the notion of families so as to allow for that. More precisely, we define

$$Fam(\Gamma) := Fam_{type}(\Gamma) \mathbin{\dot\cup} Fam_{prop}(\Gamma)$$

where $Fam_{type}(\Gamma)$ is the set of families of specifications over Γ as defined in Section 4.3.2 and $Fam_{prop}(\Gamma)$ is defined as the set of propositions in context Γ_{set}, i.e. the set of terms P with $\Gamma_{set} \vdash P : Prop$. These are called propositional families over Γ. Now we must show that even with this extended notion of families we still have a syntactic category with attributes supporting all the structure we are interested in. The proof of this is absolutely straightforward and tedious. An element of $Fam_{prop}(\Gamma)$ behaves like an ordinary family of specifications having the extensional unit type as $-_{set}$-component. All reference to it is omitted. So for example if $P \in Fam_{prop}(\Gamma)$, i.e. $\Gamma_{set} \vdash P : Prop$ then the comprehension $\Gamma \cdot P$ is defined by $(\Gamma \cdot P)_{set} = \Gamma_{set}$ and $(\Gamma \cdot P)_{pred}[\gamma] = \gamma\colon \Gamma_{pred}[\gamma], q\colon Prf(P)$. Also if τ is a propositional family over $\Gamma \cdot \sigma$ then the product $\Pi(\sigma, \tau)$ is a propositional family over Γ.

A more systematic way of eliminating extensional unit types consists of exhibiting an interpretation of type theory with them in type theory without them. This can be done by constructing a syntactic category with attributes whose base category is the same as that of the term model—the category of contexts and context morphisms—and where a family over Γ is either a type in context Γ or a formal constant $\mathbf{1}_E$. The relevant operations may then be extended to this constant in the obvious way, for instance we define $\Gamma \cdot \mathbf{1}_E := \Gamma$ and $p(\mathbf{1}_E) := id_\Gamma$.

4.5.5 Proof irrelevance

In the deliverables model any two proofs of a proposition are indistinguishable because proof types have $\mathbf{1}_E$ as $-_{set}$-component and observations only take the $-_{set}$-component into account. To make this statement precise we first need to define a semantic analogue to Leibniz equality.

Definition 4.5.6. Fix a loose model of the Calculus of Constructions and let $\sigma \in Fam(\Gamma)$ and $M, N \in Sect(\sigma)$. *Leibniz equality* of M and N, written $L\_Eq(M, N)$, is the proposition over Γ defined by

$$\forall_{\sigma \to \mathbf{Prop}}(\forall_{\mathbf{Prf}\{\overline{PM}\}}(\overline{PN^+})) : \Gamma \to \mathbf{Prop}$$

where we have used the following abbreviations

$$\begin{aligned} \sigma \to \mathbf{Prop} &:= \Pi(\sigma, \mathbf{Prop}\{!_{\Gamma\cdot\sigma}\}) \in Fam(\Gamma) \\ PM &:= App_{\sigma^+,(\sigma\to\mathbf{Prop})^+}(v_{\sigma\to\mathbf{Prop}}, M^+) \in Sect(\mathbf{Prop}\{!_{\Gamma\cdot\sigma\to\mathbf{Prop}}\}) \\ PN &:= App_{\sigma^+,(\sigma\to\mathbf{Prop})^+}(v_{\sigma\to\mathbf{Prop}}, N^+) \in Sect(\mathbf{Prop}\{!_{\Gamma\cdot\sigma\to\mathbf{Prop}}\}) \end{aligned}$$

Syntactically, the thus defined Leibniz equality corresponds to the usual one.

$$\Gamma \vdash \forall P{:}\sigma \to Prop.(P\ M) \Rightarrow (P\ N) : Prop$$

(To be precise, this gets interpreted as $Hd(L\_Eq([\![M]\!], [\![N]\!]))$.) By mimicking the usual syntactic manipulations in the semantics we can show that $\mathbf{Prf}\{L\_Eq(M, M)\}$ is always inhabited, and that if $\mathbf{Prf}\{L\_Eq(M, N)\}$ is inhabited, then whenever $\mathbf{Prf}\{\overline{App_{\sigma,\mathbf{Prop}}(P, M)}\}$ is inhabited for some $P \in Sect(\sigma \to \mathbf{Prop})$ then so is $\mathbf{Prf}\{\overline{App_{\sigma,\mathbf{Prop}}(P, N)}\}$.

Let us now examine what Leibniz equality means in the deliverables model. If $\sigma \in Fam(\Gamma)$ and $M, N \in Sect(\sigma)$ in this model then we have

$$L\_Eq(M, N)_{fun}[\gamma] = \forall P{:}\sigma_{set}[\gamma] \to Prop.(P\ (M_{fun}[\gamma])) \Rightarrow (P\ (N_{fun}[\gamma]))$$

whereas $L\_Eq(M, N)_{resp}$ is some uninteresting proof of tt—the $-_{pred}$-component of **Prop**. So $L\_Eq(M, N)$ simply states actual Leibniz equality of the $-_{fun}$-components of M and N. Given the definition of **Prf** in the model we see that $Sect(\mathbf{Prf}\{L\_Eq(M, N)\})$ is nonempty iff M_{fun} and N_{fun} are Leibniz equal in context $\Gamma_{set}, \Gamma_{pred}$. We thus obtain the following proposition expressing proof irrelevance in the deliverables model.

Proposition 4.5.7. *Let $\Gamma \in \mathbf{C}$ be a context of specifications and $A : \Gamma \to \mathbf{Prop}$ and $M, N \in Sect(\mathbf{Prf}\{A\})$. Then the set $Sect(\mathbf{Prf}\{\overline{L\_Eq(M, N)}\})$ is nonempty.*

Proof. Since the $-_{set}$-component of the family $\mathbf{Prf}\{A\}$ is $\mathbf{1}_E$, both M_{fun} and N_{fun} must equal $\star$. So by the above considerations the section Pr_Ir with $Pr\_Ir_{fun} = \star$ and $Pr\_Ir_{resp}[\gamma{:}\, \Gamma_{set}, p{:}\, \Gamma_{pred}[\gamma]] = \lambda P{:}\, \mathbf{1}_E \to Prop \lambda x{:}\, Prf(P\, \star).x$ defines the desired section.

However, we do not have $M = N$ because the second components (which are not observed by propositional equality) may differ. We can also obtain a model if we identify morphisms and sections with equal first component. In the more general setting of partial equivalence relations instead of predicates this will be done in the next chapter. Here we prefer not to identify propositionally equal terms because this would thwart the application to modules described in Section 4.6.2.2 below. There is no fundamental reason for this choice, though.

Corollary 4.5.8. *The deliverables model permits the interpretation of a type theory extending the Calculus of Constructions by the rule*

$$\frac{\Gamma \vdash A : Prop \qquad \Gamma \vdash M, N : Prf(A)}{\Gamma \vdash Pr\_Ir(A, M, N) : Prf(M \stackrel{L}{=} N)} \ \textsc{Pr-Ir}$$

Proof. We interpret Pr_Ir by the section defined in the proof above. It remains to show that the choice of this section is stable under substitution, but this follows immediately from its uniform syntactic definition.

4.5.6 Universes

If the underlying type theory has universes so has the deliverables model. In particular since the underlying type theory has an impredicative universe (*Prop*) we can define such a universe in $\mathbf{D}$ which astonishingly is different from $\mathbf{Prop}$. We therefore call it *Set* and write *El* instead of *Prf*. We define

$$\begin{aligned}
&Set_{set} = \Sigma X{:}\, Prop.X \to Prop\\
&Set_{pred}[x{:}\, Set_{set}] = tt\\
&El_{set}[X{:}\, Set_{set}] = X.1\\
&El_{pred}[X{:}\, Set_{set}, x{:}\, X.1] = X.2\ x
\end{aligned}$$

It is now possible to define a $\forall$-operator for this universe. Moreover, the universe *Set* will be closed under the subset types defined below. Since we do not make further use of *Set*, we do not give the details here. Notice that since the $-_{set}$-component of an *El*-type is not $\mathbf{1}_E$, proof irrelevance does not hold for *Set*. In this sense $\mathbf{D}$ allows us to use the impredicative universe *Prop* both for proof-irrelevant logic and for impredicative quantification. One may compare this to the two impredicative universes *Spec* and *Prop* in Paulin-Mohring's thesis work [91].

4.6 Subset types

Given PR-IR we can encode subset comprehension using Σ-types, more precisely if $\Gamma \vdash \sigma$ and $\vdash P : \sigma \to Prop$ then the type $\vdash \Sigma x{:}\,\sigma.Prf(P\ x)$ can be seen as the subset of σ containing those elements for which P holds. The first projection $\vdash \pi : \Sigma x{:}\,\sigma.Prf(P\ x) \to \sigma$ where $\pi = \lambda u : \Sigma x{:}\,\sigma.Prf(P\ x).(u.1)$ is now provably injective in the sense that the proposition

$$u, v{:}\,\Sigma x{:}\,\sigma.Prf(P\ x) \vdash (\pi\ u) \stackrel{L}{=} (\pi\ v) \Rightarrow u \stackrel{L}{=} v$$

is provable using Σ-elimination (R^{Σ}) and PR-IR. Let us now look at how this particular Σ-type gets interpreted in the deliverables model. If $\sigma \in Fam(\Gamma)$ and $P : \Gamma \cdot \sigma \to Prop$ then we have

$$\begin{aligned}
&\Sigma(\sigma, Prf\{P\})_{set}[\gamma{:}\,\Gamma_{set}] = \Sigma x{:}\,\sigma_{set}.\mathbf{1}_E \\
&\Sigma(\sigma, Prf\{P\})_{pred}[\gamma{:}\,\Gamma_{set}, u{:}\,\Sigma(\sigma, Prf\{P\})_{set}[\gamma]] = \\
&\qquad R^{\Sigma}_{\dots}([x{:}\,\sigma_{set}, y{:}\,\mathbf{1}_E]\sigma_{pred}[x] \wedge P[\gamma, x]\ ,\ u)
\end{aligned}$$

That is, the $-_{set}$-component of the Σ-type is almost the $-_{set}$-component of σ and the $-_{pred}$-component is the conjunction of σ_{pred} and P. It makes sense to replace "is almost" by "is" and to define the following subset type former inside $\mathbf{D}$.

Definition 4.6.1. Let $\sigma \in Fam(\Gamma)$ and $P : \Gamma \cdot \sigma \to Prop$. The *subset type* $\{\sigma \mid P\} \in Fam(\Gamma)$ is defined by

$$\begin{aligned}
&\{\sigma \mid P\}_{set}[\gamma{:}\,\Gamma_{set}] = \sigma_{set}[\gamma] \\
&\{\sigma \mid P\}_{pred}[\gamma{:}\,\Gamma_{set}, x{:}\,\sigma_{set}[\gamma]] = \sigma_{pred}[\gamma, x] \wedge P_{fun}[\gamma, x]
\end{aligned}$$

Now if $M \in Sect(\sigma)$ and $H \in Prf\{P \circ \overline{M}\}$ then we define a section $(M)_H$ of $\{\sigma \mid P\}$ by

$$\begin{aligned}
((M)_H)_{fun} &= M_{fun} \\
((M)_H)_{resp} &= pair(M_{resp}, P_{resp})
\end{aligned}$$

where here *pair* is the (defined) introduction operator corresponding to $\wedge$. Conversely, if $M \in Sect(\{\sigma \mid P\})$ then we have a section $wit(M)$ of σ ("witness") given by $wit(M)_{fun} = M_{fun}$ and $wit(M)_{resp} = fst(M_{resp})$. Every section of the form $wit(M)$ "satisfies" P in the sense that there is a section $cor(M)$ of $Prf\{P \circ \overline{M}\}$ ("correctness") given by $cor(M)_{fun} = \star$ and $cor(M)_{resp} = snd(M_{resp})$ where *fst* and *snd* are the projections corresponding to $\wedge$.

Proposition 4.6.2. *The deliverables model permits the interpretation of a type theory extending the Calculus of Constructions by a subset type governed by the following rules:*

$$\frac{\vdash \sigma \qquad x{:}\sigma \vdash P : Prop}{\vdash \{x{:}\sigma \mid P\}} \; \{\}\text{-FORM}$$

$$\frac{\vdash M : \sigma \qquad \vdash H : Prf(P[M])}{\vdash (M)_H : \{x{:}\sigma \mid P\}} \; \{\}\text{-INTRO}$$

$$\frac{\vdash M : \{x{:}\sigma \mid P\}}{\vdash wit(M) : \sigma} \; \{\}\text{-WIT}$$

$$\frac{\vdash M : \{x{:}\sigma \mid P\}}{\vdash cor(M) : Prf(P[wit(M)])} \; \{\}\text{-COR}$$

$$\frac{}{\vdash wit((M)_H) = M : \sigma \qquad \vdash cor((M)_H) = H : Prf(P[M])} \; \{\}\text{-BETA}$$

Proof. The subset type and its associated combinators are interpreted by the semantic operations defined above. {}-BETA and stability under substitution are immediate from the definition.

The above proposition could also be "proved" in a purely syntactical way by interpreting $\{x{:}\sigma \mid P\}$ as $\Sigma x{:}\sigma.Prf(P)$. The point is that the more economical encoding above also interprets them.

An extended example for the use of subset types will be given in Section 6.1.

4.6.1 Subset types without impredicativity

The above development of subset types and proof irrelevance does not hinge on impredicative propositions; one could carry out the same development in the core type theory TT and would then obtain a model for a type theory with two different sorts: types and propositions. So instead of $\Gamma \vdash \sigma$ we would have to write $\Gamma \vdash \sigma\ type$ and we would also have a judgement $\Gamma \vdash \sigma\ prop$. These "propositions" would support the same type formers as the "types" and moreover we would have an inclusion rule

$$\frac{\Gamma \vdash \sigma\ prop}{\Gamma \vdash \sigma\ type}$$

Semantically, the "propositions" would be interpreted as families with extensional unit type in the $-_{set}$-component. In the absence of Leibniz equality we would then define a propositional equality using the identity type from the syntax. The propositions then would be at most single-valued with respect to this equality. The resulting model is then quite close to the subset interpretation of Martin-Löf type theory.

4.6.2 A non-standard rule for subset types

In Section 4.5.4.1 we sketched a refinement of the deliverables model in which families could have no $-_{set}$-component, which then was meant to be $\mathbf{1}_E$. We will now consider a refinement under which there are families without $-_{pred}$-component, which is understood to be *tt*. This will allow us to derive a "non-standard" rule for subset types in Proposition 4.6.6 below which shows that every function on a subset type is in fact defined on the whole of the underlying $-_{set}$-component. So we define a new model with

$$Fam(\Gamma) := Fam_{type}(\Gamma) \dot{\cup} Fam_{nprop}(\Gamma)$$

where $Fam_{type}(\Gamma)$ is the set of families of specifications as defined before, whereas $Fam_{nprop}(\Gamma)$ is the set of (syntactic) types in context Γ_{set}. The elements of this set are called *non-propositional families of specifications over* Γ. If $\sigma \in Fam_{nprop}(\Gamma)$ then we define $(\Gamma \cdot \sigma)_{set}$ as $\Gamma_{set}, x{:}\sigma$ and $(\Gamma \cdot \sigma)_{pred} = \Gamma_{pred}$. Substitution in non-propositional families is simply syntactic substitution. A section of σ is a term $\Gamma_{set} \vdash M : \sigma$. The corresponding morphism $\overline{M}$ is given by $\overline{M}_{fun}[\gamma] = (\gamma, M)$ and $\overline{M}_{resp}[\gamma, p] = p$ which makes sense since the comprehension of σ does not add anything to the $-_{pred}$-component. We call the resulting structure the *deliverables model with non-propositional families*.

Proposition 4.6.3. *The deliverables model with non-propositional families is a loose model of the Calculus of Constructions and supports natural numbers and* Σ*-types. Moreover, Prop and* **N** *are non-propositional, and* $\Pi(\sigma, \tau)$ *is non-propositional iff* τ *is, and* $\Sigma(\sigma, \tau)$ *is non-propositional iff both* σ *and* τ *are.*

Proof. The model equations for non-propositional families are readily checked by straightforward expansion of the definitions. **N** and *Prop* are defined as their syntactic companions viewed as non-propositional families. If $\tau \in Fam_{nprop}(\Gamma \cdot \sigma)$ then $\Pi(\sigma, \tau) \in Fam_{nprop}(\Gamma)$ is defined as

$$\gamma{:}\, \Gamma_{set} \vdash \Pi x{:}\, \sigma_{set}[\gamma].\tau_{set}[\gamma, x]$$

If $\sigma \in Fam_{nprop}(\Gamma)$ and $\tau \in Fam(\Gamma \cdot \sigma)$ we define the dependent product by

$$\begin{aligned}
&\Pi(\sigma,\tau)_{set}[\gamma{:}\,\Gamma_{set}] := \Pi x{:}\,\sigma_{set}.\tau_{set}[x] \\
&\Pi(\sigma,\tau)_{pred}[\gamma{:}\,\Gamma_{set}, f{:}\,\Pi(\sigma,\tau)_{set}[\gamma]] := \forall x{:}\,\sigma_{set}.\tau_{pred}[\gamma, x, f\ x]
\end{aligned}$$

The dependent sum of two non-propositional families is defined as in the syntax; the dependent sum of an ordinary family and a non-propositional one is defined as in Section 4.5.2 where all references to the (non-existing) $-_{pred}$-part of the non-propositional family are dropped. The other components and the verifications are straightforward.

Next we define a syntactical counterpart to non-propositional families which will serve to define the desired rule for subset types.

Definition 4.6.4. Consider the Calculus of Constructions with natural numbers, Σ-types, and subset types. The set of *non-propositional* pre-types is defined by the following clauses.

- **N** and *Prop* are non-propositional.
- If σ and τ are non-propositional then so are $\Sigma(\sigma, \tau)$ and $\Pi(\rho, \tau)$ for any pre-type ρ.

We write $\Gamma \vdash \sigma$ *Nprop* if $\Gamma \vdash \sigma$ and σ is a non-propositional pre-type. In this situation σ is called a non-propositional type.

Proposition 4.6.5. *The notion of non-propositional types is stable under equality and substitution, more precisely if $\Gamma \vdash \sigma = \tau$ and $\Gamma \vdash \sigma$ Nprop then also $\Gamma \vdash \tau$ Nprop and $B \vdash \sigma[f]$ Nprop for every syntactic substitution $B \vdash f \Rightarrow \Gamma$.*

Proof. The only type equality rule which is not a congruence rule is PROP-EQ. Therefore, if $\Gamma \vdash \sigma = \tau$ then either both σ and τ have the same outermost type former or both types contain an instance of *Prf* in which case $\Gamma \vdash \sigma$ *Nprop* does not hold. We then use induction. The same argument applies to substitution.

Proposition 4.6.6. *The following rules can be interpreted in the deliverables model with non-propositional families.*

$$\frac{\begin{array}{l} x{:}\sigma \vdash \tau[x]\ \mathit{Nprop} \\ x{:}\sigma \vdash P[x] : \mathit{Prop} \\ p{:}\{x{:}\sigma \mid P[x]\} \vdash M[p] : \tau[\mathit{wit}(p)] \\ \vdash N : \sigma \end{array}}{\vdash \mathit{extend}_{\sigma,P,\tau}(M,N) : \tau[N]} \quad \{\}\text{-ELIM-NONPROP}$$

$$\frac{\vdash H : \mathit{Prf}(P[M])}{\vdash \mathit{extend}_{\sigma,P,\tau}(M,N) = M[(N)_H] : \tau[N]} \quad \{\}\text{-ELIM-NONPROP-COMP}$$

Proof. By induction along the definition of the interpretation function we show that if $\Gamma \vdash \sigma$ *Nprop* then $[\![\Gamma \mid \sigma]\!] \in \mathit{Fam}_{nprop}([\![\Gamma]\!])$. It remains to interpret the two rules on the semantic level. Thus let $\sigma \in \mathit{Fam}(\Gamma)$, $\tau \in \mathit{Fam}_{nprop}(\Gamma)$, and $P : \Gamma \cdot \sigma \to \mathit{Prop}$. Let $w : \Gamma \cdot \{\sigma \mid P\} \to \Gamma \cdot \sigma$ be the morphism $q(p(\{\sigma \mid P\}), \sigma) \circ \overline{\mathit{wit}(v_{\{\sigma \mid P\}})}$. We have $w_{fun}[\gamma, s] = (\gamma, s)$ and $w_{resp}[\gamma, p, q] = (p, \mathit{fst}(q))$. Now let $M \in \mathit{Sect}(\tau\{w\})$ and $N \in \mathit{Sect}(\sigma)$. We must construct a section $\mathit{extend}_{\sigma,P,\tau}(M,N)$ of $\tau\{\overline{M}\}$. We define it by

$$\mathit{extend}_{\sigma,P,\tau}(M,N)[\gamma{:}\Gamma_{set}] = M_{fun}[\gamma, N_{fun}[\gamma]]$$

Notice that no $-_{resp}$-component is required because τ is non-propositional. The equation {}-ELIM-NONPROP-COMP now holds by definition.

The non-standard rule allows one to extend a function $p{:}\{x{:}\sigma \mid P[x]\} \vdash M[p] : \tau[\mathit{wit}(p)]$ defined on a subset type to the whole of σ, hence the name of the operator symbol.

In the ordinary deliverables model we could still interpret the rule {}-ELIM-NONPROP by showing that non-propositional types get interpreted as families with a $-_{pred}$-component which is universally valid (equivalent to *tt*). One would then use this proof to interpret the $-_{resp}$-component of *extend*. But first, some care would have to be taken in order to retain stability under substitution, and second the equation {}-ELIM-NONPROP-COMP would only hold up to Leibniz equality, because the $-_{resp}$-components of the two sections might disagree.

The "computation" rule {}-ELIM-NONPROP-COMP is suspicious from a rewriting point of view, for if we direct it from left to right we would have to "guess" a proof $H : \mathit{Prf}(P[M])$. If on the other hand we choose to direct it from right to left there is no way of eliminating instances of *extend*. Such rules do however make sense as a clarification of the semantically determined definitional equality induced by a syntactic model (here **D**) discussed in the Introduction. Below in Section 4.7 we make that more precise.

4.6.2.1 Internal program extraction. The rule {}-ELIM-NONPROP may be used to perform an internal version of program extraction. For example if we have defined a function $f : Even \to Even$ where $Even$ is $\{x{:}\,\mathbf{N} \mid \exists m{:}\,\mathbf{N}.n \stackrel{L}{=} 2m\}$, then we obtain a function $\hat{f}$ from $\mathbf{N}$ to $\mathbf{N}$ by

$$\hat{f} = \lambda n{:}\,\mathbf{N}.extend([x{:}\,Even]wit(f\ x), n)$$

which is valid since $\mathbf{N}$ is non-propositional. Now using the rule {}-ELIM-NONPROP-COMP we obtain

$$x{:}\,Even \vdash \hat{f}\ wit(x) = wit(f\ x) : \mathbf{N}$$

which states that $\hat{f}$ is the "underlying algorithm" of f.

In a similar vein, the non-standard rule for subset types admits the following generalisation of Proposition 4.1.1 above.

Proposition 4.6.7. *Consider a type theory with subset types and the extend-operator. Let* $\Gamma \vdash P : Prop$ *and* $\Gamma, p{:}\,Prf(P) \vdash M : \sigma$ *for some non-propositional type* σ*, e.g.* $\mathbf{N}$*. Then there exists a term* $\Gamma \vdash N : \sigma$ *with* $\Gamma, p{:}\,Prf(P) \vdash M = N : \sigma$*.*

Proof. Consider the type $\{n{:}\,\mathbf{N} \mid P\}$ (where n does not occur in P). We have

$$\Gamma\ , x{:}\,\{n{:}\,\mathbf{N} \mid P\} \vdash M[p := cor(x)] : \sigma$$

Therefore, by {}-ELIM-NONPROP

$$\Gamma \vdash extend_{\sigma,P,\sigma}([x{:}\,\{n{:}\,\mathbf{N} \mid P\}]M[p := cor(x)], 0) : \sigma$$

Let N be this term. By rule {}-ELIM-NONPROP-COMP we get

$$\Gamma, p{:}\,Prf(P) \vdash M = N : \sigma$$

as required.

We remark that the non-propositional families and the families of the form $Prf\{s\}$ are *independent* in the sense of [87]. In our terminology this means, that for $s : \Gamma \to Prop$ the non-propositional families over Γ are in 1-1 correspondence to the non-propositional families over $\Gamma \cdot Prf\{s\}$ and that for non-propositional $\sigma \in Fam_{nprop}(\Gamma)$ the weakening map

$$-\{p(Prf\{s\})\} : Sect(\sigma) \to Sect(\sigma\{p(Prf\{s\})\})$$

is a bijection. The non-standard operator *extend* can be seen as an internalisation or a syntactic version of Moggi's concept of independence. From equation {}-ELIM-NONPROP-COMP we can only deduce that the above mapping is surjective. To deduce bijectivity in a purely syntactic way we could introduce an

η-like equation: $extend([u\colon\{x\colon\sigma \mid P\}M[wit(u)], N) = M[N]$ if $x\colon\sigma \vdash M[x] : \sigma$. This is a case in point for the type theory with semantically defined definitional equality described in Section 4.7.

In the set-theoretic model (Example 2.4.3), where subset types can be interpreted as ordinary subsets, one can interpret *extend* as $extend_{\sigma,P,\tau}(M, N) = M(N)$ if $P(N)$ holds and c_τ otherwise where $c_\tau \in \tau$ is a "default element" defined by induction on the definition of "non-propositional". However, this understanding is not constructive, whereas the interpretation in **D** is.

4.6.2.2 Application to modules. We can also use subset types and the deliverables model for modules and higher-order functors, where it is important that the typing component of a functor does not depend upon its implementation component. Although the account of modules we end up with is essentially the same as the one given by Moggi and others in [87, 42] we found it interesting to exhibit the relationship with subset types and deliverables and to rephrase their constructions in terms of syntactic models.

For what follows we need a type theory in which *Prop* (the name "*Set*" would be more appropriate) contains the natural numbers and other datatypes of interest. This means that there is a constant $\Gamma \vdash \hat{\mathbf{N}} : Prop$ with $Prf(\hat{\mathbf{N}}) = \mathbf{N}$. In the deliverables model we can then also define such a constant by putting $\hat{\mathbf{N}}_{fun}[\gamma] = \hat{\mathbf{N}}$ and $\hat{\mathbf{N}}_{resp}$ returns a proof of *tt*. Now in the model $Prf(\hat{\mathbf{N}})$ is not $\mathbf{N}$ because any family of the form $Prf\{s\}$ has $-_{set}$-part $\mathbf{1}_E$, but we can interpret the following rules, where we abbreviate $Prf\{\hat{\mathbf{N}}\}$ by $\mathbf{N}$.

$$\frac{\Gamma \vdash}{\Gamma \vdash 0 : \mathbf{N}}\ \textsc{N-Intro-0} \qquad\qquad \frac{\Gamma \vdash M : \mathbf{N}}{\Gamma \vdash Suc(M) : \mathbf{N}}\ \textsc{N-Intro-Suc}$$

$$\frac{\begin{array}{l}\Gamma\,,\, x\colon\mathbf{N} \vdash S : Prop\\ \Gamma \vdash M_z : Prf(S[x := 0])\\ \Gamma\,,\, x : \mathbf{N}\,,\, p : Prf(S) \vdash M_s : Prf(S[x := Suc(x)])\\ \Gamma \vdash N : \mathbf{N}\end{array}}{\Gamma \vdash R^{\mathbf{N}}_S(M_z, M_s, N) : Prf(S[x := N])}\ \textsc{N-Elim}$$

Now we may view *Prop* as the universe of datatypes, and types like $Prop \to Prop$ as "kinds". We may then use the subset type to give signatures. For example a signature for sets with a binary operation would be[4]

$$BIN = \{X\colon Prop \mid X \Rightarrow X \Rightarrow X\}$$

The signature for "monads" as used in functional programming [86, 113] would be

$$\begin{array}{rl} MON = & \{T\colon Prop \to Prop \mid \\ & \quad (\forall X\colon Prop.X \Rightarrow (T\ X))\ \wedge \\ & \quad (\forall X, Y\colon Prop.(X \Rightarrow (T\ Y))\ \Rightarrow\ (T\ X) \Rightarrow (T\ Y))\} \end{array}$$

[4] Recall that for $X, Y\colon Prop$ we have $Prf(X \Rightarrow Y) = Prf(X) \to Prf(Y)$ and that $Prf(X \wedge Y)$ together with *pair*, *fst*, and *snd* is a binary product without surjective pairing.

If $S : BIN$ then $wit(S) : Prop$ and $cor(S) : Prf(S) \to Prf(S) \to Prf(S)$, so an element of type BIN consists of a datatype and a binary operation on it. For example we have $(\hat{\mathbf{N}})_{\lambda x.\lambda y.x+y} : BIN$. We may therefore call elements of a signature "structures matching the signature". If S is a structure then we call $wit(S)$ its *type component* and $cor(S)$ its *implementation* component.

Now we still have that any two "proofs" (elements) of a "proposition" (datatype) are "Leibniz-equal", for example the type

$$\vdash Prf(\forall P : Prf(\mathbf{N}) \to Prop.(Prf(P\ 0)) \to Prf(P\ (Suc(0))))$$

is inhabited in the model (by PR-IR). But the reason for this is that in the model a datatype (a "proposition") can never depend upon values of a datatype and so every P the quantification ranges over must be constant. Thus by working in $\mathbf{D}$ it is possible to typecheck signatures at compile time without performing any computations on datatypes. Using the deliverables model with non-propositional families this can be internalised in the following sense. If $F : BIN \to BIN$ is a "functor" mapping structures matching BIN to themselves, then just as in Section 4.6.2.1 we can define $\hat{f} : Prop \to Prop$ where $\hat{f}\ X$ is the "type component" of f applied to a structure matching BIN with type component X.

It should be pointed out that now that we have used up the two components of a specification for types and implementation respectively, there is no possibility of including logical information anymore. This could be done in a similar model where a family has three components: type, implementation, and proof. This might form the basis of a type-theoretic semantics for a modular specification languages like Extended ML [97] in which programs and specifications can coexist. The internal language of such a model, i.e. the type theory interpretable by it should resemble one of the higher-order type theories considered in [56].

4.7 Reinterpretation of the equality judgement

We have now described a syntactic model of type theory in which subset types and proof irrelevance can be interpreted. We have argued that type theory extended with these additional constructs can be understood as a macro language for working under the deliverables approach described in Section 4.2 above. Upon application of the semantic interpretation function to derivations in the extended type theory one obtains a corresponding derivation under the deliverables approach as explained by the semantic interpretation of the various type and term formers. Rather than actually computing the deliverables interpretation of some derivation one may also look at the new equality judgements induced by this interpretation. We therefore propose a new type theory in which all the equality rules are replaced by rules making reference to semantic equality.

Definition 4.7.1. Let $[\![\ldots]\!]$ denote the semantic function associated to **D** and take $x = y : X$ to mean that x, y, and X are defined, and x, y are equal elements of the set X. The type theory **D** is defined by the following clauses.

- The pre-constructions of **D** are the same as those of ordinary type theory extended with *Pr_Ir* and the operators associated to subset types.
- The typing rules are the same as those of ordinary type theory.
- There are only the following three equality rules:

$$\frac{[\![\Gamma \mid M]\!] = [\![\Gamma \mid N]\!] \in \mathit{Sect}([\![\Gamma \mid \sigma]\!])}{\Gamma \vdash M = N : \sigma}\ \textsc{Eq-Term}$$

$$\frac{[\![\Gamma \mid \sigma]\!] = [\![\Gamma \mid \tau]\!] \in \mathit{Fam}([\![\Gamma]\!])}{\Gamma \vdash \sigma = \tau}\ \textsc{Eq-Type}$$

$$\frac{[\![\Gamma]\!] = [\![\Delta]\!] \in \mathit{Ob}(\mathbf{C})}{\vdash \Gamma = \Delta}\ \textsc{Eq-Cont}$$

Since the interpretation function is recursive and equality in the deliverables model is decidable by decidability of equality in the Calculus of Constructions, we find that the new type theory defined above also has decidable typing and equality judgements. In this way we can make sense out of equality rules like {}-Elim-Nonprop-Comp which now appear as special cases of Eq-Term.

This semantic definition of equality also applies to the two syntactic models for extensionality and quotient types we study in the next chapter.

4.8 Related work

A subset type former very similar to ours has also been introduced by Martin-Löf and justified by a very similar model construction—the *subset interpretation of type theory* (described in [88]). The main difference is that our model accounts for the full Calculus of Construction, i.e. higher-order logic, and more subtly that Martin-Löf identifies morphisms with equal first components and families with equal first and equivalent second components. This last identification makes semantic equality undecidable. A difference in presentation is that Martin-Löf's subset interpretation is not described in terms of categorical semantics. In this simple case this is only a matter of taste; however, for the more involved setoid models we look at in the next chapter, we found the use of the abstract semantics unavoidable.

Subset types in an extensional context without unicity of typing (in the sense of Proposition 2.1.7) have been studied by Salvesen and Smith [96]. They consider a subset introduction rule which allows one to conclude $M : \{x{:}\sigma \mid P[x]\}$ provided $M : \sigma$ and $P[M]$ *true*. They do not have a rule which gives $P[M]$ from $M : \{x{:}\sigma \mid P[x]\}$, but only allow to conclude C *true* if $x{:}\sigma, p{:}\mathit{Prf}(P[x]) \vdash C$ *true* and x does not occur in C. They ask the question under what conditions on P this does allow $P[M]$ to be derived and find that this is the case for *stable formulae*, i.e. those for which $((P \Rightarrow \mathit{ff}) \Rightarrow \mathit{ff}) \Rightarrow P$ *true*.

This work does not really compare to ours because in an intensional setting the questions answered in *loc. cit.* either make no sense or have a trivial answer.

The idea that types endowed with predicates can be organised into a model of type theory also appears in [13]. There, however, only semi-cartesian closure is shown, thus it follows that "deliverables" form a model of the simply typed lambda calculus. In *loc. cit.* also "relativised specifications" and "second order deliverables" are considered. A *relativised specification* relative to a specification $\vdash \sigma$ and $x{:}\sigma \vdash P[x] : \mathit{Prop}$ consists of a type τ and a relation $x{:}\sigma, y{:}\tau \vdash R[x,y] : \mathit{Prop}$. If (τ, R) and (τ', R') are two relativised specifications over (σ, P) then a *second-order deliverable* from (τ, R) to (τ', R') consists of a function $x{:}\sigma \vdash f[x] : \tau \to \tau'$ and a proof $x{:}\sigma, p{:}\mathit{Prf}(P[x]) \vdash F : \mathit{Prf}(\forall y{:}\tau. R[x,y] \Rightarrow R'[x, f[x]\ y])$. In our model a relativised specification is a special case of a family in $\mathit{Fam}(\top \cdot (\sigma, P))$ (where in general τ may depend upon σ). A second order deliverable then appears as a morphism from $p((\tau, R))$ to $p((\tau', R'))$ in the slice category over $\top \cdot (\sigma, P)$. It is easily seen that the relativised specifications form a full submodel of $\mathbf{D}$. The setoid model $\mathbf{S}_0$ presented in the next chapter has indeed the property that the type component of families does not depend on the context.

Many of the ideas underlying the model $\mathbf{D}$ are also present in the thesis work of C. Paulin-Mohring [91]. She describes a type theory with two impredicative universes *Spec* and *Prop* which is then interpreted in an amalgamation of pure Calculus of Constructions and system F_ω. It is explained in great detail how the F_ω component of the interpretation (it corresponds to our $-_{fun}$- and $-_{set}$-components) can be seen as a program extracted from a constructive proof. An analogy is drawn to the notion of *realisability* in proof theory and various extensions are proposed such as an interpretation of well-founded recursion using a fixpoint combinator on the F_ω level. Also a subset type is introduced, albeit without the non-standard rule from Section 4.6.2. The main differences to our construction are that categorical models are not used and that the "$-_{set}$-, $-_{fun}$-components" in Paulin-Mohring's model do not contain type dependency, very much like our model $\mathbf{S}_0$ to be introduced in the next chapter.

Summing up, we can say that the constructions leading to the deliverables model were all known, but that some new facets have been presented such as the organisation using categorical models and the non-standard rule for subset types. As was said before, our main purpose in presenting the deliverables model is to exemplify the use of categorical models for syntactic translations.

5. Extensionality and quotient types

In this chapter we study models for quotient types and the related constructs of functional and propositional extensionality. The general method is to construct models in which types are interpreted as types together with partial equivalence relations. Propositional equality at some type is then the associated partial equivalence relation.

As we discuss below in Section 5.4 this idea has been considered by several authors with various aims. What is new here is that the concept of types with partial equivalence relations ("setoids") is studied in the context of intensional dependent type theory, which makes it necessary to define setoids depending on setoids. In the sequel we shall essentially give three different answers to this question. The first and simplest one (Section 5.1) only provides a restricted notion of type dependency because in the model dependency arises only at the level of the relations, not at the level of the types themselves. The second one (Section 5.2)—the groupoid model—supports all the type and term-formers we know of and should be considered the "correct" answer, but unfortunately it is not definable in intensional type theory. However, it answers the question of independence of the uniqueness of identity and provides valuable insights into the nature of propositional equality. The third one (Section 5.3), finally, is an attempt to overcome the syntactic problems with the groupoid model and the limitations of the first setoid model. It provides slightly more type dependency, but—as we shall see—has other disadvantages, so that the definitive answer to the question of dependent setoids remains open.

We encounter two operations the soundness of which relies on the difference between extensional propositional equality and intensional definitional equality: a choice operator for quotient types (Section 5.1.7) and an axiom identifying propositional equality on a universe with the type of isomorphisms (Section 5.2.4).

5.1 The setoid model

We consider an extension of the Calculus of Constructions by dependent sums, natural numbers, and possibly some datatypes. No further universes are required. Our aim is to construct a model in which quotient types with respect to Leibniz equality can be interpreted and in which the extensional constructs of proof irrelevance, functional extensionality, and propositional extensionality are available. In view of the analysis in Section 1.2 we then also have proof irrelevance which we shall, however, derive directly. The model we construct has the remarkable property that Leibniz equality behaves like the identity type in the sense that a substitution operator for dependent types in the sense of Section 3.2.3.1 may be defined for it. If one looks closer at the model this is not so astonishing because the only source for type dependency in this model is the dependency of proof types on propositions, with respect to which Leibniz equality is substitutive by definition.

In the model types will be interpreted as types together with *Prop*-valued partial equivalence relations. Type dependency is modelled only at the level of the relations, i.e. a family indexed over some type σ is a type τ (not depending on σ) and for each $x{:}\,\sigma$ a partial equivalence relation on τ compatible with the relation on σ. By analogy to Bishop's definition of sets as assemblies together with an equality relation [9] we call the pairs of types and partial equivalence relations *setoids*.

We use the name *target type theory* to refer to the version of the Calculus of Constructions the model is built upon, and *source type theory* to refer to the internal language of the model which contains all the type and term formers from the target type theory together with the desired extensional constructs and the "strong" Leibniz equality. Definitional equality in the source type theory is defined semantically via interpretation in the model according to Section 4.7. The model we construct is called the *setoid model* $\mathbf{S}_0$. Its components are called contexts, families, morphisms, etc. of setoids. The source type theory itself will also be called $\mathbf{S}_0$.

Large parts of this section have been previously published in [48]. The extension to universes (Section 5.1.8) and the discussion of choice principles and Church's thesis (Section 5.1.4.2 and 5.1.4.3) appear here for the first time.

5.1.1 Contexts of setoids

A *context of setoids* is a pair $\Gamma = (\Gamma_{set}, \Gamma_{rel})$ where Γ_{set} is a (syntactic) context (i.e. $\Gamma_{set} \vdash$) and Γ_{rel} is a proposition in context $\gamma{:}\,\Gamma_{set}, \gamma'{:}\,\Gamma_{set}$, i.e. $\gamma{:}\,\Gamma_{set}, \gamma'{:}\,\Gamma_{set} \vdash \Gamma_{rel}[\gamma, \gamma'] : \mathit{Prop}$, in such a way that

$$\gamma, \gamma'{:}\,\Gamma_{set}\ ,\ \mathit{Prf}(\Gamma_{rel}[\gamma, \gamma']) \vdash \Gamma_{rel}[\gamma', \gamma]\ \mathit{true} \qquad \textsc{Sym}$$
$$\gamma, \gamma', \gamma''{:}\,\Gamma_{set}\ ,\ \mathit{Prf}(\Gamma_{rel}[\gamma, \gamma'])\ ,\ \mathit{Prf}(\Gamma_{rel}[\gamma', \gamma'']) \vdash \Gamma_{rel}[\gamma, \gamma'']\ \mathit{true} \qquad \textsc{Trans}$$

Here we just write $\mathit{Prf}(\Gamma_{rel}[\gamma, \gamma'])$ in a context instead of $p : \mathit{Prf}(\Gamma_{rel}[\gamma, \gamma'])$ for some variable p since this variable does not appear in the conclusion of the judgement. We shall adopt this abbreviation in the sequel as well. The axioms Sym and Trans thus express that Γ_{rel} is a partial equivalence relation on Γ_{set}. Two contexts of setoids Γ, Δ are equal if their components are definitionally equal, i.e. if $\vdash \Gamma_{set} = \Delta_{set}$ and $\gamma, \gamma'{:}\,\Gamma_{set} \vdash \Gamma_{rel}[\gamma, \gamma'] = \Delta_{rel}[\gamma, \gamma'] : \mathit{Prop}$. The implicit witnesses for Sym and Trans are not compared. It is important that the relation is not assumed reflexive for otherwise we could not interpret propositions, see Section 5.1.4.

Examples. The *empty context of setoids* is given by $\top = (\diamond, tt)$ where SYM and TRANS are validated by the canonical proof of tt. Another example is the context *Int* given by $Int_{set} = p\colon \mathbf{N} \times \mathbf{N}$ and

$$Int_{rel}[p, p'\colon Int_{set}] := p.1 + p'.2 \stackrel{L}{=} p.2 + p'.1$$

where in order to establish symmetry and transitivity we use some elementary lemmas about addition on natural numbers.

5.1.1.1 Morphisms. Let Γ, Δ be contexts of setoids. A *morphism* from Γ to Δ is a (syntactic) context morphism f from Γ_{set} to Δ_{set} with

$$\gamma, \gamma'\colon \Gamma_{set}\ ,\ Prf(\Gamma_{rel}[\gamma, \gamma']) \vdash \Delta_{rel}[f[\gamma], f[\gamma']]\ true \qquad \text{RESP}$$

In this situation we write $f : \Gamma \to \Delta$ as usual. Equality of morphisms is definitional equality in the target type theory. Clearly, the composition of two morphisms is a morphism again and the identity is a morphism. Moreover, the unique context morphism into $\top$ trivially satisfies RESP so that we have

Proposition 5.1.1. *The contexts of setoids and their morphisms form a category* $\mathbf{C}$ *with terminal object.*

Notice that we do not identify "provably equal" morphisms, i.e. f and g with $\gamma\colon \Gamma_{set}, Prf(\Gamma_{rel}[\gamma, \gamma]) \vdash \Delta_{rel}[f[\gamma], g[\gamma]]\ true$. Doing so would render the semantic equality undecidable and therefore the source type theory would be undecidable. The difference between actual semantic equality and provable equality is also important for the choice operator we consider in Section 5.1.7 which would be unsound if the two were identified. The latter equality will turn out to be captured by Leibniz equality (Lemma 5.1.6).

5.1.1.2 Families of setoids. Let Γ be a context of setoids. A family of setoids indexed over Γ is a pair $\sigma = (\sigma_{set}, \sigma_{rel})$ where σ_{set} is a type *in the empty context* $(\diamond \vdash \sigma_{set})$ and

$$\gamma\colon \Gamma_{set}\ ,\ s, s'\colon \sigma_{set} \vdash \sigma_{rel}[\gamma, s, s'] : Prop$$

is a family of relations on σ_{set} indexed by Γ_{set} such that

$$\begin{array}{ll}\gamma\colon \Gamma_{set}\ ,\ s, s'\colon \sigma_{set}\ , & \text{SYM}\\ \quad Prf(\Gamma_{rel}[\gamma, \gamma])\ ,\ Prf(\sigma_{rel}[\gamma, s, s']) \vdash \sigma_{rel}[\gamma, s', s]\ true & \end{array}$$

$$\begin{array}{ll}\gamma\colon \Gamma_{set}\ ,\ s, s', s''\colon \sigma_{set}\ , & \text{TRANS}\\ \quad Prf(\Gamma_{rel}[\gamma, \gamma])\ ,\ Prf(\sigma_{rel}[\gamma, s, s'])\ ,\ Prf(\sigma_{rel}[\gamma, s', s'']) \vdash & \\ \qquad \sigma_{rel}[\gamma, s, s'']\ true & \end{array}$$

$$\begin{array}{ll}\gamma, \gamma'\colon \Gamma_{set}\ ,\ s, s'\colon \sigma_{set}\ , & \text{COMP}\\ \quad Prf(\Gamma_{rel}[\gamma, \gamma'])\ ,\ Prf(\sigma_{rel}[\gamma, s, s']) \vdash \sigma_{rel}[\gamma', s, s']\ true & \end{array}$$

Two families are equal if their two components $-_{set}$ and $-_{rel}$ are definitionally equal.

Thus, a family of setoids consists of a Γ_{set}-indexed family of partial equivalence relations on one and the same type σ_{set} with the property expressed by COMP ("compatibility") that the relations over related elements in Γ_{set} are equivalent (by virtue of symmetry of Γ_{rel}). It is important that symmetry and transitivity are relativised to "existing" $\gamma{:}\,\Gamma_{set}$, i.e. those with $\Gamma_{rel}[\gamma,\gamma]$, since otherwise most of the type formers including quotient types would not go through. Also intuitively this restriction makes sense since σ_{set} and σ_{rel} are meant to be undefined or unconstrained outside the "existing" part of Γ. An example of a family is again given by $Int_{set} = \mathbf{N} \times \mathbf{N}$ and $Int_{rel}[\gamma, p, p'] = p.1 + p'. \stackrel{L}{=} p.2 + p'.1$. Here the family σ_{rel} does not actually depend on Γ_{set}. Examples where there is such a dependency will arise later in Section 5.1.4 when we interpret an impredicative universe of propositions. The set of families of setoids over Γ is denoted by $Fam(\Gamma)$.

5.1.1.3 Substitution and comprehension. Let B, Γ be contexts of setoids, $f : B \to \Gamma$ and $\sigma \in Fam(\Gamma)$. A family $\sigma\{f\} \in Fam(B)$ is defined by $\sigma\{f\}_{set} = \sigma_{set}$ and

$$\sigma\{f\}_{rel}[\beta{:}\, B_{set}\ ,\ s, s'{:}\, \sigma\{f\}_{set}] = \sigma_{rel}[f[\beta], s, s']$$

The comprehension of σ is defined by $(\Gamma \cdot \sigma)_{set} = \Gamma_{set}, s{:}\, \sigma_{set}$ and

$$(\Gamma \cdot \sigma)_{rel}[(\gamma, s)\ ,\ (\gamma', s')] = \Gamma_{rel}[\gamma, \gamma'] \wedge \sigma_{rel}[\gamma, s, s']$$

The morphism $p(\sigma) : \Gamma \cdot \sigma \to \Gamma$ is defined by $p(\sigma)[\gamma, s] = \gamma$; the relation is preserved by virtue of $\wedge$-elimination. Finally, the morphism $q(f, \sigma) : B{\cdot}\sigma\{f\} \to \Gamma{\cdot}\sigma$ is given by $q(f,\sigma)[\beta, s] = (f[\beta], s)$. This preserves the relation since f does. It is also readily checked that the relevant square formed out of f, p, and q commutes (and in fact is a pullback).

5.1.1.4 Sections. Let Γ be a context of setoids and $\sigma \in Fam(\Gamma)$. A *section* of σ is a term $\Gamma_{set} \vdash M{:}\,\sigma_{set}$ which respects the relations, that is

$$\gamma, \gamma'{:}\, \Gamma_{set}\ ,\ Prf(\Gamma_{rel}[\gamma, \gamma']) \vdash \sigma_{rel}[\gamma, M[\gamma], M[\gamma']]\ true \qquad \text{RESP}$$

Equality of these sections is again definitional equality in the target type theory. If M is a section then its associated context morphism is defined by $\overline{M}[\gamma] = (\gamma, M[\gamma])$ which is a morphism by virtue of RESP. Conversely, if $(f, M) : B \to \Gamma \cdot \sigma$ is a morphism then $Hd((f, M)) := M$ is a section of $\sigma\{f\}$. To see RESP let $\beta, \beta'{:}\, B_{set}$ and $p{:}\, B_{rel}[\beta, \beta']$. We must show that in this context $\sigma_{rel}[f[\beta], M[\beta], M[\beta']]$ *true*. But this is merely the last component of condition RESP for $(f, M) : B \to \Gamma \cdot \sigma$ by definition of $(\Gamma \cdot \sigma_{rel})$. Summing up, we have the following:

Proposition 5.1.2. *Contexts, families, and sections of setoids form a syntactic category with attributes—the setoid model* $\mathbf{S}_0$.

The dependent type theory with extensional constructs modelled by this syntactic category with attributes in the sense of Section 4.7 will also be called $\mathbf{S}_0$.

5.1.2 Implementing the setoid model $\mathbf{S}_0$ in Lego

If we want to use Lego to check the equations of $\mathbf{S}_0$ we must deal with of the extensional nature of the judgements of the form $\Gamma \vdash P\ true$. We have used an approach whereby semantic objects are rendered as pairs consisting of the actual part which figures in the model and a proof that the required judgements hold. For example, we define

```
[Con = <<set:Type(0), rel:set->set->Prop>>]
[CON[G:Con] =
    ({g,g'|G.set}(G.rel g g')->(G.rel g' g)) /\
    ({g,g',g''|G.set}(G.rel g g')->(G.rel g' g'')->(G.rel g g''))]
```

Now a context is an element `G` of `Con` and a proof of `(CON G)`, but these proofs are ignored when two contexts are compared. For each semantic operation we have a part which works on the first components and a part which establishes preservation of these properties. For instance, composition has two components:

```
Comp:{A,B,G|Con}(Mor B G)->(Mor A B)->(Mor A G)
COMP:{A,B,G|Con}{g|Mor B G}{f|Mor A B}
                (CON A)->(CON B)->(CON G)->
                (MOR g)->(MOR f)->(MOR (Comp g f))
```

Thus in a sense we work in the deliverables model. We cannot use the internal language of the deliverables model directly, since it only offers propositional equality of proofs, but we must compare the semantic objects using definitional equality.

5.1.3 Type formers in the setoid model

Let $\sigma \in Fam(\Gamma)$ and $\tau \in Fam(\Gamma \cdot \sigma)$. The dependent product $\Pi(\sigma, \tau)$ is defined by

$$\begin{aligned}
&\Pi(\sigma,\tau)_{set} = \sigma_{set} \to \tau_{set}\\
&\Pi(\sigma,\tau)_{rel}[\gamma{:}\,\Gamma_{set}\ ,\ u,v{:}\,\Pi(\sigma,\tau)_{set}] =\\
&\qquad \forall s,s'{:}\,\sigma_{set}.\sigma_{rel}[\gamma,s,s'] \Rightarrow \tau_{rel}[(\gamma,s)\ ,\ u\ s\ ,\ v\ s']
\end{aligned}$$

If $M \in Sect(\tau)$ then we define

$$\lambda_{\sigma,\tau}(M)[\gamma{:}\,\Gamma_{set}] = \lambda s{:}\,\sigma_{set}.M[\gamma,s]$$

and conversely, if $M \in Sect(\Pi(\sigma,\tau))$ and $N \in Sect(\sigma)$ we define

$$App_{\sigma\ ,\ \tau}(M,N)[\gamma{:}\,\Gamma_{set}] = (M[\gamma]\ N[\gamma])$$

Now by straightforward calculation we obtain:

Proposition 5.1.3. *These data endow the model $\mathbf{S}_0$ with dependent products in the sense of Definition 2.4.10.*

In very much the same way we can interpret Σ-types, natural numbers, and various inductive types present in the target type theory. Since the interpretation of these is almost forced we leave it out here. In particular we have $(\mathbf{N}_\Gamma)_{set} = \mathbf{N}$ and $(\mathbf{N}_\Gamma)_{rel}[\gamma, x, x'] = (x \stackrel{L}{=} x')$. This allows us to establish in exactly the same way as in the proof of Proposition 4.5.3 that in the empty context the natural numbers contain canonical elements only, in spite of the extensional constructs (N-canonicity).

5.1.4 Propositions

Our next goal is to identify a type of propositions (and proofs) which allows us to "externalise" the relations associated to each type and to interpret quotient types and extensionality. We use the semantic framework of Definition 2.4.31 (loose model of Constructions). Again, we denote by "**Prop**" both the family in $Fam(\top)$, and its comprehension $\top \cdot \mathbf{Prop}$.

To simplify the exposition we assume as in Chapter 4 that the target type theory supports an extensional unit type, i.e. there is a type $\Gamma \vdash \mathbf{1}_E$ in every context, and a term $\Gamma \vdash \star : \mathbf{1}_E$ together with the equation $\Gamma \vdash M = \star : \mathbf{1}_E$ for every $\Gamma \vdash M : \mathbf{1}_E$. We shall use the extensional unit type as the underlying type of propositional types. It is possible to get rid of the extensional unit type in the same way as in Section 4.5.4.1, by defining families of setoids as being either of the form defined in Section 5.1.1.2, or to consist simply of a proposition in context Γ_{set} compatible with Γ_{rel} (in which case it is silently understood that the $-_{set}$-component is the extensional unit type). Comprehension, substitution, and the type formers have then to be defined by case distinction.

Now we are ready to define the required ingredients to interpret propositions. The underlying type of $\mathbf{Prop} \in Fam(\top)$ in $\mathbf{S}_0$ is just the syntactic type of propositions.

$$\mathbf{Prop}_{set} = Prop$$

In order to identify equivalent propositions we define the relation as bi-implication.

$$\mathbf{Prop}_{rel}[p, q{:}\,Prop] = (p \Leftrightarrow q)$$

This is clearly symmetric and transitive and so a family over $\top$ has been defined. Note that COMP degenerates to a tautology in the case of families over $\top$.

The family $\mathbf{Prf} \in Fam(\top \cdot \mathbf{Prop})$ is given by

$$\mathbf{Prf}_{set}[p{:}\,Prop] = \mathbf{1}_E$$
$$\mathbf{Prf}_{rel}[p{:}\,Prop\ ,\ x, x'{:}\,\mathbf{1}_E] = p$$

where $\mathbf{1}_E$ is the extensional unit type. The relation on **Prf** is trivially symmetric and transitive; for COMP we use the relation on **Prop**. More precisely, if $\mathbf{Prop}_{rel}[p, q]$ and $\mathbf{Prf}_{rel}[p, x, y]$, i.e. p, then also q, thus $\mathbf{Prf}_{rel}[q, x, y]$.

Next we define universal quantification. If $\sigma \in Fam(\Gamma)$ and $S : \Gamma{\cdot}\sigma \to \mathbf{Prop}$ then we put

$$\forall_\sigma(S)[\gamma{:}\,\Gamma_{set}] = \forall x{:}\,\sigma_{set}.\sigma_{rel}[\gamma, x, x] \Rightarrow S[\gamma, x]$$

For property RESP assume $\gamma, \gamma' : \Gamma_{set}$ and $\Gamma_{rel}[\gamma, \gamma']$. We must show that $\forall_\sigma(S)[\gamma] \Leftrightarrow \forall_\sigma(S)[\gamma']$. So assume $\forall_\sigma(S)[\gamma]$ and $x:\sigma_{set}$ with $\sigma_{rel}[\gamma', x, x]$. Using SYM and COMP we get $\sigma_{rel}[\gamma, x, x]$, hence $S[\gamma, x]$ by assumption. Property RESP for S using $\Gamma_{rel}[\gamma, \gamma']$ and $\sigma_{rel}[\gamma, x, x]$ gives $S[\gamma, x] \Leftrightarrow S[\gamma', x]$ and thus $S[\gamma', x]$ as required. The other direction is symmetric. We see that the relativisation to "existing" $x:\sigma_{set}$ is necessary to prove RESP.

For the abstraction let $M \in Sect(\mathbf{Prf}\{S\})$. We are forced to define

$$\lambda_{\sigma,S}(M)[\gamma:\Gamma_{set}] \doteq \star$$

To see that this is indeed a section of $\mathbf{Prf}\{\forall_\sigma(S)\}$ assume $\gamma, \gamma' : \Gamma_{set}$ and $\Gamma_{rel}[\gamma, \gamma']$. We must show that

$$\mathbf{Prf}\{\forall_\sigma(S)\}_{rel}[\gamma, \lambda_{\sigma,S}(M)[\gamma] \, , \lambda_{\sigma,S}(M)[\gamma']]$$

which equals $\forall x:\sigma_{set}.\sigma_{rel}[\gamma, x, x] \Rightarrow S[\gamma, x]$ by definition of $\mathbf{Prf}_{rel}$ and $\forall$. Now assuming $x:\sigma_{set}$ and $\sigma_{rel}[\gamma, x, x]$ we have that $(\Gamma \cdot \sigma)_{rel}[(\gamma, x) \, , \, (\gamma', x)]$, hence $S[\gamma, x]$ by RESP for M.

Finally, the evaluation morphism is given by

$$ev_{\sigma,S}[\gamma:\Gamma_{set} \, , \, x:\sigma_{set} \, , \, u:\mathbf{1}_E] = (\gamma, x, \star)$$

We must show that this defines a morphism from $\Gamma \cdot \sigma \cdot \mathbf{Prf}\{\forall_\sigma(S) \circ p(\sigma)\}$ to $\Gamma \cdot \sigma \cdot \mathbf{Prf}\{S\}$. Assume $\gamma, \gamma' : \Gamma_{set}$, $x, x' : \sigma_{set}$ with $\Gamma_{rel}[\gamma, \gamma']$ and $\sigma_{rel}[\gamma, x, x']$; furthermore assume $\forall x:\sigma_{set}.\sigma_{rel}[\gamma, x, x] \Rightarrow S[\gamma, x]$ (meaning that the two variables of unit type corresponding to $\mathbf{Prf}\{\forall_\sigma(S) \circ p(\sigma)\}$ are "related" according to the definition of $\mathbf{Prf}_{rel}$). We must show that $S[\gamma, x]$. But this follows since by SYM and TRANS for Γ we have $\Gamma_{rel}[\gamma, \gamma]$ and thus $\sigma_{rel}[\gamma, x, x]$ by SYM and TRANS for σ.

The equations relating evaluation and abstraction follow straightforwardly from the properties of the extensional unit type. Also the stability under substitution of all components follows readily by expanding the definitions.

Proposition 5.1.4. *The model* $\mathbf{S}_0$ *is a loose model of the Calculus of Constructions.*

5.1.4.1 Proof irrelevance and extensionality. Next we explore which additional propositions are provable in $\mathbf{S}_0$. First, we have a non-provability result:

Proposition 5.1.5 (Consistency). *The family*

$$ff := \mathbf{Prf}\{\forall_{\mathbf{Prop}}(id_{\mathbf{Prop}})\} \in Fam(\top)$$

has no sections.

Proof. We have $ff_{set} = \mathbf{1}_E$ and $ff_{rel}[x, y:\mathbf{1}_E] = \forall p: Prop.\mathbf{Prop}_{rel}[p, p] \Rightarrow p$. A section of ff consists of an element of $\mathbf{1}_E$ (necessarily $\star$) such that $\forall p: Prop.(p \Leftrightarrow p) \Rightarrow p$ (expanding $\mathbf{Prop}_{rel}$). But this is not possible by consistency of the syntax.

Next we look at Leibniz equality in the model. If $\sigma \in Fam(\Gamma)$ and $M, N \in Sect(\sigma)$ then let $L\_Eq(M,N) : \Gamma \to \mathbf{Prop}$ be the denotation of Leibniz equality as given in Definition 4.5.6. By unfolding its definition we find that in $\mathbf{S}_0$

$$\begin{aligned}
&L_Eq(M,N)[\gamma{:}\,\Gamma_{set}] = \\
&\qquad \forall P{:}\,\sigma_{set} \to Prop. \\
&\qquad\quad (\forall x, x'{:}\,\sigma.\sigma_{rel}[\gamma, x, x'] \Rightarrow ((P\ x) \Leftrightarrow (P\ x'))) \Rightarrow \\
&\qquad\qquad \forall y{:}\,\mathbf{1}_E.(P\ M[\gamma]) \Rightarrow (P\ N[\gamma])
\end{aligned}$$

So (modulo the trivial quantification over $\mathbf{1}_E$) M and N are Leibniz equal in the model if they are indistinguishable by observations which respect the relation on σ.

Lemma 5.1.6. *Let $\sigma \in Fam(\Gamma)$ and $M, N \in Sect(\sigma)$. The family $\mathbf{Prf}\{L\_Eq(M,N)\}$ over Γ has a section iff*

$$\gamma{:}\,\Gamma_{set}\ ,\ Prf(\Gamma_{rel}[\gamma,\gamma]) \vdash \sigma_{rel}[\gamma, M[\gamma], N[\gamma]]\ true$$

Proof. Assume a section of $\mathbf{Prf}\{L\_Eq(M,N)\}$. By definition this means that if $\Gamma_{rel}[\gamma,\gamma']$ for some $\gamma,\gamma'{:}\,\Gamma_{set}$ then $M[\gamma]$ and $N[\gamma]$ (not $N[\gamma']$!) are indistinguishable by observations respecting σ_{rel}. Now

$$\lambda x{:}\,\sigma_{set}.\sigma_{rel}[\gamma, M[\gamma], x]$$

is such an observation as can be seen using Sym and Trans for σ and $\Gamma_{rel}[\gamma,\gamma]$ by virtue of $\Gamma_{rel}[\gamma,\gamma']$ and Sym, Trans for Γ. Therefore, we can deduce $\sigma_{rel}[\gamma, M[\gamma], N[\gamma]]$ provided we can show $\sigma_{rel}[\gamma, M[\gamma], M[\gamma]]$, but this follows from Resp for M. Conversely, if $\sigma[\gamma, M[\gamma], N[\gamma]]$ and $P : \sigma \to Prop$ respects σ_{rel} then obviously $(P\ M[\gamma]) \Rightarrow (P\ N[\gamma])$ by definition of "respects".

Thus Leibniz equality "externalises" the relation associated to each family.

Proposition 5.1.7. *The following rules providing proof irrelevance as well as functional and propositional extensionality can be interpreted in $\mathbf{S}_0$.*

$$\frac{\Gamma \vdash A : Prop \qquad \Gamma \vdash M, N : Prf(A)}{\Gamma \vdash M = N : Prf(A)}\ \textsc{Pr-Ir}$$

$$\frac{\Gamma \vdash P : Prop \qquad \Gamma \vdash Q : Prop \qquad \Gamma \vdash H : Prf(P \Leftrightarrow Q)}{\Gamma \vdash Bi\_Imp(P,Q,H) : Prf(P \stackrel{L}{=} Q)}\ \textsc{Bi-Imp}$$

$$\frac{\begin{array}{l}\Gamma \vdash U, V : \Pi x{:}\,\sigma.\tau \\ \Gamma, x{:}\,\sigma \vdash H : Prf(U\ x \stackrel{L}{=} V\ x)\end{array}}{\Gamma \vdash Ext(H) : Prf(U \stackrel{L}{=} V)}\ \textsc{Ext}$$

Proof. We identify syntactic objects with their denotations in $\mathbf{S}_0$. Rule Pr-Ir is immediate because both M and N equal $\star$. For rule Bi-Imp assume $P, Q \in Sect(\mathbf{Prop}\{!_\Gamma\})$. The assumption H gives rise to a proof of $P[\gamma] \Leftrightarrow Q[\gamma]$ for every $\gamma : \Gamma_{set}$ with $\Gamma_{rel}[\gamma,\gamma]$. But by Lemma 5.1.6 this implies that there exists a section of $\mathbf{Prf}\{L\_Eq(P,Q)\}$. The proof for Ext is similar.

Now we present a somewhat unexpected feature of Leibniz equality, namely that it behaves like Martin-Löf's identity type in the sense that a Leibniz principle for dependent types can be interpreted from which the definability of Martin-Löf's elimination rule [88] follows using PR-IR and the encoding of J in terms of *Subst* and *IdUni* given in Section 3.2.3.1.

Proposition 5.1.8. *For each $\sigma \in Fam(\Gamma)$ and $\tau \in Fam(\Gamma \cdot \sigma)$ and $M, N \in Sect(\sigma)$ and $P \in L\_Eq(M,N)$ and $U \in Sect(\tau\{\overline{M}\})$ there exists a well-determined section $Subst_{\sigma,\tau}(P,U) \in Sect(\tau\{\overline{N}\})$ in such a way that*

$$Subst_{\sigma,\tau}(Refl(M), U) = U$$

where $Refl(M)$ is the canonical section of $\mathbf{Prf}\{L\_Eq(M,M)\}$ *corresponding to reflexivity. Moreover, this operator Subst is stable under substitution in all its arguments.*

Proof. We define $Subst_{\sigma,\tau}(P,U)$ simply as U. That this is a section of $\tau\{\overline{N}\}$ follows from Lemma 5.1.6 applied to P and rule COMP for the family τ. The other properties are trivially satisfied.

Obviously, this means that we can interpret rules LEIBNIZ and LEIBNIZ-COMP from Section 3.2.3.1 with $Id_\sigma(M,N)$ replaced by $Prf(M \stackrel{L}{=} N)$. As long as we do not use induction over definable families the semantic formulation of closure properties as in Proposition 5.1.8 above and the more syntactic one in Proposition 5.1.7 are equivalent and it is a mainly a matter of presentation which one is being used.

Proposition 5.1.8 is important because it states that $\mathbf{S}_0$ has the same strength as TT_I and thus is conservative over extensional type theory (TT_E) by Theorem 3.2.5.

5.1.4.2 Axiom of choice. We may start with a target type theory in which the "internal axiom of choice" [5], i.e. the following schema of propositions

$$\frac{\Gamma \vdash \sigma \qquad \Gamma \vdash \tau}{\Gamma \vdash (\forall x{:}\,\sigma.\exists y{:}\,\tau.R[x,y]) \to (\exists f{:}\,\sigma \to \tau.\forall x{:}\,\sigma.R[x, f\ x])\ true}\ \text{IAC}$$

holds.[1] Then in the source type theory this axiom is nevertheless not in general valid because in its translation the witness f is required to respect the equivalence relations present on σ and τ which cannot be guaranteed for the witness produced by IAC in the target type theory. In fact one can show by mimicking Diaconescu's argument (see [62, II.7]) that IAC together with EXT and BI-IMP implies the principle of the excluded middle, $\forall p{:}\,Prop.p \vee (p \Rightarrow ff)$; if this holds in the source type theory then it must hold in the target type theory. This in turn is not necessarily the case even if IAC is present in the target type theory. The fact that the axiom of choice may fail because a choice function fails to preserve extensional equality is well-known and is e.g. discussed in [110, 36]. See also the example with real numbers in Section 5.1.7.1 below.

[1] IAC is not valid in the Calculus of Constructions, but it does hold in its set-theoretic interpretation (see Example 2.4.6).

We conjecture, however, that in the particular case where the type σ in IAC is $\mathbf{N}$ ("countable choice") or any other type with Leibniz equality as the relation we can actually interpret IAC in the source type theory provided we have it for the target type theory. The reason is that this relation is always preserved. See also Remark 5.1.11 below. The same goes for the principle of *dependent choice*, see e.g. [32, Section 4.4.3]:

$$\forall x{:}\,\sigma.\exists y{:}\,\sigma.R[x,y] \rightarrow \forall x{:}\,\sigma.\exists f{:}\,\mathbf{N} \rightarrow \sigma.(f\ 0 \stackrel{L}{=} x) \wedge (\forall n{:}\,\mathbf{N}.R[f\ n, f(Suc(n))])$$

where again the choice function has domain $\mathbf{N}$. The internal axiom of choice must not be confused with the following "proof-relevant" version of the axiom of choice, which is obviously inhabited:

$$\Pi x{:}\,\sigma.\Sigma y{:}\,\tau.Prf(P[x,y]) \rightarrow \Sigma f{:}\,\sigma \rightarrow \tau.\Pi x{:}\,\sigma.Prf(P[x,y])$$

The difference is that elements of a Σ-type are distinct if they have different witnesses, whereas all proofs of an existential statement are identified.

5.1.4.3 Church's thesis. Another principle which gets lost when passing from the target type theory to the source type theory is Church's thesis in the following formulation:

$$CT = \exists F{:}\,(\mathbf{N} \rightarrow \mathbf{N}) \rightarrow \mathbf{N}.\forall f{:}\,\mathbf{N} \rightarrow \mathbf{N}.computes[F\ f, f]$$

where

$$computes[e{:}\,\mathbf{N}, f{:}\,\mathbf{N} \rightarrow \mathbf{N}] = \forall n{:}\,\mathbf{N}.\exists z{:}\,\mathbf{N}.(f\ n \stackrel{L}{=} U[z]) \wedge T[e,n,z]$$

and T is Kleene's T-predicate and U is the output extraction function. Informally, $computes[e,f]$ expresses that e is a (code for a) program which computes f, and CT states that there exists a functional which associates a program to every function $f{:}\,\mathbf{N} \rightarrow \mathbf{N}$ associates a program.

If CT holds in the target type theory[2] then it need not hold in the source type theory, because there is no reason why the witness F should preserve pointwise equality of functions. Even worse, in the source type theory one can actually prove $\neg CT$ by mimicking the proof in [110] that CT together with the axiom of choice and functional extensionality is inconsistent in HA_ω. Intuitively, the reason is that in the presence of functional extensionality F must yield equal results when applied to extensionally equal functions, so one can check whether a function is e.g. constantly zero by examining whether F applied to it equals $F(\lambda x{:}\,\mathbf{N}.0)$.[3]

Let us point out that the following weaker formulation of CT

[2] We do not know whether there exists a model of the Calculus of Constructions in which CT is valid, but this seems rather likely since CT is consistent in higher-order intuitionistic arithmetic (HA_ω) using a realisability model, which interprets types as *subsets* of ω rather than partial equivalence relations[32, 107].

[3] Per Martin-Löf has told the author that this phenomenon was one of the reasons for him to reject equality reflection (ID-DEFEQ), which gives functional extensionality and thus makes (a TT_E-version of) CT inconsistent.

$$CT' = \forall f{:}\,\mathbf{N} \to \mathbf{N}.\exists e{:}\,\mathbf{N}.computes[e, f]$$

does hold in the source type theory if it holds in the target type theory because it gets translated into basically the same formula. In the presence of IAC this weaker version implies CT, so that $\mathbf{S}_0$ together with CT' provides a type theory in which the internal axiom of choice is inconsistent (provided CT' is consistent in the target type theory.).

We also believe that CT' can be consistently added to TT_I or TT_E if we translate the existential quantification using the *squash type* former to be introduced in Section 5.3.5, rather than into a Σ-type.

5.1.5 Quotient types

We now turn to the interpretation of quotient types in the model. Their syntax is given by the following rules:

$$\frac{\Gamma \vdash \sigma \qquad \Gamma\,,\, s, s'{:}\,\sigma \vdash R[s, s'] : Prop}{\Gamma \vdash \sigma/R}\ \textsc{Q-Form}$$

$$\frac{\Gamma \vdash M{:}\,\sigma}{\Gamma \vdash [M]_R{:}\,\sigma/R}\ \textsc{Q-Intro}$$

$$\frac{\begin{array}{lll}\Gamma \vdash \tau & \Gamma\,,\, s{:}\,\sigma \vdash M[s] : \tau & \Gamma \vdash N : \sigma/R\\ \multicolumn{3}{l}{\Gamma\,,\, s, s'{:}\,\sigma\,,\, p{:}\,Prf(R[s, s']) \vdash H : Prf(M[s] \stackrel{L}{=} M[s'])}\end{array}}{\Gamma \vdash plug_R\, N\ in\ M\ using\ H : \tau}\ \textsc{Q-Elim}$$

$$\frac{}{\Gamma \vdash plug_R\, [N]_R\ in\ M\ using\ H = M[N] : \tau}\ \textsc{Q-Comp}$$

$$\frac{\Gamma \vdash M, N : \sigma \qquad \Gamma \vdash H : Prf(R[M, N])}{\Gamma \vdash Qax_R(H) : Prf([M]_R \stackrel{L}{=} [N]_R)}\ \textsc{Q-Ax}$$

$$\frac{\begin{array}{l}\Gamma\,,\, x{:}\,\sigma/R \vdash P[x] : Prop\\ \Gamma\,,\, s{:}\,\sigma \vdash H : Prf(P[[s]_R])\\ \Gamma \vdash M : \sigma/R\end{array}}{\Gamma \vdash Qind_R(H, M) : Prf(P[M])}\ \textsc{Q-Ind}$$

This syntax is a formalisation of usual mathematical practice. Rule Q-FORM allows for the formation of a type σ/R from a type σ and a relation R. We do not require R to be an equivalence relation. Using Q-INTRO one constructs elements ("classes") of σ/R from "representatives". The rule Q-ELIM allows one to construct functions on the quotient type by definition on representatives. The axiom Q-AX states that classes of related elements are equal; the axiom Q-IND states that σ/R consists of "classes" only.

For example, we may define a type of integers as $Int := \mathbf{N} \times \mathbf{N}/R_{Int}$ where $R_{Int}[u, v{:}\,\mathbf{N} \times \mathbf{N}] = (u.1 + v.2 \stackrel{L}{=} u.2 + v.1)$. The elimination rule now allows us to define functions like addition on the integers and the induction principle together with the equations permits us to derive properties of these functions from their implementations. In examples we assume that various such functions such as $+$, $|-|$, and $-$ (negation) have been defined.

5.1.5.1 Comparison to quotient types in $\mathbf{TT}_I$. Notice that Q-IND resembles a special case of the dependent elimination rule Q-I-ELIM from Section 3.2.6.1; the additional proviso in Q-I-ELIM stating that H preserves R is trivially satisfied in this special situation because of proof irrelevance.

Using Proposition 5.1.8 and the fact that the quotienting relation never appears in the conclusion of a rule we can simulate the rules for quotients given in Section 3.2.6.1 in the following way. If $s, s'{:}\sigma \vdash \rho[s, s']$ we put $R[s, s'] := \exists p{:}\rho[s, s'].tt$ and $\sigma/\rho := \sigma/R$. We also put $[-]_\rho = [-]_R$. Obviously we have $R[s, s']$ *true* if $\rho[s, s']$ *true* and conversely, if $R[s, s']$ *true* and $p{:}\rho[s, s'] \vdash P[s, s']$ *true* for some $s, s'{:}\sigma \vdash P[s, s'] : Prop$ then $P[s, s']$ *true* by $\exists$-elimination. Therefore, the rules Q-I-FORM, Q-I-INTRO, Q-I-AX are validated. For Q-I-ELIM assume $x{:}\sigma/R \vdash \tau[x]$ and $x{:}\sigma \vdash M[x] : \tau[[x]_R]$ such that

$$x, x'{:}\sigma\ , p{:}\rho[x, x'] \vdash H : Prf(Subst_{\sigma/\rho,\tau}(Qax_\rho(p), M[x]) \stackrel{L}{=} M[x'])$$

We put $\tau' = \Sigma x{:}\sigma/R.\tau[x]$ and

$$M'[s{:}\sigma] = pair([s]_R, M[s]) : \tau'$$

Now if $R[s, s']$ then we can find $H : M'[s] \stackrel{L}{=} M'[s']$ using R^Σ, $\exists$-elimination, and equality reasoning. Therefore the premises to rule Q-ELIM are satisfied. Let $N : \sigma/R$ and put $U = plug_R\, N\ in\ M'\ using\ H : \tau'$. The second projection $U.2$ has type $\tau[U.1]$ rather than $\tau[N]$ as we would need it in order to simulate Q-I-ELIM. But using Q-IND (over N) we can construct $P : Prf((plug_R\, N\ in\ M'\ using\ H).1 \stackrel{L}{=} N)$, and thus have $Subst_{\sigma/R,\tau}(P, U.2) : \tau[N]$ as required. The computation rule Q-I-COMP follows from Q-COMP and the semantic interpretation of *Subst* as the identity.

5.1.6 Interpretation of quotient types in $\mathbf{S_0}$

Let $\sigma \in Fam(\Gamma)$ and $R : \Gamma \cdot \sigma \cdot \sigma^+ \to \mathbf{Prop}$ be a "relation" over σ. Viewed internally R is a term of type $\gamma{:}\Gamma_{set}\ , x, x'{:}\sigma_{set} \vdash Prop$ such that

$$\begin{array}{l}\gamma{:}\Gamma_{set}\ , x, y{:}\sigma_{set}\ , \gamma'{:}\Gamma_{set}\ , x', y'{:}\sigma_{set}\ , \\ \quad Prf(\Gamma_{rel}[\gamma, \gamma'] \wedge \sigma_{rel}[\gamma, x, x'] \wedge \sigma_{rel}[\gamma, y, y']) \vdash \\ \qquad R[x, y] \Leftrightarrow R[x', y']\ true\end{array}$$

We say that R "respects" σ_{rel}. Our aim is to define a new family σ/R on which the relation is given by R. The underlying type remains unchanged by quotienting.

$$(\sigma/R)_{set}[\gamma] = \sigma_{set}[\gamma]$$

Now since R is not guaranteed to be symmetric and transitive, we have to take the symmetric and transitive closure of R, moreover we must ensure compatibility with σ_{rel} — the relation already present on σ. It turns out that the right choice for $(\sigma/R)_{rel}$ is the following higher-order encoding of symmetric, transitive closure.

$$\begin{array}{ll}
(\sigma/R)_{rel}[\gamma{:}\,\Gamma_{set}\ ,\ s, s'{:}\,\sigma_{set}] = & \\
\quad \forall R'{:}\,\sigma_{set} \to \sigma_{set} \to Prop. & \\
\qquad (\forall x, x'{:}\,\sigma_{set}.(R'\ x\ x') \Rightarrow (R'\ x'\ x)) & (\sigma) \\
\qquad \Rightarrow (\forall x, x', x''{:}\,\sigma_{set}.(R'\ x\ x') \Rightarrow (R'\ x'\ x'') \Rightarrow (R'\ x\ x'')) & (\tau) \\
\qquad \Rightarrow (\forall x, x'{:}\,\sigma_{set}.\sigma_{rel}[\gamma, x, x'] \Rightarrow (R'\ x\ x')) & (\rho) \\
\qquad \Rightarrow (\forall x, x'{:}\,\sigma_{set}.R[\gamma, x, x'] \Rightarrow \sigma_{rel}[\gamma, x, x] \Rightarrow \sigma_{rel}[\gamma, x', x'] & \\
\qquad\qquad \Rightarrow (R'\ x\ x')) & (\kappa) \\
\qquad \Rightarrow (R'\ s\ s') &
\end{array}$$

In other words, $(\sigma/R)_{rel}[\gamma, -, -]$ is the least partial equivalence relation on σ_{set} which contains $\sigma_{rel}[\gamma, -, -]$ and $R[\gamma, -, -]$ restricted to the domain of $\sigma_{rel}[\gamma, -, -]$. We shall call such a relation *suitable*, so $(\sigma/R)_{rel}$ is the least suitable relation.

5.1.6.1 Equivalence classes. Assume $M \in Sect(\sigma)$. We want to construct a section $[M]_R$ of σ/R. We put

$$[M]_R[\gamma] = M[\gamma]$$

so $[M]_R$ behaves just like M. We must prove RESP for $[M]_R$. Let $\gamma, \gamma'{:}\,\Gamma_{set}$ and $\Gamma_{rel}[\gamma, \gamma']$. Moreover, let $R'{:}\,\sigma_{set} \to \sigma_{set} \to Prop$ be a suitable relation. We must show that $R'[M[\gamma], M[\gamma']]$. But since M is a section we have $\sigma_{rel}[\gamma, M[\gamma], M[\gamma']]$ so we are done since suitable relations contain $\sigma_{rel}[\gamma, -, -]$.

For Q-AX assume $M, N \in Sect(\sigma)$ and $H \in Sect(Prf\{R \circ \overline{N^+} \circ \overline{M}\})$, i.e.

$$\gamma, \gamma'{:}\,\Gamma_{set}\ ,\ Prf(\Gamma_{rel}[\gamma, \gamma']) \vdash R[\gamma, M[\gamma], N[\gamma]]\ true$$

We want to show that $Sect(Prf\{L\_Eq([M]_R, [N]_R)\})$ is non-empty. By Lemma 5.1.6 and the definition of $[-]_R$ it suffices to show that

$$\gamma{:}\,\Gamma_{set}\ ,\ Prf(\Gamma_{rel}[\gamma, \gamma]) \vdash \sigma/R_{rel}[\gamma, M[\gamma], N[\gamma]]\ true$$

In this context we can show $\sigma_{rel}[\gamma, M[\gamma], M[\gamma]]$ and $\sigma_{rel}[\gamma, N[\gamma], N[\gamma]]$ using RESP. Thus for every suitable relation $R'{:}\,\sigma_{set} \to \sigma_{set} \to Prop$ we have $(R'\ M[\gamma]\ N[\gamma])$ by (κ) which gives the desired result.

5.1.6.2 Lifting. Now we want to define functions on the quotient family σ/R, from functions on σ which respect R. More precisely, suppose we are given $\tau \in Fam(\Gamma)$ and $M \in Sect(\tau\{p(\sigma)\})$ and

$$\begin{array}{l}
H \in Sect(Prf\{L_Eq(\\
\qquad M\{p(\sigma^+) \circ p(Prf\{R\})\}, \\
\qquad M\{q(p(\sigma), \sigma) \circ p(Prf\{R\})\} \\
\quad)\})
\end{array}$$

(The two sections inside L_Eq are instances of M applied to one of the σ-variables in $\Gamma \cdot \sigma \cdot \sigma^+ \cdot Prf\{R\}$. Thus informally H states that M maps R-related elements to Leibniz-equal elements.) Finally assume $N \in Sect(\sigma/R)$. We want to construct a section $plug_R\ N\ in\ M\ using\ H$ of τ from this. We define

$$(plug_R\ N\ in\ M\ using\ H)[\gamma{:}\,\Gamma_{set}] = M[\gamma\ ,\ N[\gamma]]$$

Let $\gamma, \gamma' : \Gamma_{set}$ and $\Gamma_{rel}[\gamma, \gamma']$. Since $N \in Sect(\sigma/R)$ we have

$$(\sigma/R)_{rel}[\gamma, N[\gamma], N'[gamma']]$$

Now consider the particular relation

$$\begin{aligned} R' = \lambda x, x' &: \sigma_{set}. \\ &\sigma_{rel}[\gamma, x, x] \wedge \\ &\sigma_{rel}[\gamma, x', x'] \wedge \\ &\tau_{rel}[\gamma, M[\gamma, x], M[\gamma', x']] \end{aligned}$$

Now the result follows provided we can show that R' is suitable. This requires a somewhat lengthy calculation in which H is used to establish that R' extends R.

5.1.6.3 Induction. Finally, we interpret the induction principle *Qind*. Assume $P : \Gamma \cdot \sigma/R \to \mathbf{Prop}$ and let an element of

$$Sect(Prf\{P \circ q(p(\sigma), \sigma/R) \circ \overline{[v_\sigma]_{R+}}\})$$

be given. In other words assume

$$\gamma, \gamma' : \Gamma_{set} \text{ , } Prf(\Gamma_{rel}[\gamma, \gamma']) \text{ , } s, s' : \sigma_{set} \text{ , } Prf(\sigma_{rel}[\gamma, s, s']) \vdash P[s] \; true$$

Now, if $M \in Sect(\sigma/R)$ then we want to find a section of $Prf\{P \circ \overline{M}\}$, i.e. we must show

$$\gamma, \gamma' : \Gamma_{set} \text{ , } Prf(\Gamma_{rel}[\gamma, \gamma']) \vdash P[M[\gamma]] \; true$$

In this context we find $\sigma/R_{rel}[\gamma, M[\gamma], M[\gamma]]$ using SYM and TRANS and RESP. Now putting $s = s' = M[\gamma]$ the result follows if we can show (the stronger condition) $\sigma_{rel}[\gamma, M[\gamma], M[\gamma]]$. But this follows by using the suitable relation

$$R' := \lambda x, x' : \sigma_{set}.\sigma_{rel}[\gamma, x, x] \wedge \sigma_{rel}[\gamma, x', x']$$

Proposition 5.1.9. *The setoid model* $\mathbf{S}_0$ *supports the interpretation of quotient types.*

5.1.6.4 Effectiveness of quotient types. Here we want to address the question as to whether the converse to Q-AX is also true, i.e. whether we can conclude $R[M, N]$ from $[M]_R \stackrel{L}{=} [N]_R$. In general, this cannot be the case because together with Q-AX this implies that R is an equivalence relation. Quotient types are called *effective* [57, 85] if this condition is already sufficient. It turns out that in $\mathbf{S}_0$ all quotient types are effective and that this is a purely syntactic consequence of the rules for quotient types and rule BI-IMP.

Proposition 5.1.10. *Fix a type theory with quotient types and* BI-IMP. *Let* $\Gamma \vdash \sigma$ *and* $\Gamma \text{ , } x, y : \sigma \vdash R[x, y] : Prop$ *be an equivalence relation, i.e. in context* Γ *we have*

$$\begin{aligned} &\forall x : \sigma. R[x, x] \; true && (\textit{reflexivity}) \\ &\forall x, y : \sigma. R[x, y] \Rightarrow R[y, x] \; true && (\textit{symmetry}) \\ &\forall x, y, z : \sigma. R[x, y] \Rightarrow R[y, z] \Rightarrow R[x, z] \; true && (\textit{transitivity}) \end{aligned}$$

Then for each $\Gamma \vdash U, V : \sigma$ *with* $[U]_R \stackrel{L}{=} [V]_R$ *we have* $R[U, V]$.

Proof. Consider the term $M[y:\sigma] = R[U,y] : Prop$. We have

$$\Gamma\ ,\ y,y':\sigma\ ,\ R[y,y'] \vdash M[y] \stackrel{L}{=} M[y']\ true$$

by symmetry and transitivity and BI-IMP. Let H denote the proof of this. Now put

$$M'[z:\sigma/R] = plug_R\ z\ in\ M\ using\ H : Prop$$

By Q-COMP $M'[[U]_R]$ equals $R[U,U]$ and $M'[[V]_R]$ equals $R[U,V]$. The former is true by reflexivity of R and both are Leibniz equal by assumption on U and V. Therefore $R[U,V]$ is true as well.

Remark 5.1.11 (Quotient types without impredicativity). Remark 4.6.1 on the role of impredicativity for subset types applies *mutatis mutandis* also to $\mathbf{S}_0$. We have used the impredicative Calculus of Constructions as a target type theory (and as the basis of the source type theory) mainly for convenience because it offers encodings of the logical connectives and equality so that their existence does not have to be verified in the model. Also effectiveness of quotients can then be derived in the source type theory. However, if for certain reasons one wants to stay within a predicative framework such as Martin-Löf type theory one can still interpret a type theory with a *sort* rather than a type of propositions. One reason why one might want to do this is that Martin-Löf type theory as a target type theory satisfies the internal axiom of choice so that one could hope to interpret the principles of countable choice and dependent choice in the source type theory as indicated in Section 5.1.4.2 above. The details are left to future research.

5.1.7 A choice operator for quotient types

Our next goal consists of finding a way of getting hold of a representative for a given element of a quotient type. The idea is that since in the model an element M of a quotient type is nothing but an element of the underlying type, albeit with a weaker RESP requirement, it should under certain circumstances be possible to view M as an element of this underlying type. In other words we seek to interpret a rule of the form

$$\frac{\Gamma \vdash M : \sigma/R \qquad \text{“certain proviso”}}{\Gamma \vdash choice(M) : \sigma}\ \text{Q-CHOICE}$$

and the following two equality rules.

$$\frac{\Gamma \vdash choice([M]_R) : \sigma \qquad \Gamma \vdash M : \sigma}{\Gamma \vdash choice([M]_R) = M : \sigma}\ \text{Q-CHOICE-COMP}$$

$$\frac{\Gamma \vdash M : \sigma/R \qquad \Gamma \vdash choice(M) : \sigma}{\Gamma \vdash [choice(M)]_R = M : \sigma/R}\ \text{Q-CHOICE-AX}$$

There is no error in rule Q-CHOICE-COMP; we explain in Section 5.1.7.1 how it can be consistent with Q-AX. Semantically, we want to interpret the *choice* operator as the identity, i.e. if $M \in Sect(\sigma/R)$ then we want to put $choice(M)[\gamma{:}\,\Gamma_{set}] = M[\gamma]$. Now from $M \in Sect(\sigma/R)$ we can deduce that if $\Gamma_{rel}[\gamma, \gamma']$ for some $\gamma, \gamma'{:}\,\Gamma_{set}$ then $\sigma/R_{rel}[\gamma, M[\gamma], M[\gamma']]$. But in order to conclude $choice(M) = M \in Sect(\sigma)$ we need to have (the stronger) $\sigma_{rel}[\gamma, M[\gamma], M[\gamma']]$. One situation in which we can deduce the latter from the former occurs when from $\Gamma_{rel}[\gamma, \gamma']$ we can conclude that $M[\gamma]$ and $M[\gamma']$ are actually Leibniz equal, because then we can reason like in Section 5.1.6.3 above. Now this situation in turn occurs for example when $\Gamma_{rel}[\gamma, \gamma']$ entails that γ and γ' are Leibniz equal themselves. This motivates the following definition.

Definition 5.1.12. The set of *non-quotiented* types is defined by the following clauses.

- $\mathbf{N}$ (and other inductive types) and $\Pi x_1{:}\,\sigma_1 \ldots \Pi x_n{:}\,\sigma_n.Prf(M)$ for $n \geq 0$ are non-quotiented.
- If σ and τ are non-quotiented, so is $\Sigma x{:}\,\sigma.\tau$.

A (syntactic) context Γ is non-quotiented if it is made up out of non-quotiented types only.

Notice that if $\vdash \Gamma = \Delta$ and Γ is non-quotiented then so is Δ.

Proposition 5.1.13. *Let Γ be a non-quotiented syntactic context and let $(\Gamma_{set}, \Gamma_{rel})$ be its interpretation in the setoid model. We have*

$$\gamma, \gamma'{:}\,\Gamma_{set}\ ,\ \Gamma_{rel}[\gamma, \gamma'] \vdash \gamma.i \stackrel{L}{=} \gamma'.i\ true$$

Proof. Easy induction on the definition of "non-quotiented".

Now we could semantically justify the application of *choice* in non-quotiented contexts, i.e. the above "certain proviso" would be that Γ is non-quotiented. With such a rule we would, however lose the syntactic weakening and substitution properties which are implicit in e.g. the typing rule for application. For example if $\vdash M : \sigma/R$ we can infer $\vdash choice(M) : \sigma$ because the empty context is non-quotiented, but we would not have $x{:}\,\sigma/R \vdash choice(M) : \sigma$ although x does not occur in M. Therefore, we close the rule up under arbitrary substitution and weakening and thus arrive at the following definitive rule for *choice*:

$$\frac{\Gamma \vdash M : \sigma/R \qquad \begin{array}{l}\text{There exists a non-quotiented context } \Delta \text{ and}\\ \text{a type } \tau \text{ and a term } N \text{ with } \Delta \vdash N : \tau \text{ and a}\\ \text{syntactic context morphism } \Gamma \vdash f \Rightarrow \Delta \text{ such}\\ \text{that } M \equiv N[f] \text{ and } \sigma/R \equiv \tau[f].\end{array}}{\Gamma \vdash choice(M) : \sigma}\ \text{Q-CHOICE}$$

Recall that $\equiv$ means syntactic identity and not just definitional equality so that the side condition is decidable since the possible substitutions f are bounded by the size of M. The condition is e.g. satisfied for $z{:}\,Int \vdash -|z| : Int$ with $\Delta = n{:}\,\mathbf{N}$ and $f[z{:}\,Int] = |z|$, but it does not hold for $z{:}\,Int \vdash z + z : Int$

Proposition 5.1.14. *The above rules* Q-CHOICE, Q-CHOICE-COMP, *and* Q-CHOICE-AX *can be soundly interpreted in the setoid model.*

Proof. We interpret $choice(M)$ like M. Suppose that $\Gamma \vdash choice(M) : \sigma$ by Q-CHOICE then $M \equiv N[f]$ for some (syntactic) substitution $f : \Gamma \Rightarrow \Delta$. We identify these syntactic objects with their interpretations. If $\Gamma_{rel}[\gamma, \gamma']$ then $f[\gamma]$ and $f[\gamma']$ are actually Leibniz equal by Proposition 5.1.13 using the assumption that Δ is non-quotiented. So $M[\gamma]$ and $M[\gamma']$ are Leibniz equal and thus related in σ_{rel} by the same argument as the one used in Section 5.1.6.3. Thus we can assert $M \in Sect(\sigma)$. Since both *choice* and $[-]_R$ are interpreted as the identity, the rules Q-CHOICE-COMP and Q-CHOICE-AX are validated.

5.1.7.1 Discussion. The choice operator is quite unusual and seems paradoxical at first so that a few examples explaining its use are in order. Let $-1 := [(0,1)]_{R_{\mathrm{Int}}}$ and $-1' := [(2,3)]_{R_{\mathrm{Int}}}$ where $1, 2, 3$ are abbreviations for the corresponding numerals of type $\mathbf{N}$ and R_{Int} is as defined in Section 5.1.5. Now since $0+3 \stackrel{L}{=} 1+2$ we have $R_{Int}[(0,1)\ ,\ (2,3)]$, and thus $-1 \stackrel{L}{=} -1'$. On the other hand, $choice(-1) = (0,1)$ and $choice(-1') = (2,3)$. It seems as if we could conclude $(0,1) \stackrel{L}{=} (2,3)$ which would be a contradiction. Now $-1 \stackrel{L}{=} -1'$ means

$$\forall P{:}\, Int \to Prop.(P\ \ -1) \Rightarrow (P\ \ -1')$$

In order to get a contradiction from this, we would have to instantiate with $P = \lambda z{:}\, Int.choice(-1) \stackrel{L}{=} choice(z)$. But this expression is not well-typed since $z{:}\, Int \vdash choice(z)$ (the context $z{:}\, Int$ contains a quotient type) is not. Thus, in some sense the *choice* operator does not respect Leibniz equality. As any term former it does of course respect definitional equality, so in an extensional setting where the two equalities are identified *choice* would be unsound.

The main usage of the *choice* operator is that it permits to recover the underlying implementation of a function between quotients internally. For example, if we have defined some function $F : Int \to Int$ in non-quotiented context Γ (so in particular F may not just be a variable) then we may form $\Gamma, u{:}\,\mathbf{N}\times\mathbf{N} \vdash F\ [u]_{R_{\mathrm{Int}}} : Int$, apply *choice* and abstract from u to obtain $F' := \lambda u{:}\,\mathbf{N}\times\mathbf{N}.choice(F\ [u]_{R_{\mathrm{Int}}}) : (\mathbf{N}\times\mathbf{N}) \to (\mathbf{N}\times\mathbf{N})$ Now using Q-CHOICE-AX we obtain $\Gamma\ ,\ u{:}\,\mathbf{N}\times\mathbf{N} \vdash F\ [u]_{R_{\mathrm{Int}}} = [F'\ u]_{R_{\mathrm{Int}}} : Int$ and moreover—since R_{Int} is an equivalence relation— $\Gamma \vdash \forall u, v{:}\,\mathbf{N}\times\mathbf{N}.R_{Int}[u,v] \Rightarrow R_{Int}[F'\ u, F'\ v]$ using the above and Proposition 5.1.10.

Yet another application of *choice* arises from the use of quotient types to model "non-constructive" types like the real numbers. Assume that we have defined a type of real numbers *Real* as a quotient of a suitable subset type *RealRep* of $\mathbf{N} \to \mathbf{N}$ (thought of as of decimal or continued fraction expansions), the quotienting relation being that the difference is a fundamental sequence defined in the usual ε-δ style. Now since different real numbers can be arbitrarily close together, there can be no non-constant function from the quotient type *Real* to a ground type like $\mathbf{N}$ and, more generally, all functions from *Real* to *Real* must be continuous. This has been brought forward as a serious argument against the use of quotienting in a constructive setting, since if one has defined a real number one cannot even put one's hands on its first digit! Using the choice operator one can at least solve this last problem. Assume that we have defined a real number R in the empty context, say e or π. We may then form

$\vdash choice(R) : \mathbf{N} \to \mathbf{N}$ and from this extract desired intensional information like the first digit or other things. However, it is still not possible to write a non-constant *function* from the reals to $\mathbf{N}$. Notice that this lack is not particular to the setoid model, but directly inherited from the target type theory which does not provide functions from $\mathbf{N} \to \mathbf{N}$ to $\mathbf{N}$ which respect the "book"-equality for real numbers. Similarly, in the setoid model we have no non-constant function from *Prop* to $\mathbf{N}$ because this is so in the target type theory.

The constructive real numbers furnish another illustrative example due to Troelstra [107] for the failure of the axiom of choice because choice functions need not preserve equality. Consider the function $f[x] = x^3 - 3x$. Although f is surjective it has no continuous right inverse. In $\mathbf{S}_0$ surjectivity, that is $\forall a{:}\, Real.\exists x{:}\, Real.f[x] \stackrel{L}{=} a$, follows using *Qind* from $\forall \hat{a}{:}\, RealRep.\exists x{:}\, Real.f[x] \stackrel{L}{=} [\hat{a}]_{R_{\mathrm{Real}}}$ where $Real = RealRep/R_{Real}$. This in turn is readily established by case distinction on $\hat{a}$. But since this case distinction does not preserve equality of real numbers we cannot lift this operation to a function on the real numbers themselves.

5.1.8 Type dependency and universes

As announced earlier, the model $\mathbf{S}_0$ does not provide any type dependency other than the one induced by propositions. In particular we have neither universes nor "large eliminations" [2]. Formally, this can be seen as follows. Assume that the target type theory contains an empty type $\mathbf{0}$ and an operator $?_\sigma$ such that $?_\sigma(M) : \sigma$ if $M : \mathbf{0}$. Then in the setoid model $\mathbf{S}_0$ we also have this empty type. Now if we had a universe containing the natural numbers and the empty type then we could define a family of types $n{:}\,\mathbf{N} \vdash \sigma[n]$ such that $\sigma[0] = \mathbf{0}$ and $\sigma[Suc(n)] = \mathbf{N}$. Now remember that the $-_{set}$-component is left unchanged upon substitution. So in view of the first equation ($\sigma[0] = \mathbf{0}$) there would have to be a function $?_\tau : \sigma_{set} \to \tau$ for every type τ which contradicts the second equation if $\tau = \mathbf{0}$.

A major application of such families of types is that in their presence Peano's fourth axiom [88, 100] is derivable. On the level of propositions we are still able to interpret this axiom, that is we have $0 \stackrel{L}{=} Suc(0) \Rightarrow ff$, provided this holds in the target type theory. The difference is that $Prf(ff)$ is weaker than the empty type because in the presence of an element of $Prf(ff)$ every proposition is true, but not every type is inhabited.

Universes are also used for modularisation and structuring. For instance, using a universe it is possible to define a type of monoids or groups and to identify the construction of the free group over a monoid as a function. If we want to use universes in this way then we are not interested in quotienting the universe itself or types derived from it like $U \to U$. So it should be enough to allow for quotienting of *small types*, i.e. those of the form $El(M)$ for some $M : U$. We will now look at an extension to the setoid model which admits a universe restricted in the sense that quotient types and inductive types and elimination of them is only allowed within the universe.

In this model, types (in the empty context for now) are triples $\sigma = (\sigma_{type}, \sigma_{set}, \sigma_{rel})$ where σ_{type} is a type, σ_{set} is a family of types indexed by σ_{type} and finally σ_{rel} is a partial equivalence relation on each $\sigma_{set}[s]$ for $s\colon\sigma_{type}$. Note that it is not possible to compare by σ_{rel} to elements $x\colon\sigma_{set}[s]$ and $x'\colon\sigma_{set}[s']$ unless $s, s'\colon\sigma_{type}$ are *definitionally* equal. Now the universe has $-_{type}$ component U, whereas types of the form $El(M)$ have trivial $-_{type}$ component $\mathbf{1}$. Quotienting will then only be possible for such "small" types. Similarly, the families in elimination rules like the one for the natural numbers or the one for quotient types will only range over small types. So in particular we cannot define the above family σ over $\mathbf{N}$, because this would require primitive recursion with result type U. Let us now study the model in more detail.

Contexts are defined as triples $\Gamma = (\Gamma_{type}, \Gamma_{set}, \Gamma_{rel})$ where Γ_{type} is a syntactic context. Γ_{set} is a syntactic telescope over Γ_{type}. The third component Γ_{rel} is a proposition

$$g\colon\Gamma_{type}\ , \gamma, \gamma'\colon\Gamma_{set}[g] \vdash \Gamma_{rel}[g, \gamma, \gamma'] : Prop$$

such that Γ_{rel} defines a partial equivalence relation, i.e. we have

$$\begin{array}{ll} g\colon\Gamma_{type}\ , \gamma, \gamma'\colon\Gamma_{set}[g]\ , & \\ \quad Prf(\Gamma_{rel}[g, \gamma, \gamma']) \vdash \Gamma_{rel}[g, \gamma', \gamma]\ true & \textsc{Sym} \\ g\colon\Gamma_{type}\ , \gamma, \gamma', \gamma''\colon\Gamma_{set}[g]\ , & \\ \quad Prf(\Gamma_{rel}[g, \gamma, \gamma'])\ , Prf(\Gamma_{rel}[g, \gamma', \gamma'']) \vdash \Gamma_{rel}[g, \gamma, \gamma'']\ true & \textsc{Trans} \end{array}$$

As usual, equality between such contexts is given by definitional equality of all three components. So a context can be seen as a context of deliverables equipped with a partial equivalence relation on each fibre.

A morphism between two contexts Γ, Δ is a pair $f = (f_{type}, f_{set})$ such that

$$\begin{array}{l} g\colon\Gamma_{type} \vdash f_{type}[g] : \Delta_{type} \\ g\colon\Gamma_{type}\ , \gamma\colon\Gamma_{set}[g] \vdash f_{set}[g, \gamma] : \Delta_{set}[f_{type}[g]] \end{array}$$

and moreover

$$\begin{array}{ll} g\colon\Gamma_{type}\ , \gamma, \gamma'\colon\Gamma_{set}[g]\ , & \\ \quad Prf(\Gamma_{rel}[g, \gamma, \gamma']) \vdash \Delta_{rel}[f_{type}[g], f_{set}[g, \gamma], f_{set}[g, \gamma']]\ true & \textsc{Resp} \end{array}$$

A family over context Γ is now given by a triple $\sigma = (\sigma_{type}, \sigma_{set}, \sigma_{rel})$ where

$$\begin{array}{l} g\colon\Gamma_{type} \vdash \sigma_{type}[g] \\ g\colon\Gamma_{type}, s\colon\sigma_{type}[g] \vdash \sigma_{set}[g, s] \\ g\colon\Gamma_{type}\ , s\colon\sigma_{type}[g]\ , \gamma\colon\Gamma_{set}[g]\ , x, x'\colon\sigma_{set}[g, s] \vdash \sigma_{rel}[g, s, \gamma, x, x'] : Prop \end{array}$$

in such a way that σ_{rel} is symmetric and transitive and compatible with Γ_{rel}, i.e.

$$g{:}\,\Gamma_{type}\ ,\ s{:}\,\sigma_{type}[g]\ ,\ \gamma{:}\,\Gamma_{set}[g]\ ,\ x,x'{:}\,\sigma_{set}[g,s]\ , \qquad \textsc{Sym}$$
$$Prf(\Gamma_{rel}[g,\gamma,\gamma])\ ,\ Prf(\sigma_{rel}[g,\gamma,s,x,x'])\vdash\sigma_{rel}[g,\gamma,s,x',x]\ \mathit{true}$$

$$g{:}\,\Gamma_{type}\ ,\ s{:}\,\sigma_{type}[g]\ ,\ \gamma{:}\,\Gamma_{set}[g]\ ,\ x,x',x''{:}\,\sigma_{set}[g,s]\ , \qquad \textsc{Trans}$$
$$Prf(\Gamma_{rel}[g,\gamma,\gamma])\ ,\ Prf(\sigma_{rel}[g,\gamma,s,x,x'])\ ,\ Prf(\sigma_{rel}[g,\gamma,s,x',x''])$$
$$\vdash\sigma_{rel}[g,\gamma,s,x,x'']\ \mathit{true}$$

$$g{:}\,\Gamma_{type}\ ,\ s{:}\,\sigma_{type}[g]\ ,\ \gamma,\gamma'{:}\,\Gamma_{set}[g]\ ,\ x,x'{:}\,\sigma_{set}[g,s]\ , \qquad \textsc{Comp}$$
$$Prf(\Gamma_{rel}[g,\gamma,\gamma'])\ ,\ Prf(\sigma_{rel}[g,\gamma,s,x,x'])\vdash\sigma_{rel}[g,\gamma',s,x,x']\ \mathit{true}$$

We notice that since σ_{set} does not depend on Γ_{set} the last axiom Comp can be formulated in the same simple way as before for ordinary setoids. Indeed, in this setup dependency and quotienting are entirely separated. Again, we have the relativisation of symmetry and transitivity to "existing" elements of Γ_{set}.

The remaining structure including Π-types is forced by the construction, so we leave it to the reader.

More interesting is the definition of the universe in the empty context. We shall require the target type theory to support a universe (U, El). In particular, this universe may be $(Prop, Prf)$ in which case we obtain an impredicative universe in the model different from the semantic universe of propositions which exists too. The universe is defined by

$$\begin{aligned}
&U_{type} = U\\
&U_{set}[X{:}\,U] = El(X)\to El(X)\to Prop\\
&U_{rel}[X{:}\,U][R,R'{:}\,El(X)\to El(X)\to Prop] =\\
&\quad(\forall x,x'{:}\,El(X).(R\ x\ x')\Rightarrow(R\ x'\ x)\wedge\\
&\quad(\forall x,x',x''{:}\,El(X).(R\ x\ x')\Rightarrow(R\ x'\ x'')\Rightarrow(R\ x\ x''))\wedge\\
&\quad(\forall x,x'{:}\,El(X).(R\ x\ x')\Leftrightarrow(R'\ x\ x'))
\end{aligned}$$

This means that a morphism f from Γ to $\top\cdot U$ assigns to $g{:}\,\Gamma_{type}$ an element $f_{type}[g] : U$ and to $\gamma{:}\,\Gamma_{set}[g]$ a relation $El(f_{type}[g])\to El(f_{type}[g])\to Prop$ in such a way that if $\Gamma_{rel}[g,\gamma,\gamma]$ then this relation is symmetric and transitive and moreover if $\Gamma_{rel}[g,\gamma,\gamma']$ then the relations associated to γ and γ' are equivalent. In other words such a morphism induces a family over Γ. This family is formally obtained by substituting f into the generic family $El\in Fam(\top\cdot U)$ defined by

$$\begin{aligned}
&El_{type}[X{:}\,U] = \mathbf{1}\\
&El_{set}[X{:}\,U, u{:}\,\mathbf{1}] = El(X)\\
&El_{rel}[X{:}\,U\ ,\ u{:}\,\mathbf{1}\ ,\ R{:}\,X\to X\to Prop\ ,\ x,x'{:}\,El(X)] = R\ x\ x'
\end{aligned}$$

It is easy to check that this indeed defines a family, because symmetry and transitivity are relativised to those $R{:}\,X\to X\to Prop$ for which $U_{rel}[X,R,R]$ holds.

The universe thus obtained is closed under all the type formers supported by the setoid model $\mathbf{S}_0$ including quotient types provided the universe of the target type theory has the required closure properties. We leave the details to future work.

5.2 The groupoid model

We have seen in the last section that the setoid model $\mathbf{S}_0$ is of rather limited use when one wants to interpret genuine type dependency other than the one arising from proofs depending on propositions. For instance, we were unable to define a family of types over the natural numbers by primitive recursion. The reason was that the set-part of a family did not depend on the set-part of its context, only the relations were made dependent. In order to overcome these limitations an obvious change is to make the $-_{set}$-parts dependent, too. So a setoid depending on a context of setoids Γ would consist of a type in context Γ_{set}, i.e. $\Gamma_{set} \vdash \sigma_{set}$ and an equivalence relation on each of the $\sigma[\gamma]$, i.e. $\Gamma_{set}\ ,\ x, x' : \sigma_{set} \vdash \sigma_{rel}$ together with terms *refl*, *sym*, *trans* having the appropriate types. Now in order to define context extension one needs a relation on the context $\Gamma, x : \sigma_{set}$. This means that there must be a way of comparing elements in $\sigma_{set}[\gamma]$ and $\sigma_{set}[\gamma']$ if γ and γ' are related. A natural way of achieving this consists of assuming functions mediating between the two, i.e. a term

$$\gamma, \gamma' : \Gamma_{set}\ ,\ p : \Gamma_{rel}[\gamma, \gamma']\ ,\ x : \sigma_{set}[\gamma] \vdash \sigma_{reindex}[p, x] : \sigma_{set}[\gamma']$$

One would then expect that $\sigma_{reindex}$ sends related elements to related elements and so forth. It is because the proof p of $\Gamma_{rel}[\gamma, \gamma']$ is used computationally in $\sigma_{reindex}$ that all the relations have to be proper dependent types and not just families of propositions. Another way, which we pursue in Section 5.3, consists of defining the relations σ_{rel} between elements in different fibres $\sigma_{set}[\gamma]$ and $\sigma_{set}[\gamma']$. The component $\sigma_{reindex}$ is, however, still needed.

Now it turns out that in order to obtain a model of type theory along these lines one needs to impose certain equational constraints on these proofs of relatedness and also on the witnesses *refl*, *sym*, *trans* which cannot be guaranteed in a purely syntactic model. The construction can, however, be carried through in an extensional set-theoretic framework and is of interest for the following three reasons.

First, the model shows that from the identity elimination rule J alone it is not possible to prove that any two elements of an identity type are propositionally equal, that is it provides the proof promised in Section 3.1.2 that the additional elimination rule *IdUni* introduced there is indeed necessary.

Second, the model gives a semantic view on the dichotomy between definitional and propositional equality. Definitional equality is the ambient set-theoretic equality in the model, whereas propositional equality gets interpreted as generalised isomorphism. This establishes in particular the soundness of a rule which defines propositional equality between elements of a universe as isomorphism.

Thirdly, we consider the model as the "correct" definition of dependent setoids as opposed to the somewhat *ad hoc* formulation in Section 5.3 below. This latter construction is motivated and clarified by the interpretation we shall give now.

Since in the model types are interpreted as *groupoids*, that is small categories with isomorphisms only, we call the model *the groupoid interpretation of type theory*. Much of the material presented here has been published in a joint paper with Streicher [49]. A similar model has independently been studied by Lamarche [61] without, however, noticing the implications on propositional equality. In [49] the model was described in rather abstract terms using the notion of fibration of groupoids. Here we give an elementary description avoiding categorical language as much as possible. We use informal Martin-Löf type theory to denote sets and families of sets as well as elements. In particular we write $\Pi x \in X.S(x)$ for the cartesian product of a family of sets S indexed over a set X, and we use $\Sigma x \in X.S(x)$ for the disjoint union of all sets in an X-indexed family S. We denote pairing by (x, y) and projections by $x.1$ and $x.2$. Also we use $\lambda x \in A. \ldots$ and $M\ N$ for set-theoretic functional abstraction and application, respectively.

In the following we define groupoids and families of groupoids and show how they give rise to a syntactic category with attributes which interprets the standard type formers including universes and the intensional identity type, but which nevertheless violates uniqueness of identity. Functional extensionality will be available, though.

5.2.1 Groupoids

Definition 5.2.1. A *groupoid* is a small category in which all morphisms are isomorphisms.

In elementary terms a groupoid X consists of a set X_{set} and a "proof-relevant" relation on it, i.e. for each $x, y \in X_{set}$ a set $X_{rel}(x, y)$ which is provably an equivalence relation. This means that there are functions

$$\begin{aligned}
&\mathit{refl} \in \Pi x \in X_{set}.X_{rel}(x,x)\\
&\mathit{sym} \in \Pi x,y \in X_{set}.X_{rel}(x,y) \rightarrow X_{rel}(y,x)\\
&\mathit{trans} \in \Pi x,y,z \in X_{set}.X_{rel}(y,z) \times X_{rel}(x,y) \rightarrow X_{rel}(x,z)
\end{aligned}$$

Notice the order of arguments to *trans* which differs from the one used for transitivity of *Id* or of setoids in Section 5.1. We have adopted the applicative order here to be in line with usual practice in category theory.

Moreover, these functions must satisfy the following equations.

$$\begin{aligned}
&\mathit{trans}(p, \mathit{trans}(q, r)) = \mathit{trans}(\mathit{trans}(p, q), r)\\
&\mathit{trans}(p, \mathit{refl}) = p = \mathit{trans}(\mathit{refl}, p)\\
&\mathit{trans}(p, \mathit{sym}(p)) = \mathit{refl}\\
&\mathit{trans}(\mathit{sym}(p), p) = \mathit{refl}
\end{aligned}$$

where we have omitted the X-arguments. If X is a groupoid we refer to its components by X_{set}, X_{rel}, X_{refl}, X_{sym}, X_{trans}. If the groupoid we are referring to is clear from the context we use the following abbreviations

$$\begin{aligned}
&\iota := X_{refl}(x)\\
&pq := X_{trans}(p, q)\\
&p^{-1} := X_{sym}(p)
\end{aligned}$$

We also sometimes leave out the $-_{set}$ and $-_{rel}$ subscripts.

Examples. For each set X we have the discrete groupoid $\nabla(X)$ defined by $\nabla(X)_{set} = X$ and $\nabla(X)_{rel}(x,y) = \{\star\}$ if $x = y$ and $\emptyset$ otherwise.

If X is a set and G is a group we define a groupoid $G \otimes X$ by $(G \otimes X)_{set} = X$ and $(G \otimes X)_{rel}(x,y) = G$ if $x = y$ and $\emptyset$ otherwise. Here *refl*, *sym*, and *trans* are defined in the obvious way using the group structure.

Every topological space $\mathcal{X}$ gives rise to a groupoid with underlying set being the set of points $|\mathcal{X}|$ of $\mathcal{X}$ and with the set of morphisms between points x and y being the set of paths from x to y quotiented by *homotopy*, i.e. two paths are identified if there exists a continuous mapping sending one to the other. For example the groupoid thus associated to the surface of a ball is the discrete one element groupoid $\nabla(|\mathcal{X}|)$ and the groupoid associated to the surface of a torus is the groupoid $\mathbf{Z} \otimes |\mathcal{X}|$.

Every type $\Gamma \vdash \sigma$ in intensional type theory TT (without extensional constructs) gives rise to a groupoid $\mathcal{G}(\sigma)$ in the following way. The underlying set $\mathcal{G}(\sigma)_{set}$ is the set of terms $\Gamma \vdash M : \sigma$ and the set of morphisms $\mathcal{G}(\sigma)_{rel}(M,N)$ between terms $\Gamma \vdash M, N : \sigma$ is the set of proofs $\Gamma \vdash P : Id_\sigma(M,N)$ quotiented by propositional equality, i.e. $\Gamma \vdash P : Id(M,N)$ and $\Gamma \vdash Q : Id(M,N)$ are identified if $\Gamma \vdash Id(P,Q)$ is inhabited. Identity, composition, and inverses are given by $Refl_\sigma$, $Trans_\sigma$, and Sym_σ which are readily seen to lift to the quotient using *Resp*. The proof of the groupoid equations is a nice application of the elimination rule ID-ELIM-J.

Remark 5.2.2. The definition of groupoids also makes perfect sense in intensional type theory replacing "set" by "type" or "context". The problem is that not many groupoids exist if we ask the defining equations to hold in terms of definitional equality. Indeed we see below why this definition would not be closed under most type formers, notably function space and Π-type.

Definition 5.2.3. Let X and Y be groupoids. A *morphism of groupoids* from X to Y is a functor from X to Y when viewed as categories.

In elementary terms such a morphism f consists of a function $f_{fun} : X_{set} \to Y_{set}$ and a "proof" that f_{fun} respects the relations, i.e. a function

$$f_{resp} : \Pi x, y \in X_{set}.X_{rel}(x,y) \to Y_{rel}(f_{fun}(x), f_{fun}(y))$$

Moreover, this proof must respect *refl* and *trans*, i.e.

$$f_{resp}(\iota) = \iota$$

$$resp(pq) = resp(p)resp(q)$$

From this it follows that *sym* is also preserved. We omit the $-_{fun}$ and $-_{resp}$ subscripts if no confusion can arise.

Example. If $\Gamma \vdash \sigma, \tau$ are types in intensional type theory and $\Gamma, x{:}\sigma \vdash F[x] : \tau$ is a "function" from σ to τ then F induces a morphism of groupoids $\mathcal{G}(F)$ from $\mathcal{G}(\sigma)$ to $\mathcal{G}(\tau)$ by $\mathcal{G}(F)_{fun}(M) := F[M]$ and $\mathcal{G}(F)_{resp}(P) = Resp_{\sigma,\tau}(F, P)$. Again, the required equations are readily verified using ID-ELIM-J.

Proposition 5.2.4. *Groupoids with morphisms of groupoids form a cartesian closed category.*

Proof. Composition and identities are their set-theoretic companions taken for the element and *resp*-part separately. The terminal object **1** is the discrete one-element groupoid denoted $\nabla(\{\star\})$; the cartesian product is the product category. We give the explicit construction of the exponential $X \Rightarrow Y$ of two groupoids X and Y.

$$(X \Rightarrow Y)_{set} = \text{"the set of morphisms from } X \text{ to } Y\text{"}$$

$$(X \Rightarrow Y)_{rel}(f, g) = \text{"the set of natural transformations from } f \text{ to } g\text{"}$$

Thus an element of $(X \Rightarrow Y)_{rel}(f, g)$ is a function

$$\mu \in \Pi x \in X_{set}.Y_{rel}(f(x), g(x))$$

such that for each $p \in X_{rel}(x, y)$ the equation

$$\mu(x)g(p) = f(p)\mu(y)$$

holds. The groupoid structure is given by

$$\begin{aligned} &\iota(x) = \iota \\ &pq(x) = p(x)q(x) \\ &p^{-1}(x) = (p(x))^{-1} \end{aligned}$$

If f is a groupoid morphism from $Z \times X$ to Y its abstraction $\lambda(f)$ is a groupoid morphism from Z to $X \Rightarrow Y$. If $z \in Z_{set}$ then $\lambda(f)(z)$ is the functor which sends $x \in X_{set}$ to $f(z, x)$ and $p \in X_{rel}(x, x')$ to $f(\iota, p)$. Now if $q \in Z_{rel}(z, z')$ then $\lambda(f)(q)$ is a natural transformation from $\lambda(f)(z)$ to $\lambda(f)(z')$. Its component at $x \in X_{set}$ is $f(q, \iota)$. The naturality condition amounts to checking that $f(q, \iota)f(\iota, p) = f(\iota, p)f(q, \iota)$ which follows since f preserves transitivity (composition) in $X \times Y$. It remains to verify the functor laws for $\lambda(f)$. We have

$$\lambda(f)(\iota)(x) = f(\iota, \iota) = \iota$$

$$\lambda(f)(pq)(x) = f(pq, \iota) = f((p, \iota)(q, \iota)) = f(p, \iota)f(q, \iota)$$

We leave it to the reader to define the evaluation morphism and to check the β and η equations.

The exponential of groupoids cannot be defined syntactically in intensional type theory, because there is no way to restrict the type-theoretic function space in such a way that only those functions are included which preserve reflexivity and transitivity up to definitional equality. An encoding in extensional type theory should, however, be possible.

The assignment $\nabla(-)$ of groupoids to sets extends to a functor between these categories and as such has a left adjoint Δ which sends a groupoid G to the quotient set $G_{set}/\!\sim$ where $g \sim g' \Leftrightarrow G_{rel}(g, g') \neq \emptyset$.

Definition 5.2.5. Let Γ be a groupoid. A *family of groupoids indexed over Γ* is a functor from Γ viewed as a category to the category of groupoids. The set[4] of families over a groupoid Γ is denoted $Fam(\Gamma)$.

In elementary terms a family of groupoids over Γ consists of a groupoid $\sigma(\gamma)$ for each element $\gamma \in \Gamma_{set}$. Moreover, these groupoids must be compatible with each other, i.e. there is a reindexing function

$$\sigma_{reindex} \in \Pi\gamma, \gamma' \!\in \Gamma_{set}.\Gamma_{rel}(\gamma, \gamma') \to \sigma(\gamma)_{set} \to \sigma(\gamma')_{set}$$

and a "proof" that this operation respects the relations, i.e. a function

$$\begin{aligned}\sigma_{resp} \in \Pi\gamma, \gamma' \!\in \Gamma_{set}.\Pi p \!\in \Gamma_{rel}(\gamma, \gamma').\Pi u, v \!\in \sigma(\gamma)_{set}.\\ \sigma(\gamma)_{rel}(u, v) \to \sigma(\gamma')_{rel}(\sigma_{reindex}(p, u), \sigma_{reindex}(p, v))\end{aligned}$$

So far we have the notion of dependent setoid sketched in the introduction to this section. But now again we impose further equational constraints. Both $\sigma_{reindex}$ and σ_{resp} must respect the groupoid structure, i.e.

$$\sigma_{reindex}(\iota, u) = u$$

$$\sigma_{reindex}(pq, u) = \sigma_{reindex}(p, \sigma_{reindex}(q, u))$$

$$\sigma_{resp}(\iota, p) = p$$

$$\sigma_{resp}(pq, r) = \sigma_{resp}(p, \sigma_{resp}(q, r))$$

$$\sigma_{resp}(p, \iota) = \iota$$

$$\sigma_{resp}(p, qr) = \sigma_{resp}(p, q)\sigma_{resp}(p, r)$$

If σ is a family of groupoids indexed over Γ we refer to its components by σ_{set}, σ_{rel}, σ_{refl}, σ_{sym}, σ_{trans}. We thus allow both $\sigma_{rel}(\gamma\ , s, s')$ and $\sigma(\gamma)_{rel}(s, s')$ to denote relatedness of s and s' in the fibre over $\gamma \in \Gamma_{set}$. If the family under consideration is clear from the context, we abbreviate $\sigma_{reindex}(p, u)$ by $p \cdot u$ and also $\sigma_{resp}(p, q)$ by $p \cdot q$ in accordance with the convention to use the same name for object and morphism part of a functor (For $p \in \Gamma_{rel}(\gamma, \gamma')$ the pair $(\sigma_{reindex}(p, -), \sigma_{resp}(p, -))$ is a functor from $\sigma(\gamma)$ to $\sigma(\gamma')$.).

[4] This is meant to be a "naive" set. In order to avoid size problems one might require the groupoids taken on by the functor to be small with respect to some universe.

Examples. Every family of sets $\{S_x\}_{x\in X}$ induces a family of (discrete) groupoids indexed over $\nabla(X)$ in the obvious way. Moreover if X and Y are groupoids we define the constant family $Const(X,Y)$ over X by

$$\begin{array}{ll} Const(X,Y)_{set}(x) = Y_{set} & \text{independent of } x \in X_{set} \\ Const(X,Y)_{rel}(x,y,y') = Y_{rel}(y,y') & \\ Const(X,Y)_{reindex}(p,y) = y & \\ Const(X,Y)_{resp}(p,q) = q & \end{array}$$

We also note that families of groupoids over the terminal groupoid **1** are in 1-1 correspondence to (non-dependent) groupoids.

A family of types $\Gamma, x{:}\sigma \vdash \tau[x]$ in TT *almost* gives rise to a family of groupoids over $\mathcal{G}(\sigma)$. Indeed, if $\Gamma \vdash M : \sigma$ is an element of $\mathcal{G}(\sigma)_{set}$ then we obtain a groupoid $\mathcal{G}(\tau)(M) := \mathcal{G}(\tau[M])$. Moreover, if $\Gamma \vdash P : Id_\sigma(M,M')$ then $Subst_{\sigma,\tau}(P,-)$ induces a morphism of groupoids from $\mathcal{G}(\tau[M])$ to $\mathcal{G}(\tau[N])$. However, we do not have $Subst\,(Trans\,(P,Q),M) = Subst\,(Q, Subst\,(P,M))$ as required by the second law for σ_{resp}, but only the corresponding propositional equality. Also the *Subst*-functions do not lift to functions defined on $\mathcal{G}(\sigma)_{rel}$. Nevertheless it is an instructive exercise to verify the equations for families of groupoids in this case up to propositional equality.

More examples arise from the interpretation of the type constructors which we are going to describe.

5.2.1.1 Comprehension and Substitution. Let Γ be a groupoid and $\sigma \in Fam(\Gamma)$. We define the comprehension of σ denoted $\Gamma \cdot \sigma$ as the following groupoid:

$$\begin{array}{l} (\Gamma\cdot\sigma)_{set} = \Sigma\gamma\in \Gamma_{set}.\sigma_{set}(\gamma) \\ (\Gamma\cdot\sigma)_{rel}((\gamma,u)\ ,\ (\gamma',v)) = \Sigma p\in \Gamma_{rel}(\gamma,\gamma').\sigma_{rel}(\gamma'\ ,\ p\cdot u, v) \\ (\Gamma\cdot\sigma)_{refl}((\gamma,u)) = (\Gamma_{refl}(\gamma), \sigma_{refl}(\gamma,u)) \\ (p,q)^{-1} = (p^{-1}, q^{-1}) \\ (p,q)(p',q') = (pp', q(p\cdot q')) \end{array}$$

Notice that in order for these definitions to typecheck the equations imposed on families are required. For example, for $(\Gamma_{refl}(\gamma), \sigma_{refl}(\gamma,u)) \in (\Gamma\cdot\sigma)_{rel}((\gamma,u)\ ,\ (\gamma,u))$ it is required that $\sigma_{refl}(\gamma,u)\in\sigma_{rel}(\gamma\ ,\ \Gamma_{refl}\cdot u, u)$ and this follows from $\Gamma_{refl}\cdot u = u$.

It remains to check that $\Gamma\cdot\sigma$ is indeed a groupoid. We only verify associativity of transitivity.

$$\begin{array}{l} ((p,q)(p',q'))(p'',q'') = (pp', q(p\cdot q'))(p'',q'') = \\ (pp'p'', q(p\cdot q')(pp'\cdot q'')) = (pp'p'', q(p\cdot(q'(p'\cdot q'')))) = \\ (p,q)(p'p'', q'(p'\cdot q'')) = (p,q)((p',q')(p'',q'')) \end{array}$$

The canonical projection $p(\sigma)$ is the first projection from $(\Gamma\cdot\sigma)_{set}$ to Γ_{set} which is obviously a morphism of groupoids. The reader familiar with category theory will have noticed that this defininition of comprehension is a special case of the so-called "Grothendieck construction". Now let Γ, Δ be groupoids,

$\sigma \in Fam(\Delta)$ and $f : \Gamma \to \Delta$. The composition $\sigma \circ f$ is a family of groupoids over Γ which defines the substitution $\sigma\{f\}$. In more explicit terms we have

$$\sigma\{f\}_{set}(\gamma \in \Gamma_{set}) = \sigma_{set}(f_{fun}(\gamma))$$

$$\begin{aligned}
&\sigma\{f\}_{rel}(\gamma \in \Gamma_{set}\ ,\ u, v \in \sigma\{f\}_{set}(\gamma)) = \\
&\qquad \sigma_{rel}(f_{fun}(\gamma), u, v) \\
&\sigma\{f\}_{reindex}(\gamma, \gamma' \in \Gamma_{set}, p \in \Gamma_{rel}(\gamma, \gamma')\ ,\ u \in \sigma\{f\}_{set}(\gamma)) = \\
&\qquad f(p) \cdot u \in \sigma\{f\}_{set}(y)
\end{aligned}$$

The other components are defined accordingly. It is an immediate consequence of the definition that this substitution operation satisfies the split property $\sigma\{f \circ g\} = \sigma\{f\}\{g\}$ and $\sigma\{id_\Delta\} = \sigma$.

Finally, we have the morphism $q(f, \sigma)$ from $\Gamma \cdot \sigma\{f\}$ to $\Delta \cdot \sigma$ whose function part sends (γ, u) to $(f_{fun}(\gamma), u)$, and it is readily seen that the following diagram is a pullback in the category of groupoids:

$$\begin{array}{ccc}
\Gamma \cdot \sigma\{f\} & \xrightarrow{q(f,\sigma)} & \Delta \cdot \sigma \\
{\scriptstyle p(\sigma\{f\})}\downarrow & & \downarrow{\scriptstyle p(\sigma)} \\
\Gamma & \xrightarrow[f]{} & \Delta
\end{array}$$

This shows that groupoids and families of groupoids form a category with attributes in the sense of Remark 2.4.7. For technical reasons it is, however, appropriate to separate sections from their associated context morphisms as follows.

Sections. Let $\sigma \in Fam(\Gamma)$. A *section* of σ is a pair $M = (M_{el}, M_{resp})$ where $M_{el} \in \Pi\gamma \in \Gamma_{set}.\sigma_{set}(\gamma)$ and $M_{resp} \in \Pi\gamma, \gamma' \in \Gamma_{set}.\Pi p \in \Gamma_{rel}(\gamma, \gamma').\sigma_{rel}(\gamma'\ ,\ p \cdot M_{el}(\gamma),\ M_{el}(\gamma'))$ in such a way that the following two equations hold for $\gamma, \gamma', \gamma'' \in \Gamma_{set}$ and $q \in \Gamma_{rel}(\gamma, \gamma')$ and $q \in \Gamma_{rel}(\gamma', \gamma'')$:

$$M_{resp}(refl(\gamma)) = refl(M_{el}(\gamma)) \in \sigma_{rel}(\gamma, M_{el}(\gamma), M_{el}(\gamma))$$

$$M_{resp}(pq) = M_{resp}(p)(p \cdot M_{resp}(q)) \in \sigma_{rel}(\gamma'', (pq) \cdot M_{el}(\gamma), M_{el}(\gamma''))$$

If $M \in Sect(\sigma)$ we define the associated context morphism $\overline{M} : \Gamma \to \Gamma \cdot \sigma$ by $\overline{M}_{fun}(\gamma) := (\gamma, M_{el}(\gamma))$ and $\overline{M}_{resp}(p) := (p, M_{resp}(p))$. This obviously defines a 1-1 correspondence between the set of sections and the set of right inverses to $p(\sigma)$. We can now conclude:

Proposition 5.2.6. *Groupoids and families of groupoids form a syntactic category with attributes.*

5.2.2 Interpretation of type formers

We now investigate the closure properties of this model with respect to various type formers. It turns out that the groupoid model supports almost all the set-theoretic type formers, so in particular those studied in the monograph [88]. Here we restrict ourselves to the interpretation of sum and product types, the identity type, natural numbers, and a universe.

5.2.2.1 Interpretation of the dependent sum. Let σ be a family over Γ and τ a family over $\Gamma \cdot \sigma$. We want to construct a family over Γ which interprets the Σ-type $\Sigma s : \sigma.\tau(s)$. It is a dependent version of the comprehension groupoid from Section 5.2.1.1. It is defined by

$$\Sigma(\sigma, \tau)_{set}(\gamma \in \Gamma_{set}) = \Sigma s \in \sigma_{set}(\gamma).\tau_{set}(\gamma, s)$$

$$\begin{aligned}
&\Sigma(\sigma, \tau)_{rel}(\gamma \in \Gamma_{set}\ ,\ u, v \in \Sigma(\sigma, \tau)_{set}(\gamma)) = \\
&\qquad \Sigma p \in \sigma_{rel}(\gamma, u.1, v.1).\tau_{rel}((\gamma, v.1)\ ,\ (\iota, p) \cdot u.2, v.2)
\end{aligned}$$

$$\Sigma(\sigma, \tau)_{refl}(\gamma \in \Gamma_{set}, u \in \Sigma(\sigma, \tau)_{set}(\gamma)) = (\iota, \iota)$$

$$(p, q)^{-1} = (p^{-1}, q^{-1})$$

$$(p, q)(p', q') = (pp', q(p \cdot q'))$$

$$p \cdot (s, t) = (p \cdot s, (p, \iota) \cdot t) \qquad \text{Reindexing on elements.}$$

$$p \cdot (q, r) = (p \cdot q, (p, \iota) \cdot r) \qquad \text{Reindexing on morphisms.}$$

The verifications are essentially the same as in Section 5.2.1.1 and are left to the reader.

Projections and pairing. If M is a section of $\Sigma(\sigma, \tau)$ then taking the first/second projection in the *el*- and the *resp*-part separately yields two sections, one of σ denoted $M.1$ and one of $\tau\{\overline{M.1}\}$ denoted $M.2$. On the other hand, componentwise pairing turns sections N_1 of σ and N_2 of $\tau\{\overline{N_1}\}$ into a section (N_1, N_2) of $\Sigma(\sigma, \tau)$. These operations are inverse to each other, i.e. $(N_1, N_2).1 = N_1$, $(N_1, N_2).2 = N_2$ and $(M.1, M.2) = M$. Compatibility with substitution is straightforward.

The groupoid model thus admits extensional Σ-types in the sense of Proposition 2.4.23.

5.2.2.2 Dependent product of groupoids. Let σ, τ, Γ be as before. For each $\gamma \in \Gamma_{set}$ let $\tau(\gamma)$ denote the family of groupoids over the groupoid $(\sigma_{set}(\gamma), \sigma_{rel}(\gamma))$ defined by

$$\tau(\gamma)_{set}(s \in \sigma_{set}(\gamma)) = \tau_{set}((\gamma, s))$$
$$\tau(\gamma)_{rel}(s \in \sigma_{set}(\gamma)) = \tau_{rel}((\gamma, s))$$

In fact $\tau(\gamma)$ is just a special instance of substitution along the morphism from the one point groupoid to Γ determined by $\gamma \in \Gamma_{set}$. Using this notation we define the dependent product of τ along σ as a family over Γ by

$$\Pi(\sigma, \tau)_{set}(\gamma \in \Gamma) = \text{“The set of sections of } \tau(\gamma)\text{”}$$

$$\Pi(\sigma, \tau)_{rel}(\gamma \in \Gamma\ ,\ M, N \in \Pi(\sigma, \tau)_{set}(\gamma)) =$$
$$\text{“The set of natural transformations from } M \text{ to } N\text{”}$$

Natural transformations are defined in the same way as for the exponential of non-dependent groupoids. Also the fibre-wise groupoid structure is like in this example. A bit trickier is the definition of reindexing. For $p \in \Gamma_{rel}(\gamma, \gamma')$ and $M \in \Pi(\sigma, \tau)_{set}(\gamma)$ we put

$$(p \cdot M)_{el}(s \in \sigma_{set}(\gamma')) = (p, \iota) \cdot M(p^{-1} \cdot s)$$

Here $M(p^{-1} \cdot s)$ is an element of $\tau_{set}(\gamma, p^{-1} \cdot s)$. Now since $p \cdot (p^{-1}s) = s$ the pair (p, ι) lies in $(\Gamma \cdot \sigma)_{rel}((\gamma, p^{-1} \cdot s), (\gamma', s))$. So the definition makes sense. The morphism part is similar:

$$(p \cdot M)_{resp}(q \in \sigma_{rel}(\gamma', s, s')) = (p, \iota) \cdot M(p^{-1} \cdot q)$$

Finally we must define the *resp* part of reindexing. So let μ be a natural transformation from M to M' in $\Pi(\sigma, \tau)_{set}(\gamma)$. We must define a natural transformation from $p \cdot M$ to $p \cdot M'$. Its component at $s \in \sigma_{set}(\gamma')$ is

$$(p, \iota) \cdot \mu(p^{-1} \cdot s)$$

which is an element of $\tau_{rel}((\gamma', s)\ ,\ (p \cdot M)(s), (p \cdot M')(s))$. The verifications are tedious but straightforward.

Abstraction and application. We want to show that the groupoid model admits products in the sense of Proposition 2.4.13; so we have to define abstraction and the evaluation morphism. In the above situation let M be a section of τ. Its abstraction $\lambda_{\sigma,\tau}(M)$ is a section of $\Pi(\sigma, \tau)$. Its element part at $\gamma \in \Gamma_{set}$ is the section of $\tau(\gamma)$ sending $s \in \sigma_{set}(\gamma)$ to $M_{el}(\gamma, s)$ and $q \in \sigma_{rel}(\gamma, s, s')$ to $M_{resp}(\iota, q)$. Its *resp*-part sends $p \in \Gamma_{rel}(\gamma, \gamma')$ to a natural transformation from $p \cdot \lambda_{\sigma,\tau}(M)(\gamma)$ to $\lambda_{\sigma,\tau}(M)(\gamma')$. Its component at $s \in \sigma_{set}(\gamma')$ is $M(p, \iota)$ which is an element of

$$\tau_{rel}((\gamma', s)\ ,\ (p \cdot \lambda_{\sigma,\tau}(M)(\gamma))(s), M(\gamma', s))$$

The evaluation morphism

$$ev_{\sigma,\tau} : \Gamma \cdot \sigma \cdot \Pi(\sigma, \tau)\{p(\sigma)\} \to \Gamma \cdot \sigma \cdot \tau$$

sends $(\gamma, s, M) : (\Gamma \cdot \sigma \cdot \Pi(\sigma, \tau)\{p(\sigma)\})_{set}$ to $(\gamma, M_{el}\ s)$ and acts similarly on morphisms. The verifications are left to the reader.

5.2.2.3 The identity groupoid. Now we come to the most important part of the model construction. Due to the particular properties of groupoids and families of groupoids it becomes possible to define an identity type which is different from the categorical equaliser, and hence is truly intensional. We use the semantic framework of Definition 2.4.20.

Let σ be a family of groupoids over Γ. We form the groupoid $\Gamma \cdot \sigma \cdot \sigma^+$ which corresponds to the context $\gamma{:}\,\Gamma\ ,\ s, s'{:}\,\sigma(\gamma)$. Over this groupoid we define the family $Id(\sigma)$ by

$$Id(\sigma)((\gamma, s, s')) = \nabla(\sigma_{rel}(\gamma, s, s'))$$

Thus a proof that $s, s' \in \sigma_{set}(\gamma)$ are propositionally equal is an element of $\sigma_{rel}(\gamma, s,\ s')$. Two such proofs are related only if they are actually equal. Thus propositional and definitional equality on identity types coincide in the groupoid model. Coming back to the example of topological spaces and homotopy we would thus consider two points of a topological space as "propositionally equal" if there exists a path linking the two, and two such paths are propositionally equal if they can be continuously transformed into one another.

It remains to define the reindexing functions for $Id(\sigma)$. If $(p, q, r) \in (\Gamma \cdot \sigma \cdot \sigma^+)_{rel}((\gamma, s, s')\ ,\ (\gamma_1, s_1, s_1'))$ and $h \in Id(\sigma)_{set}(x, s, s') = \sigma_{rel}(\gamma\ ,\ s, s')$ then

$$(p, q, r) \cdot h := r(p \cdot h)q^{-1} \in \sigma_{rel}(\gamma', s_1, s_1') = Id(\sigma)_{set}(\gamma', s_1, s_1')$$

Notice that by the definition of $\Gamma \cdot \sigma \cdot \sigma^+$ we have $p \in \Gamma_{rel}(\gamma, \gamma_1)$, $q \in \sigma_{rel}(p \cdot s, s_1)$ and $r \in \sigma_{rel}(p \cdot s', s_1')$. Moreover since $h \in \sigma_{rel}(s, s')$ by definition of families we have $p \cdot h \in \sigma_{rel}(p \cdot s, p \cdot s')$. So the above definition of reindexing in $Id(\sigma)$ is "well typed".

We check the transitivity law:

$$\begin{aligned}
&(p, q, r)(p', q', r') \cdot h = (pp', q(p \cdot q'), r(p \cdot r')) \cdot h = \\
&r(p \cdot r')(pp' \cdot h)(q(p \cdot q')^{-1} = r(p \cdot r')(pp' \cdot h)(p \cdot q'^{-1})q^{-1} = \\
&r(p \cdot (r'(p' \cdot h)q'^{-1}))q^{-1} = (p, q, r) \cdot ((p', q', r') \cdot h)
\end{aligned}$$

Now if $\star \in Id(\sigma)_{rel}((\gamma, s, s')\ ,\ h, h')$—in other words if h and h' are equal—then clearly $(p, q, r) \cdot h = (p, q, r) \cdot h'$, so we put

$$(p, q, r) \cdot \star = \star$$

For this definition to work all the coherence equations imposed on families of groupoids are indeed required.

Identity introduction. Recall that $\overline{v_\sigma} : \Gamma \cdot \sigma \to \Gamma \cdot \sigma \cdot \sigma^+$ is the diagonal morphism from $\Gamma \cdot \sigma$ to $\Gamma \cdot \sigma \cdot \sigma^+$ defined on elements by $(\overline{v_\sigma})_{fun}(\gamma, s) = (\gamma, s, s)$. We must construct a morphism $Refl_\sigma\ :\ \Gamma \cdot \sigma\ \to\ \Gamma \cdot \sigma \cdot \sigma^+ \cdot Id(\sigma)$ with $p(Id_\sigma) \circ Refl_\sigma = \overline{v_\sigma}$. It is defined by

$$\begin{aligned}
(Refl_\sigma)_{fun}(\gamma, s) &:= (\gamma, s, s, \sigma_{refl}(s)) \\
(Refl_\sigma)_{resp}(p, q) &:= (p, q, q, \star)
\end{aligned}$$

This definition implicitly uses the equation

$$(p, q, q) \cdot refl\,(\sigma)(\gamma, s) = q(p \cdot \iota)q^{-1} = qq^{-1} = \iota$$

deduced from the laws for families of groupoids.

Identity elimination. Let τ be a family over the family $\Gamma \cdot \sigma \cdot \sigma^{+} \cdot Id(\sigma)$ and let M be a section of $\tau\{Refl_\sigma\}$. We construct a section $J_{\sigma,\tau}(M)$ of τ by

$$J_{\sigma,\tau}(M)_{el}((\gamma, s, s', q) \in (\Gamma \cdot \sigma \cdot \sigma^{+} \cdot Id(\sigma))_{set}) := (\iota, \iota, q, \star) \cdot M(\gamma, s)$$

This works since $(\gamma, s, s, \sigma_{refl}(s))$ and (γ, s, s', q) are related in $\Gamma \cdot \sigma \cdot \sigma^{+} \cdot Id(\sigma)$ by $(\iota, \iota, q, \star)$ because $(\iota, \iota, q) \cdot \iota = q\iota\iota = q$.

It remains to define $J_{\sigma,\tau}(M)_{resp}$. Let

$$(p, r, r', \star) \in (\Gamma \cdot \sigma \cdot \sigma^{+} \cdot Id(\sigma))_{rel}((\gamma, s, s', q), (\gamma_1, s_1, s_1', q_1))$$

or equivalently $p \in \Gamma_{rel}(\gamma, \gamma')$, $r \in \sigma_{rel}(\gamma_1, p \cdot s, s_1)$, $r' \in \sigma_{rel}(\gamma_1, p \cdot s', s_1')$, and $q_1 = r'(p \cdot q)r^{-1}$. We must exhibit an element of

$$\tau_{rel}((\gamma_1, s_1, s_1', q_1)\ ,\ (p, r, r', \star) \cdot (\iota, \iota, q, \star) \cdot M(\gamma, s)\ ,\ (\iota, \iota, q_1, \star) \cdot M(\gamma_1, s_1))$$

Now

$$M(p, r) \quad \text{relates} \quad (p, r, r, \star) \cdot M(\gamma, s) \text{ and } M(\gamma_1, s_1)$$

since M is a section. So we have

$$\begin{aligned}(\iota, \iota, r'(p \cdot q)r^{-1}, \star) \cdot M(p, r) \quad \text{relates} \quad & (p, r, r'(p \cdot q), \star) \cdot M(\gamma, s) \\ \text{and} \quad & (\iota, \iota, r'(p \cdot q)r^{-1}, \star) \cdot M(\gamma_1, s_1)\end{aligned}$$

by "multiplying" by $(\iota, \iota, r'(p \cdot q)r^{-1}, \star)$ which is the desired proof since

$$(p, r, r', \star)(\iota, \iota, q, \star) = (p, r, r'(p \cdot q), \star)$$

by definition of transitivity in $\Gamma \cdot \sigma \cdot \sigma^{+}$ and $q_1 = r'(p \cdot q)r^{-1}$ by assumption.

Proposition 5.2.7. *The groupoid model supports intensional identity types.*

5.2.2.4 Functional extensionality. It is obvious from the definitions of the dependent product and the identity type that the latter supports functional extensionality, i.e. that *Ext* constants as defined in Section 3.1.3 can be defined, and that Turner's equation (Eqn 3.2 in Section 3.1.3) holds. This is important, because we do not have uniqueness of identity in the groupoid model as shown below in Theorem 5.2.8.

5.2.2.5 The base types. The types **0**, **1**, **N** may be interpreted by the discrete groupoids $\nabla(\emptyset)$, $\nabla(\{\star\})$, $\nabla(\mathbf{N})$. These are initial, terminal, and natural numbers objects , resp., in the category of groupoids from which it follows that the introduction and elimination rules can be interpreted. One may of course construct them explicitly as well. It is also possible to define a groupoid of lists even over a non-discrete groupoid, but we have not checked whether arbitrary parametrised inductive definitions can be carried out in the groupoid model. Certainly, arbitrary inductive definitions as provided by Coquand's pattern matching [18] cannot be interpreted, for then uniqueness of identity would hold in the groupoid model contradicting Theorem 5.2.8 below. We leave it as an open problem to determine whether the groupoid model supports all "orthodox" inductive definitions, i.e. those which only provide Martin-Löf's canonical elimination rules as formalised by Dybjer [29].

5.2.2.6 Interpretation of universes. To interpret the universe we mimic the construction described in [68] of a model for the Extended Calculus of Constructions. Let κ be some inaccessible cardinal. This means that κ is regular (closed under union) and whenever a cardinal μ is strictly smaller than κ then so is 2^μ [65]. Let U be the set of all groupoids lying inside the level V_κ of the cumulative hierarchy. We interpret the universe by the discrete groupoid $\nabla(U)$. Since U is a model of ZFC it is closed under all constructions on groupoids we described, so all the universe rules can be interpreted. Using a chain of inaccessible cardinals we can also interpret universes containing each other. If we are interested in a universe closed under impredicative quantification containing only $\mathbf{0}$ and $\mathbf{1}$ we may use the discrete groupoid $\nabla(\{\emptyset, \{\star\}\})$ with $El(X) = X$.

Inside the set-theoretic groupoid model we considered so far there does not exist a non-trivial impredicative universe. However, groupoids can be defined in any locally cartesian closed category, for example the category of ω-sets [103]. Here we can take U to be the set of those ω-set groupoids for which both the set of objects and each homset are "modest sets", see *loc. cit.* We conjecture that in this way we get a non-degenerate (i.e. without proof irrelevance) model of the Calculus of Constructions. This has been worked out in some more detail in [50].

We have not looked in detail at universes admitting universe induction. However, since the category of sets is a full subcategory of the category of groupoids (via ∇), every set-theoretic construction of such a universe should carry over to the groupoid case, see for example [29].

5.2.2.7 Quotient types. The groupoid model admits the interpretation of intensional quotient types as defined in Section 3.2.6.1, at least in the case where the quotienting relation ρ is internally an equivalence relation. If $\sigma \in Fam(\Gamma)$ and $\rho \in Fam(\Gamma \cdot \sigma \cdot \sigma^+)$ is such that there exist sections of the appropriate families witnessing reflexivity, symmetry, and transitivity, then we can define a groupoid σ/ρ by $(\sigma/\rho)_{set} = \sigma_{set}$ and $(\sigma/\rho)_{rel}(\gamma, s, s') = \Delta(\rho(\gamma, s, s'))$ where Δ is the left adjoint to ∇ defined after the proof of Proposition 5.2.4. The rules for quotient types can be interpreted in this way, but due to the absence of uniqueness of identity we would like to have further rules which constrain the equality proof obtained by Qax (see rule Q-I-Ax). A special case of this problem will be considered below in Section 5.2.4. The formulation of a general quotient type former in the groupoid model of which this example would be an instance is left for future research.

5.2.3 Uniqueness of identity

We are now ready to give the promised proof that uniqueness of identity is not in general definable.

Theorem 5.2.8. *Uniqueness of identity is not uniformly definable at all types, in particular it is not definable at the type* $\vdash \Sigma x\colon U.El(x)$.

Proof. If uniqueness of identity is definable at some type $\Gamma \vdash \sigma$ then by the soundness theorem 2.5.6 and the fact that all groupoids $Id(\sigma)(\gamma, s, s')$ are discrete, every groupoid $[\![\Gamma \mid \sigma]\!](\gamma)$ for $\gamma \in [\![\Gamma]\!]$ must be a pre-order, i.e. every set $([\![\Gamma \mid \sigma]\!](\gamma))_{rel}(x, x')$ has at most one element. But the interpretation of $\Sigma x{:}\, U.El(x)$ does not have this property since in particular it has the element $(\mathbf{Z} / 2\mathbf{Z} \otimes \{\star\}, \star)$ which has two different endomorphisms.

We remark that in view of Propositions 3.1.1 and 3.1.2 the interpretations of all types definable from $\mathbf{N}$ using Π, Σ, *Id* must be pre-orders, in fact they are discrete, which may also be seen directly from the definition of these type formers in the groupoid model.

Since uniqueness of identity can be defined using pattern-matching (Section 3.1.2.2) we obtain the following important corollary:

Corollary 5.2.9. *Pattern-matching is a non-conservative extension of Martin-Löf type theory.*

5.2.4 Propositional equality as isomorphism

The lack of uniqueness of identity in the syntax has been considered as a drawback, which can be cured by the introduction of the *IdUni* constants. There may, however, be circumstances under which the existence of more than one proof of identity is actually useful. One such is provided by the following extension to TT with one universe (U, El) which may be soundly interpreted in the groupoid model. For reasons of exposition we assume functional extensionality and an explicit constant *Subst* as in TT_I although it is definable from J. Let

$$\begin{aligned} Iso[x, y{:}\, U] := {} & \Sigma f{:}\, El(x) \to El(y).\Sigma f' : El(y) \to El(x). \\ & Id_{El(x)\to El(x)}(f' \circ f, id) \times Id_{El(y)\to El(y)}(f \circ f', id) \end{aligned}$$

be the type of propositional isomorphisms between $El(x)$ and $El(y)$. If $F : Iso[X, Y]$ let F also stand for the first component $F.1 : El(X) \to El(Y)$ and let F^{-1} stand for the second component $F.2.1 : El(Y) \to El(X)$. Now consider the following new introduction rule for propositional equality on the universe:

$$\frac{\Gamma \vdash X, Y : U \qquad \Gamma \vdash F : Iso[X, Y]}{\Gamma \vdash UnivId(X, Y, F) : Id_U(X, Y)} \quad \textsc{Univ-Id}$$

Moreover, we introduce a definitional equality describing the reaction of *Subst* to *UnivId*:

$$\frac{\Gamma \vdash F : Iso[X, Y] \qquad \Gamma \vdash M : El(X)}{\Gamma \vdash Subst_{U,El}(X, Y, UnivId(F), M) = F\ M : El(Y)} \quad \textsc{Univ-Id-Eq1}$$

So if we use *Subst* to rewrite an element M of $El(X)$ as an element of $El(Y)$ then this equals the result of applying the isomorphism F to M. Clearly, this violates uniqueness of identity: this rule makes the operator *UnivId* injective so that if any two proofs of a propositional equality were identified then also any two isomorphisms between types in the universe would be identified, which is a contradiction as soon as the universe contains a code for a nontrivial type such as $\mathbf{N}$. Astonishingly however, the above extension is sound in pure TT.

Theorem 5.2.10. *The above extension of TT can be soundly interpreted in the groupoid model.*

Proof. Let U be the groupoid with underlying set U_{set} the set of all small sets (i.e. with cardinality smaller than some inaccessible cardinal) and with $U_{rel}(x, y)$ the set of bijections between x and y. The identity bijection provides reflexivity; symmetry and transitivity are given by taking inverses and composition, respectively. We define a family of groupoids $El \in Fam(U)$ by $El(x) = \nabla(x)$ and if $p : x \cong y$ is a bijection between x and y, i.e. $p \in U_{rel}(x, y)$ and $e \in El_{set}(x) = x$, then we put $p \cdot x := p(x)$. The El_{resp}-component is trivial since both $El(x)$ and $El(y)$ are discrete. It is readily verified that this defines a family of groupoids. Now we claim that this provides a model for the above extension to TT. Clearly we interpret the universe as U and the El-operator as El. Now if $X, Y \in U_{set}$ then the interpretation of *Iso* is the set of bijections from x to y because $El(X)$ and $El(Y)$ are discrete and thus propositional equality and semantic identity coincide. By definition any such bijection is an element of $U_{rel}(X, Y)$ and thus we interpret *UnivId* as the identity. Finally, it may be seen from the interpretation of J and the definition of *Subst* that the latter is interpreted exactly by $-_{reindex}$ so that the equation Univ-Id-Eq is validated.

Furthermore, we can justify the following η-like definitional equality

$$\frac{}{\Gamma \vdash UnivId(X, X, id[X]) = Refl_U(X) : Id_U(X, X)} \text{ Univ-Id-Eq2}$$

where $id[X] : Iso[X, X]$ is the identity function viewed as an isomorphism between X and X. The equations Univ-Id-Eq1 and Univ-Id-Eq2 together establish that the canonical map $Id_U(X, Y) \to Iso[X, Y]$ is an isomorphism.

We do not know whether this extension has many applications as such, but one might imagine more general extensions which allow to define quotient types with proof-relevant equality relation of which the above universe would just be a special case. For example, one could then factor a slice category by isomorphism and in this way obtain a pullback operation which is "split" up to *Id*. If one succeeds in formulating the interpretation theory for dependent types inside type theory and with propositional equality rather than set-theoretic extensional equality, one could in this way address the problem of interpreting type theory in non-split structures [24, 47].

5.3 A dependent setoid model

Our aim in this section is to give a version of truly dependent setoids which, as opposed to the groupoid model, *can* be presented as a syntactic model within TT. As we shall see, the model we end up with has other drawbacks, so that it cannot really be considered superior to $\mathbf{S}_0$. Indeed, the drawbacks and the high complexity of the following interpretation can be seen as further evidence for the canonicity of $\mathbf{S}_0$.

As we have argued, the groupoid model is not definable in intensional type theory because the preservation of symmetry and transitivity by context morphisms cannot be guaranteed as soon as one has dependent products or function spaces. A natural attempt would be to drop these requirements and to see how far one gets. Then, however, symmetry and transitivity for *contexts* may not be used in the definition of any type former, for then stability under substitution would fail. To see why, consider a (contrived) type former in the groupoid model which sends $\sigma \in Fam(\Gamma)$ to $B(\sigma) \in Fam(\Gamma)$ (for all Γ) given by

$$B(\sigma)_\Diamond = \sigma_\Diamond \qquad \text{for } \Diamond \in \{set, rel, refl, sym, trans\}.$$

$$B(\sigma)_{reindex}(p, s) = \sigma_{reindex}(\Gamma_{sym}(\Gamma_{sym}(p)), s)$$

Now, if $f : B \to \Gamma$ and $\sigma \in Fam(\Gamma)$ then we have $B(\sigma)\{f\} = B(\sigma\{f\})$, i.e. stability under substitution of B, if and only if f_{resp} preserves B_{sym}.

Now, if no type (and term) former may use symmetry and transitivity for contexts, we can just as well omit them from the definition of contexts. It turns out that the use of reflexivity in certain term formers is unavoidable, notably in the interpretation of *Subst* —the elimination operator for the identity type. Fortunately, using a trick we *can* ensure the preservation of reflexivity by all context morphisms as we see below in Section 5.3.1.2.

However, we cannot ensure that reflexivity is preserved by reindexing, i.e. that for all families $\sigma \in Fam(\Gamma)$ and $s : \sigma_{set}(\gamma)$ we have

$$\sigma_{reindex}(\Gamma_{refl}(\gamma), s) = s$$

The lack of this preservation property entails that certain definitional equalities, notably Σ-COMP and NAT-COMP-SUC, only hold up to (semantical) propositional equality in the model $\mathbf{S}_1$ so that we can only interpret a weakened type theory where these equations are replaced by propositional axioms in the style of ID-UNI-I or EXT-FORM. This is not a principal problem because the model $\mathbf{S}_1$ supports uniqueness of identity and functional extensionality so that by Theorem 3.2.5 and the remark at the end of Section 3.2.6 it has the same expressive power as TT_I, namely the one of extensional type theory TT_E.

Let us now describe the details. The target type theory , i.e. the type theory in which the syntactic model is formulated, is TT, the core type theory with dependent sums and products, identity types, and natural numbers.

Parts of the material have been previously published as [46]. The definition of squash types in Section 5.3.5 is new and the material on universes (Section 5.3.5.1) has been thoroughly revised and rewritten.

Definition 5.3.1. A *context of setoids* is a triple $\Gamma = (\Gamma_{set}, \Gamma_{rel}, \Gamma_{refl})$ where Γ_{set} is a (syntactic) context and Γ_{rel} is a (syntactic) context relative to $(\Gamma_{set}, \Gamma_{set})$, i.e.

$$\gamma : \Gamma_{set}\ , \gamma' : \Gamma_{set} \vdash \Gamma_{rel}[\gamma, \gamma']$$

and Γ_{refl} is a tuple of terms establishing reflexivity of Γ_{rel}, i.e.

$$\gamma : \Gamma_{set} \vdash \Gamma_{refl}[\gamma] : \Gamma_{rel}[\gamma, \gamma]$$

Two contexts of setoids are equal if their three components are definitionally equal.

So a context of setoids is a reflexive graph if we view Γ_{set} as the set of nodes and $\Gamma_{rel}[\gamma, \gamma']$ as the set of edges from γ to γ'.

Definition 5.3.2. Let Γ and Δ be contexts of setoids. A *morphism* from Γ to Δ is a pair $f = (f_{fun}, f_{resp})$ where

$$\gamma : \Gamma_{set} \vdash f_{fun}[\gamma] : \Delta_{set}$$

and

$$\gamma, \gamma' : \Gamma_{set}\ , p : \Gamma_{rel}[\gamma, \gamma'] \vdash f_{resp}[p] : \Delta_{rel}[f_{fun}[\gamma], f_{fun}[\gamma']]$$

such that reflexivity is preserved up to *definitional* equality, i.e.

$$\gamma : \Gamma \vdash f_{resp}[\Gamma_{refl}[\gamma]] = \Delta_{refl}[f_{fun}[\gamma]] : \Delta_{rel}[f_{fun}[\gamma], f_{fun}[\gamma]]$$

holds. Two morphisms between contexts of setoids are equal if their two components are definitionally equal.

Proposition 5.3.3. *The contexts of setoids together with their morphisms form a category* **C** *with terminal object.*

Proof. Clearly, morphisms are closed under component-wise composition and contain the identity. A terminal object $\top$ is given by $\top_{set} = \top_{rel} = \diamond$ and $\top_{refl} = ()$.

5.3.1 Families of setoids

Now we come to the most important ingredient—the definition of dependency in the model. We first introduce a useful abbreviation. If Γ is a context of setoids and $\gamma_1, \ldots, \gamma_n : \Gamma_{set}$ then

$$\Gamma_{\mathrm{conn}}[\gamma_1, \ldots, \gamma_n]$$

denotes the context consisting of all types $\Gamma_{rel}[\gamma_i, \gamma_j]$ for $i, j = 1 \ldots n, i \neq j$, i.e. the γ_i are connected by Γ_{rel}. So for example if $\gamma, \gamma' : \Gamma_{set}$ then

$$\Gamma_{\mathrm{conn}}[\gamma, \gamma'] = (\Gamma_{rel}[\gamma, \gamma']\ , \Gamma_{rel}[\gamma', \gamma])$$

Definition 5.3.4. Let Γ be a context of setoids. A *family of setoids above* Γ is a 6-tuple $\sigma = (\sigma_{set}, \sigma_{rel}, \sigma_{reindex}, \sigma_{sym}, \sigma_{trans}, \sigma_{ax})$ where

- σ_{set} is a type depending on Γ_{set}

$$\gamma : \Gamma_{set} \vdash \sigma_{set}[\gamma]$$

- σ_{rel} is a dependent relation on σ_{set}

$$\gamma, \gamma' : \Gamma_{set}\ , s : \sigma_{set}[\gamma]\ , s' : \sigma_{set}[\gamma'] \vdash \sigma_{rel}[s, s']$$

- $\sigma_{reindex}$ is the reindexing function which allows to substitute connected elements of Γ_{set} in σ_{set}

$$\gamma, \gamma' : \Gamma_{set}\ , p : \Gamma_{\mathrm{conn}}[\gamma, \gamma']\ , s : \sigma_{set}[\gamma]\ , \bar{s} : \sigma_{rel}[s, s] \vdash \\ \sigma_{reindex}[p, s, \bar{s}] : \sigma_{set}[\gamma']$$

such that σ_{rel} is symmetric and transitive and such that $\sigma_{reindex}[p, s, \bar{s}]$ is always σ_{rel}-related to s. More precisely we require terms

$$\begin{array}{ll} \gamma, \gamma' : \Gamma_{set} , & \\ p : \Gamma_{\mathrm{conn}}[\gamma, \gamma'] , & \\ s : \sigma_{set}[\gamma] , s' : \sigma[\gamma'] , & \\ q : \sigma_{rel}[s, s'] & \vdash \quad \sigma_{sym}[p, q] : \sigma_{rel}[s', s] \end{array}$$

$$\begin{array}{ll} \gamma, \gamma', \gamma'' : \Gamma_{set} , & \\ p : \Gamma_{\mathrm{conn}}[\gamma, \gamma', \gamma''] , & \\ s : \sigma_{set}[\gamma] , s' : \sigma_{set}[\gamma'] , s'' : \sigma_{set}[\gamma''] , & \\ q : \sigma_{rel}[s, s'] , q' : \sigma_{rel}[s', s''] & \vdash \quad \sigma_{trans}[p, q, q'] : \sigma_{rel}[s, s''] \end{array}$$

$$\begin{array}{ll} \gamma, \gamma' : \Gamma_{set} , & \\ p : \Gamma_{\mathrm{conn}}[\gamma, \gamma'] , & \\ s : \sigma_{set}[\gamma] , \bar{s} : \sigma_{rel}[s, s] & \vdash \quad \sigma_{ax}[p, s, \bar{s}] : \sigma_{rel}[s, \sigma_{reindex}[p, s, \bar{s}]] \end{array}$$

Two families are equal if all their six components are definitionally equal. We denote the set of families of setoids over context Γ by $Fam(\Gamma)$.

Since Γ_{rel} is not assumed to be symmetric we must make $\Gamma_{\mathrm{conn}}[\gamma, \gamma']$ a premise to $\sigma_{reindex}$ rather than merely $\Gamma_{rel}[\gamma, \gamma']$. Otherwise, we could not define type formers like the dependent product. The reflexivity assumption $\bar{s} : \sigma_{rel}[s, s]$ in σ_{ax} is needed, for otherwise from σ_{ax} one could conclude that any two elements of $\sigma_{set}[\gamma]$ are σ_{rel}-related. The function $\sigma_{reindex}$ could also be defined without this assumption.

5.3.1.1 Comprehension and substitution. Let σ be a family of setoids indexed over Γ. Its comprehension $\Gamma \cdot \sigma$ is the context of setoids defined by

$$\begin{array}{lcr} (\Gamma \cdot \sigma)_{set} & := & \gamma{:}\, \Gamma_{set} , s{:}\, \sigma_{set}[\gamma] , \bar{s}{:}\, \sigma_{rel}[s, s] \\ (\Gamma \cdot \sigma)_{rel}[(\gamma, s, p), (\gamma', s', p')] & := & p{:}\, \Gamma_{rel}[\gamma, \gamma'] , q{:}\, \sigma_{rel}[s, s']) \\ (\Gamma \cdot \sigma)_{refl}[\gamma, s, \bar{s}] & := & (\Gamma_{refl}[\gamma], \bar{s}) \end{array}$$

Thus, a typical context of setoids Γ of length 2 takes the following form.

$$\begin{array}{l} \Gamma_{set} = s_1 : (\sigma_1)_{set}, \bar{s}_1 : (\sigma_1)_{rel}[s_1, s_1], \\ \qquad s_2 : (\sigma_2)_{set}[s_1, \bar{s}_1], \bar{s}_2 : (\sigma_2)_{rel}[s_1, \bar{s}_1 , s_2, s_2] \end{array}$$

$$\begin{array}{l} \Gamma_{rel}[s_1, \bar{s}_1, s_2, \bar{s}_2 , s_1', \bar{s}_1', s_2', \bar{s}_2'] = p_1 : (\sigma_1)_{rel}[s_1, s_1'] , \\ \qquad p_2 : (\sigma_2)_{rel}[s_1, \bar{s}_1, s_1', \bar{s}_1', s_2, s_2'] \end{array}$$

$$\Gamma_{refl}[s_1, \bar{s}_1, s_2, \bar{s}_2] = (\bar{s}_1, \bar{s}_2)$$

The morphism $p(\sigma) : \Gamma \cdot \sigma \to \Gamma$ is defined by by

$$\begin{array}{lcl} p(\sigma)_{fun}[\gamma, s, \bar{s}] & := & \gamma \\ p(\sigma)_{resp}[p, q] & := & p \end{array}$$

It follows immediately from the definition of $(\Gamma \cdot \sigma)_{refl}$ that this is indeed a morphism.

Let σ be a family of setoids indexed over Δ and f be a morphism from Γ to Δ. We obtain a family of setoids over Γ denoted $\sigma\{f\}$ by pre-composing with f, i.e. by putting

$$\sigma\{f\}_{set}[\gamma : \Gamma_{set}] := \sigma_{set}[f_{fun}[\gamma]]$$

$$\sigma\{f\}_{rel}[s : \sigma_{set}[f_{fun}[\gamma]]\ ,\ s' : \sigma_{set}[f_{fun}[\gamma']]] := \sigma_{rel}[s, s']$$

$$\sigma\{f\}_{reindex}[p : \Gamma_{\text{conn}}[\gamma, \gamma']\ ,\ s : \sigma_{set}[f_{fun}[\gamma]]\ ,\ \bar{s} : \sigma_{rel}[s, s]] := \\ \sigma_{reindex}[f_{resp}[p], s, \bar{s}]$$

Here by a slight abuse of notation we have applied f_{resp} to $p : \Gamma_{\text{conn}}[\gamma, \gamma']$. This is understood component-wise. The other components of $\sigma\{f\}$ are defined similarly by pre-composition. We omit the obvious definition of the q-morphisms.

5.3.1.2 Sections.

Definition 5.3.5. If σ is a family over a context of setoids Γ then a section of σ is a pair $M = (M_{el}, M_{resp})$ where

$$\gamma : \Gamma_{set} \vdash M_{el}[\gamma] : \sigma_{set}[\gamma]$$

and

$$\gamma, \gamma' : \Gamma_{set}\ ,\ p : \Gamma_{rel}[\gamma, \gamma'] \vdash M_{resp}[p] : \sigma_{rel}[M_{el}[\gamma], M_{el}[\gamma']]$$

We denote the set of sections of σ by $Sect(\sigma)$. Two sections are equal if both components are definitionally equal.

Every section induces a context morphism from its context to the comprehension of its type. More precisely, if $M \in Sect(\sigma)$ then we define $\overline{M} : \Gamma \to \Gamma \cdot \sigma$ by

$$\begin{aligned} \overline{M}_{fun}[\gamma] &:= (\gamma, M_{el}[\gamma], M_{resp}[\Gamma_{refl}[\gamma]]) \\ \overline{M}_{resp}[p] &:= (p, M_{resp}[p]) \end{aligned}$$

We check that reflexivity is preserved:

$$\begin{aligned} &\overline{M}_{resp}[\Gamma_{refl}[\gamma]] = \\ &(\Gamma_{refl}[\gamma]\ ,\ M_{resp}[\Gamma_{refl}[\gamma]]) = \\ &\Gamma \cdot \sigma_{refl}[\gamma, M_{el}[\gamma], M_{resp}[\Gamma_{refl}[\gamma]]] = \\ &\Gamma \cdot \sigma_{refl}[\overline{M}_{fun}[\gamma]] \end{aligned}$$

So although no equational constraints are placed on sections they nevertheless induce proper morphisms. This is the "trick" used to obtain preservation of reflexivity. We have $p(\sigma) \circ \overline{M} = id_\Gamma$ as required.

Conversely, if $f : \Gamma \to \Delta \cdot \sigma$ then we construct a section $Hd(f) \in Sect(\sigma\{p(\sigma) \circ f\})$ as follows. The $-_{el}$-part of $p(\sigma) \circ f$ sends $\gamma : \Gamma_{set}$ to $\delta := f_{fun}[\gamma].1.1.2 : \Delta_{set}$. Now $f_{fun}[\gamma].1.2$ is an element of $\sigma_{set}[\delta]$. So we put

$$Hd(f)_{el}[\gamma] := f_{fun}[\gamma].1.2$$

The $-_{resp}$-component is defined analogously. We have $f = q(f, \sigma) \circ \overline{Hd(f)}$ and $M = Hd(\overline{M})$ so sections and right-inverses to canonical projections are in 1-1 correspondence. Summing up we obtain

Proposition 5.3.6. *Setoids and families of setoids form a syntactic category with attributes—the setoid model* $\mathbf{S}_1$.

A frequently asked question is whether one could not identify families as soon as they have definitionally equal $-_{set}$- and $-_{rel}$-components, but arbitrary $-_{sym}$-, $-_{trans}$-, $-_{reindex}$-, and $-_{ax}$-components. As we shall see, these components enter directly the definition of the *Subst*-operator (and the derived operators *Sym* and *Trans* witnessing symmetry and transitivity of the identity type). *A fortiori* we would then have to identify equivalent morphisms, and to make substitution respect this identification, we would also have to identify certain families. This could only be achieved using a construction like that in the proof of Theorem 3.2.5. The resulting structure would probably be a model of extensional type theory, but not a syntactic model in the sense of Section 1.3.2 for functional extensionality.

We now pass to the modelling of type formers, notably the identity type and a Π-type satisfying functional extensionality. We also look at natural numbers and Σ-types where we see that certain conversion rules are only satisfied up to propositional equality.

5.3.2 Dependent product

Let Γ be a context of setoids, σ be a family over Γ and τ be a family over $\Gamma \cdot \sigma$. We define a family $\Pi(\sigma, \tau) \in Fam(\Gamma)$ as follows. Its type component is given by

$$\Pi(\sigma,\tau)_{set}[\gamma : \Gamma_{set}] := \Pi s : \sigma_{set}[\gamma].\Pi \bar{s} : \sigma_{rel}[s,s].\tau_{set}[\gamma, s, \bar{s}]$$

The relation is defined by

$$\begin{aligned}\Pi(\sigma,\tau)_{rel}[U : \Pi(\sigma,\tau)_{set}[\gamma]\ ,\ V : \Pi(\sigma,\tau)_{set}[\gamma']] :=\\ \Pi s : \sigma_{set}[\gamma]\Pi \bar{s} : \sigma_{rel}[s,s]\Pi s' : \sigma_{set}[\gamma']\Pi \bar{s}' : \sigma_{rel}[s',s'].\\ \sigma_{rel}[s,s'] \rightarrow \tau_{rel}[U\ s\ \bar{s}, V\ s'\ \bar{s}']\end{aligned}$$

From the other components the definition of which is basically forced we single out transitivity because it displays a perhaps unexpected need for the use of reindexing. We assume the following variables:

$$\begin{aligned}&\gamma, \gamma', \gamma'' : \Gamma_{set}\ ,\ f\colon \Pi(\sigma,\tau)_{set}[\gamma], f'\colon \Pi(\sigma,\tau)_{set}[\gamma'], f''\colon \Pi(\sigma,\tau)_{set}[\gamma'']\ ,\\ &p\colon \Gamma_{\mathrm{conn}}[\gamma,\gamma',\gamma'']\ ,\ q_1\colon \Pi(\sigma,\tau)_{rel}[f,f']\ ,\ q_2\colon \Pi(\sigma,\tau)_{rel}[f',f'']\end{aligned}$$

We must show that $\Pi(\sigma,\tau)_{rel}[f, f'']$. To that end assume

$$s_1\colon\sigma_{set}[\gamma], \bar{s}_1\colon\sigma_{rel}[s_1,s_1]\ ,\ s_2\colon\sigma_{set}[\gamma''], \bar{s}_2\colon\sigma_{rel}[s_2,s_2]\ ,\ h\colon\sigma_{rel}[s_1,s_2]$$

We have to construct an element of $\tau_{rel}[f\ s_1\ \bar{s}_1\ ,\ f''\ s_2\ \bar{s}_2]$. We would like to use τ_{trans} and q_1, q_2 for this, but in order to do so we need an element of $\sigma_{set}[\gamma']$ which is related to both s_1 and s_2. Using $\sigma_{reindex}$ it is possible to obtain such element from either s_1 or s_2 using the hypothesis $p\colon \Gamma_{\mathrm{conn}}[\gamma,\gamma',\gamma'']$, for instance $\sigma_{reindex}[Q, s_1, \bar{s}_1]$ where $Q : \Gamma_{\mathrm{conn}}[\gamma,\gamma']$ is a part of p. The rest is obvious. This shows that with a more general definition of transitivity that does not include the restriction to connected elements of the context it would not have been possible to define the Π-type.

Next we define introduction and elimination for the dependent product following Definition 2.4.10. That is, if $M \in Sect(\tau)$ we construct $\lambda_{\sigma,\tau}(M) \in Sect(\Pi(\sigma,\tau))$ and conversely if $M \in Sect(\Pi(\sigma,\tau))$ and $N \in Sect(\sigma)$ we construct $App_{\sigma,\tau}(M,N) \in Sect(\tau\{\overline{N}\})$. The element parts are

$$\lambda_{\sigma,\tau}(M)_{el}[\gamma : \Gamma_{set}] := \lambda s : \sigma_{set}[\gamma].\lambda \bar{s} : \sigma_{rel}[s,s].M_{el}[\gamma, s, \bar{s}]$$

and

$$App_{\sigma,\tau}(M,N)_{el}[\gamma : \Gamma_{set}] := M_{el}[\gamma]\ N_{el}[\gamma]\ N_{resp}[\Gamma_{refl}[\gamma]]$$

The relational parts are defined similarly. Also the governing equations are readily checked. Stability under substitution may be verified by careful inspection by hand or (better) using the Lego proof checker. The fact that morphisms preserve reflexivity is crucial for this property to hold.

Proposition 5.3.7. *The setoid model* $\mathbf{S}_1$ *supports dependent products.*

5.3.3 The identity type

We are now ready to define the interpretation of the identity type in the model $\mathbf{S}_1$. The basic idea is to interpret propositional equality on a type as the associated relation endowed with a trivial relation (namely any two elements are related) so that we can interpret uniqueness of identity. In view of Section 3.2.3.1 this allows us to replace J by (the simpler combinator) *Subst*. Moreover, we can interpret functional extensionality in the sense of Section 3.1.3. *Ext* will satisfy certain definitional equations which in particular ensure that all elements of type **N** in the empty context are equal to canonical values. However, the definitional equality LEIBNIZ-COMP is only validated up to propositional equality, that is we can interpret a constant of the corresponding identity type. In [46] we sketch the definition of a more complicated identity type in $\mathbf{S}_1$ with "strict" LEIBNIZ-COMP, but since—as announced above—other computational equations only hold propositionally and there is no fix for those, we do not give this construction here.

Suppose we are given a context of setoids Γ and a family of setoids σ over Γ. We first form the context consisting of Γ and two copies of σ. In our notation this is $\Gamma \cdot \sigma \cdot \sigma^+$. The identity setoid, denoted $Id(\sigma)$, is a family over this context. We define it by

$$Id(\sigma)_{set}[(\gamma, s_1, \bar{s}_1, s_2, \bar{s}_2) : (\Gamma \cdot \sigma \cdot \sigma^+)_{set}] := \sigma_{rel}[s_1, s_2]$$

and

$$Id(\sigma)_{rel}[x, x' : (\Gamma \cdot \sigma \cdot \sigma^+)_{set}\ ,\ p : Id(\sigma)_{set}[x]\ ,\ q : Id(\sigma)_{set}[x']] = \mathbf{1}$$

i.e. any two elements of $Id(\sigma)_{set}$ are related. Clearly this is symmetric and transitive. To define the rewriter $Id(\sigma)_{reindex}$ we make use of transitivity and symmetry of σ_{rel}. We see that although the relations on contexts are not assumed to be symmetric and transitive, we do need these properties for the relations on families.

5.3.3.1 Reflexivity. The distinguished morphism $\mathit{Refl}_\sigma : \Gamma \cdot \sigma \to \Gamma \cdot \sigma \cdot \sigma^+ \cdot \mathit{Id}(\sigma)$ interpreting reflexivity is given by

$$(\mathit{Refl}_\sigma)_{fun}[(\gamma, s, \bar{s}) : (\Gamma \cdot \sigma)_{set}] = (\gamma, s, \bar{s}, s, \bar{s}, \bar{s}) : (\Gamma \cdot \sigma \cdot \sigma^+ \cdot \mathit{Id}(\sigma))_{set}$$

The $-_{resp}$ component is trivial since any two elements of the identity setoid are related (by $\star$). It is also clear from the definition that Refl_σ satisfies the required equations, i.e. is stable under substitution and equals the diagonal when composed with $p(\mathit{Id}(\sigma))$.

5.3.3.2 Identity elimination. Let $\sigma \in \mathit{Fam}(\Gamma)$ and $M, N \in \mathit{Sect}(\sigma)$. We introduce the abbreviation $\mathit{Id}(M, N)$ for the family $\mathit{Id}(\sigma)\{\overline{N^+}\}\{\overline{M}\} \in \mathit{Fam}(\Gamma)$. $\mathit{Id}(M, N)$ corresponds to the setoid of proofs that M and N are equal. We also write $\mathit{Refl}(M)$ for the section

$$\mathit{Hd}(\mathit{Refl}_\sigma \circ \overline{M}) \in \mathit{Sect}(\mathit{Id}(M, M))$$

Now assume $P \in \mathit{Sect}(\mathit{Id}(M_1, M_2))$, and $N \in \mathit{Sect}(\tau\{\overline{M_1}\})$ where $\sigma \in \mathit{Fam}(\Gamma)$, $\tau \in \mathit{Fam}(\Gamma \cdot \sigma)$, $M_1, M_2 \in \mathit{Sect}(\sigma)$ corresponding to the premises of rule LEIBNIZ. We want to define a section $\mathit{Subst}\,(P, N)$ of $\tau\{\overline{N_2}\}$ using $\tau_{reindex}$. We put

$$\begin{aligned}\mathit{Subst}\,(P, N)_{el}[\gamma] = \tau_{reindex}[\,&(\Gamma_{refl}[\gamma], P_{el}[\gamma])\ ,\ (\Gamma_{refl}[\gamma], \sigma_{sym}[\Gamma_{refl}[\gamma], \Gamma_{refl}[\gamma], P_{el}[\gamma]]),\\ &N_{el}[\gamma]\ ,\ N_{resp}[\Gamma_{refl}[\gamma]]]\end{aligned}$$

The first two arguments to $\tau_{reindex}$ constitute a proof that $(\gamma, (M_1)_{el}[\gamma])$ and $(\gamma,\ (M_2)_{el}[\gamma])$ are connected in $\Gamma \cdot \sigma$. Notice that reflexivity of Γ is required here as well as symmetry of the relation on families. In particular, the use of reflexivity (as a proof that the two first components of the two elements of $\Gamma \cdot \sigma$ are related) is crucial.

The next two arguments to $\tau_{reindex}$ are the element $N_{el}[\gamma] : \tau_{set}[\gamma, (M_1)_{el}[\gamma]]$ and a proof that it is related to itself. We hence obtain an element of $\tau_{set}[\gamma, (M_2)_{el}[\gamma]]$ as required.

The $-_{resp}$-part of Subst is defined similarly using τ_{ax}.

As announced above this eliminator does not satisfy the rule LEIBNIZ-COMP. The reason for this is that $\tau_{reindex}$ applied to a proof by reflexivity is not necessarily the identity function. However, by virtue of τ_{ax} the propositional companion to LEIBNIZ-COMP is inhabited in the model.

5.3.3.3 Extensional constructs. If $P \in \mathit{Sect}(\mathit{Id}(M, M))$ then—since the relation on $\mathit{Id}(\sigma)$ is trivial—there is a canonical section of $\mathit{Id}(\mathit{Refl}(M), P)$ sending $\gamma \in \Gamma_{set}$ to $\star$. Thus we conclude

Proposition 5.3.8. *The model* $\mathbf{S}_1$ *supports the construct of uniqueness of identity.*

Next assume that $\tau \in \mathit{Fam}(\Gamma \cdot \sigma)$, that $U, V \in \mathit{Sect}(\Pi(\sigma, \tau))$, and that

$$P \in \mathit{Sect}(\mathit{Id}(\mathit{App}_{\sigma^+, \tau^+}(U^+, v_\sigma)\ ,\ \mathit{App}_{\sigma^+, \tau^+}(V^+, v_\sigma)))$$

i.e. corresponding to the premises to rule EXT-FORM (Section 3.1.3). We want to define a canonical section $Ext_{\sigma,\tau}(U, V, P)$ of $Id(U, V)$ from this. The element part of P is basically (modulo some currying of Σ-types) a term of type

$$\gamma : \Gamma_{set} \ , \ s : \sigma_{set}[\gamma] \ , \ \bar{s} : \sigma_{rel}[s, s] \vdash \sigma_{rel}[U_{el}[\gamma, s, \bar{s}] \ , \ V_{el}[\gamma, s, \bar{s}]]$$

The element part of the section we are looking for amounts to an inhabitant of

$$\gamma : \Gamma_{set} \ , \ s : \sigma_{set}[\gamma] \ , \ \bar{s} : \sigma_{rel}[s, s] \ , \ s' : \sigma_{set}[\gamma] \ , \ \bar{s}' : \sigma_{rel}[s, s] \vdash \\ \sigma_{rel}[U_{el}[g, s, \bar{s}] \ , \ V_{el}[g, s', \bar{s}']]$$

We obtain that by using either U_{resp} or V_{resp} and transitivity of σ_{rel}. The $-_{resp}$ part is again trivial.

Proposition 5.3.9. *The model* $\mathbf{S}_1$ *supports functional extensionality.*

We discuss quotient types below in Section 5.3.5.

5.3.4 Inductive types

Simple inductive types like the unit type, the Booleans, and the natural numbers may be defined by taking the identity type from the target type theory as the relation. Unfortunately, in the case of the natural numbers we obtain the computation rule NAT-COMP-SUC (see Section 2.1.2.4) only up to propositional equality because of the reflexivity assumptions in contexts.

We show the construction of the natural numbers in $\mathbf{S}_1$ following Definition 2.4.16. Let Γ be a context of setoids.

$$\mathbf{N}_{set}[\gamma{:}\, \Gamma_{set}] := \mathbf{N}$$

$$\mathbf{N}_{rel}[\gamma, \gamma'{:}\, \Gamma_{set} \ , \ n, n'{:}\, \mathbf{N}] := Id_{\mathbf{N}}(n, n')$$

where $Id_{\mathbf{N}}$ is the identity type from the target type theory. This is obviously symmetric and transitive. The reindexing function $\mathbf{N}_{reindex}$ is simply the identity.

The section $0_\Gamma \in Sect(\mathbf{N}_\Gamma)$ and the morphisms $Suc_\Gamma : \Gamma \cdot \mathbf{N}_\Gamma \to \Gamma \cdot \mathbf{N}_\Gamma$ are defined using the derived combinator *Resp* defined in Section 3.1.1.2. For example we have

$$(Suc_\Gamma)_{fun}[\gamma{:}\, \Gamma_{set}, n{:}\, \mathbf{N}] := (\gamma, Suc(n), Resp\,([n{:}\, \mathbf{N}]Suc(n), \Gamma_{refl}[\gamma])) \\ (Suc_\Gamma)_{resp}[(\gamma, n), (\gamma', n'){:}\, (\Gamma \cdot \mathbf{N}_\Gamma)_{set} \ , \ (p, q){:}\, (\Gamma \cdot \mathbf{N}_\Gamma)_{rel}[(\gamma, n), (\gamma', n')]] := \\ (p, Resp\,([n{:}\, \mathbf{N}]Suc(n), q)) : (\Gamma \cdot \sigma)_{rel}[(\gamma, Suc(n)), (\gamma', Suc(n'))]$$

For the elimination rule $R^{\mathbf{N}}$ assume some family $\sigma \in Fam(\Gamma \cdot \mathbf{N}_\Gamma)$ and sections $M_z \in Sect(\sigma\{\overline{0}\})$ and $M_s \in Sect(\sigma\{Suc_\Gamma \circ p(\sigma)\})$. We must construct a section $R^{\mathbf{N}_\Gamma}(M_z, M_s) \in Sect(\sigma)$ from this. Assuming $\gamma{:}\, \Gamma_{set}, n{:}\, \mathbf{N}$ and $\bar{n}{:}\, Id_{\mathbf{N}}(n, n)$ we have to find an element $R^{\mathbf{N}}_\sigma(M_z, M_s)_{el}[\gamma, n, \bar{n}]$ of $\sigma_{set}[\gamma, n, \bar{n}]$. Notice the dependency of σ_{set} on the proof that n is equal to itself. As we have seen this dependency is in general unavoidable, because σ may contain term formers like *Subst* which make use of $-_{refl}$ which is in turn defined using the reflexivity assumptions like $\bar{n}$ added during context comprehension ($(\Gamma \cdot \mathbf{N})_{set}$ is $\Gamma_{set}, n{:}\,\mathbf{N}, \bar{n}{:}\,Id_{\mathbf{N}}(n,n)$).

We wish to define the desired element using $R^{\mathbf{N}}$, but in order to do so we must slightly strengthen the inductive hypothesis, because M_s expects elements of σ_{set} which are related to themselves. So we use

$$(R^{\mathbf{N}}_{[n:\mathbf{N}]\Pi\bar{n}:Id_{\mathbf{N}}(n,n).\Sigma s:\sigma_{set}[\gamma,n,\bar{n}].\sigma_{rel}[s,s]}(?_z, ?_s, n)\ \bar{n}).1$$

where the question marks are readily filled using M_z and M_s. In the case of $?_s$, i.e. the inductive step, we assume $n{:}\,\mathbf{N}$ and

$$hyp : \Pi\bar{n}{:}\, Id_{\mathbf{N}}(n,n).\Sigma s{:}\, \sigma_{set}[\gamma, n, \bar{n}].\sigma_{rel}[s,s]$$

and $\overline{sn} : Id_{\mathbf{N}}(Suc(n), Suc(n))$. We then have to construct an element of

$$\Sigma s{:}\, \sigma_{set}[\gamma, Suc(n), \overline{sn}].\sigma_{rel}[s,s]$$

There are various possibilities for obtaining this element. The easiest is probably to instantiate *hyp* with $Refl_{\mathbf{N}}(n)$ and to apply $(M_s)_{el}$. This gives an element of

$$\sigma_{set}[\gamma, Suc(n), Resp\,(Suc, Refl_{\mathbf{N}})] = \sigma_{set}[\gamma, Suc(n), Refl_{\mathbf{N}}(Suc(n))]$$

from which we obtain an element of $\sigma_{set}[\gamma, Suc(n), \overline{sn}]$. The second component, i.e. a proof that this element is related to itself, is obtained using $(M_s)_{resp}$ and the required first component using instances of *Subst* and *IdUni* (which is definable at $\mathbf{N}$).

We omit the nontrivial but essentially forced $-_{resp}$-part for $R^{\mathbf{N}}_{\sigma}(M_z, M_s)$.

Now the equation NAT-COMP-ZERO follows immediately from its syntactic companion. The other equation NAT-COMP-SUC does not even hold for the $-_{el}$-parts, because in the right-hand side, according to the definition of substitution for sections, the (omitted) $-_{resp}$-part of $R^{\mathbf{N}r}(M_z, M_s)$ applied to reflexivity appears, whereas the left-hand side does not contain this term. Of course, using induction the two can be proved propositionally equal in the target type theory and are thus propositionally equal in the source type theory so that a propositional counterpart of NAT-COMP-SUC can be interpreted. It seems that in many special cases NAT-COMP-SUC does indeed hold, for instance if σ does not depend on $\mathbf{N}$.

We are now able to establish the two important meta-properties of $\mathbf{S}_1$: N-canonicity and consistency:

Proposition 5.3.10. *If $M \in Sect(\mathbf{N}_\top)$ then $M = Suc^x(0)$ for some $x \in \omega$.*

Proof. By definition of $\mathbf{S}_1$ and in particular the interpretation of the natural numbers we must have (in the target type theory) $\diamond \vdash M_{el} : \mathbf{N}$ and $\diamond \vdash M_{resp} : Id_{\mathbf{N}}(M_{el}, M_{el})$. By Remark 2.1.6 we have $\diamond \vdash M_{el} = Suc^x(0)$ for some $x \in \omega$ and $\diamond \vdash M_{resp} = Refl_{\mathbf{N}}(M_{el}) : Id_{\mathbf{N}}(M_{el}, M_{el})$. Hence $M = Suc^x(0)$ in $\mathbf{S}_1$.

Proposition 5.3.11. *The family $Id_{\mathbf{N}}(0, Suc(0))$ over $\top$ has no sections.*

Proof. If $M \in Sect(Id_{\mathbf{N}}(0, Suc(0)))$ then in the target type theory we have $\diamond \vdash M_{el} : Id_{\mathbf{N}}(0, Suc(0))$ which is a contradiction.

5.3.4.1 Σ-types. Our definition of contexts and context morphisms is based on an equational constraint (preservation of reflexivity). So we cannot expect that they can easily be internalised in $\mathbf{S}_1$ by some sort of Σ-type. The obvious guess for a Σ-type in the setoid model has (for Γ, σ, τ as in the definition of the Π-type) as underlying set

$$\Sigma(\sigma,\tau)_{set}(\gamma : \Gamma_{set}) = \Sigma s : \sigma_{set}(\gamma).\Sigma\bar{s} : \sigma_{rel}(s,s).\tau_{set}(\gamma,s,\bar{s})$$

and as relation

$$\Sigma(S,T)_{rel}((s,\bar{s},t),(s',\bar{s}',t')) = \sigma_{rel}(s,s') \times \tau_{rel}(t,t')$$

We can now define pairing and the first projection and more generally interpret the weak Σ-elimination rule [84]. In order to define a second projection or equivalently the dependent Σ-elimination rule we must use $\tau_{reindex}$ and thus we cannot get the equation

$$(M,N).2 = N$$

but only interpret the corresponding propositional identity.

5.3.5 Quotient types

We believe that intensional quotient types obeying the syntax given in Section 3.2.6.1 are definable in $\mathbf{S}_1$, but checking all the necessary proof obligations turned out to be too complicated even with machine support. Particular instances of quotient types—especially in the empty context—can, however, easily be seen to exist in $\mathbf{S}_1$. Examples are types of integers and rationals. We therefore propose to add the required quotients as primitive when working in $\mathbf{S}_1$ until a formal verification of the quotient type former in $\mathbf{S}_1$ can be given.

As an example of how this works, we define a *squash type* former, which can be seen as a quotient by the everywhere true relation.

Proposition 5.3.12. *The setoid model* $\mathbf{S}_1$ *supports a squash type former defined by the following rules.*

$$\frac{\Gamma \vdash \sigma}{\Gamma \vdash \Box\sigma}\ \Box\text{-Form} \qquad\qquad \frac{\Gamma \vdash M : \sigma}{\Gamma \vdash \Box M : \Box\sigma}\ \Box\text{-Intro}$$

$$\frac{\Gamma \vdash M, N : \Box\sigma}{\Gamma \vdash \Box\_ax(M,N) : Id_{\Box\sigma}(M,N)}\ \Box\text{-Ax}$$

$$\frac{\Gamma \vdash \sigma,\tau \qquad \Gamma, x{:}\sigma \vdash M[x] : \tau \qquad \Gamma \vdash N : \Box\sigma \qquad \Gamma\, , x, x'{:}\sigma \vdash H : Id_\tau(M[x], M[x'])}{\Gamma \vdash plug_\Box\, N\ in\ M\ using\ H : \tau}\ \Box\text{-Elim}$$

$$\frac{\Gamma \vdash plug_\Box\, \Box N\ in\ M\ using\ H : \tau}{\Gamma \vdash \Box\_eq(M,N,H) : Id_\tau(plug_\Box\, \Box N\ in\ M\ using\ H\ ,\ M[N])}\ \Box\text{-Eq}$$

Proof. Assume $\sigma \in Fam(\Gamma)$. We define $\Box\sigma \in Fam(\Gamma)$ by $\Box\sigma_{set} := \sigma_{set}$ and

$$\Box\sigma_{rel}[\gamma, \gamma' : \Gamma_{set}, x : \sigma_{set}[\gamma], x' : \sigma_{set}[\gamma']] = \sigma_{rel}[\gamma, x, x] \times \sigma_{rel}[\gamma, x', x']$$

Furthermore, we put $\Box\sigma_{reindex} = \sigma_{reindex}$. The other components are straightforward. If $M \in Sect(\sigma)$ then $\Box M \in Sect(\Box\sigma)$ is given by $\Box M_{el} = M_{el}$ and $\Box M_{resp}$ is trivial. Finally, if $M : Sect(\tau\{p(\sigma)\})$ and $N \in Sect(\Box\sigma)$ and $H \in Sect(Id(M\{p(\sigma)^+\}, M\{p(\sigma^+)\}))$, i.e. as in the premises to rule $\Box$-ELIM, then we define $plug_{\Box}\ N\ in\ M\ using\ H \in Sect(\tau)$ by

$$plug_{\Box}\ N\ in\ M\ using\ H_{el}[\gamma : \Gamma_{set}] := M_{el}[\gamma\ ,\ N_{el}[\gamma]\ ,\ N_{resp}[\gamma, \Gamma_{refl}[\gamma]].1]$$

The $-_{resp}$ part of this section is combined using τ_{trans} out of an instance of H_{el} and M_{resp} applied to instances of N and an instance of $\sigma_{reindex}$ applied to N. One cannot use H directly because it only works for elements over one and the same $\gamma : \Gamma_{set}$, whereas for the required $-_{resp}$-part we must compare elements over different (albeit related) elements of Γ_{set}. With $\sigma_{reindex}$ being part of the definition of *plug* we cannot achieve $\Box$-EQ up to definitional equality for the same reason as in the situation of identity types where ID-COMP only holds propositionally.

We remark that an analogue to the dependent quotient elimination rule Q-I-ELIM is definable and that a definitional counterpart to $\Box$-EQ would be desirable, but cannot be interpreted in $\mathbf{S}_1$.

An application of squash types is that they permit the definition of existential quantification as

$$\exists x : \sigma.\tau[x] := \Box\Sigma x : \sigma.\tau[x]$$

For this existential quantifier the usual intuitionistic introduction and elimination rules can be defined, but with the restriction that in the elimination rule the conclusion type must be at most single valued, i.e. any two elements of it must be propositionally equal. In particular, for this existential quantifier we can prove unique choice, i.e. if $\Gamma \vdash \sigma$ then the following type is inhabited:

$$(\exists x : \sigma.\mathbf{1}) \times (\Pi x, x' : \sigma.Id(x, x')) \to \sigma$$

We also conjecture that the schemes "countable choice" and "dependent choice" from Section 5.1.4.2 (with $\forall$ replaced by Π) are inhabited.

We believe that one can define a choice operator for the squash type similar to the one in Section 5.1.7, but we have not checked the details.

5.3.5.1 Universes. We were also unable to identify a universe in $\mathbf{S}_1$ closed under Π-types. The obvious candidate would have as underlying set the type:

$$\begin{array}{ll} \Sigma S : U.\Sigma S_r : El(S) \to El(S) \to U. & \text{type and relation} \\ \quad (\Pi s, s' : El(S).El(S_r\ s\ s') \to El(S_r\ s'\ s)) \times & \text{symmetry} \\ \quad (\Pi s, s', s'' : El(S). & \\ \qquad El(S_r\ s\ s') \to El(S_r\ s'\ s'') \to El(S_r\ s\ s'')) & \text{transitivity} \end{array}$$

and the identity type as relation. Contrary to what has been stated in [46] this universe is apparently not closed under the Π-type former, unless one assumes functional extensionality in the underlying target type theory. This universe does, however, contain the empty type and a unit type so that it may e.g. be used to derive Peano's fourth axiom.

We conjecture that it is possible to add a universe to $\mathbf{S}_1$ using the method described in Section 5.2.2.6 with the same restriction on the source type theory, i.e. the extensional constructs are then only available inside the universe.

5.4 Discussion and related work

An interpretation of types, sorts or sets as sets together with (partial) equivalence relations with a corresponding relativisation of quantifiers and abstractions has been proposed by several authors with various aims.

The first was probably Gandy [35] who gives an interpretation of simple type theory (no dependency) in itself to establish relative consistency of an extensionality axiom (corresponding to our Propositions 5.1.5 and 5.3.11). In [36] he extends this result to Gödel-Bernays' formulation of axiomatic set theory. There he emphasises the semantic character of his method by using the term *inner model* for the interpretation. Gandy also recognises the value of the interpretation itself as an explanation of the nature of extensionality: "The results of this paper may be ...expressed by saying: *in ordinary mathematics it is not possible to prove the existence of non-extensional quantities*".

Feferman [32, Section 4.4.2] strengthens Gandy's result by proving *conservativity* of extensionality with respect to second-order formulae. His method does not seem to generalise easily to type dependency, which is why we must content ourselves with Conjecture 3.2.23. Luckhardt [66] applies Gandy's method to higher-order Heyting arithmetic.

Assemblies together with (partial) equivalence relations as a formalisation of the notion of a set underlies Bishop's approach to constructive analysis [9] and has also been used by Bénabou [8] where these objects are called *ensembles empiriques* (empirical sets). Bénabou's setting differs from Bishop's because the relation in empirical sets is "proof relevant" very much like the $-_{rel}$-part in groupoids, however the set of proofs that two objects are related carries a structure of topological space there. Bénabou's motivating example is the set of stars where a "proof" that two stars are "equivalent" is a person who sees them identified. The set of persons is assumed to be endowed with a structure of a topological space expressing closeness. The "relation" thus defined is symmetric and transitive and moreover has the property of *observational stability*, namely that for any two points the set of proofs that they are related is open. Or to stay within the example, if p mixes up stars x and x' and p' is close enough to p then p' identifies them, too. Bénabou then defines *functional relations* between empirical sets and proves that with this notion of morphism the empirical sets form a topos, in fact are the topos of sheaves over the topological space of observers. Bénabou's main objective seems to be of a pedagogical nature, namely to find an intuitive way of explaining sheaves and toposes.

This bears some resemblance with the Scott-Fourman theory of existence and identity [98] and the construction of a topos out of a categorical model of higher-order logic ("tripos") described in [54]. Again, a category is formed the objects of which are objects of some category endowed with an abstract equivalence relation and the morphisms are functional relations. We believe that Bénabou's construction can be cast into this abstract framework. The motivating example in [54], the *effective topos* [53] is of a similar nature.

With the exception of Bishop's approach (and ours) in all these constructions the morphisms between "setoids" are functional relations rather than functions preserving the relations. The reason for this is that functional relations provide the *axiom of unique choice* which is crucial to topos logic.[5] In the Calculus of Constructions or extensions of it like $\mathbf{S}_0$, this axiom may be phrased as follows:

$$\frac{\Gamma \vdash \sigma \qquad \Gamma \vdash P : \mathit{Prf}(\exists x{:}\,\sigma.\forall y{:}\,\sigma.x \stackrel{L}{=} y)}{\Gamma \vdash AC!(P) : \sigma}\ \text{AC!}$$

This axiom thus blurs the distinction between proofs and datatypes present in $\mathbf{S}_0$ because it allows to construct an element of type σ out of a proof. Obviously, with functional relations instead of functions one achieves unique choice because one identifies elements of a type with premises to the unique choice axiom.

We have not considered functional relations and unique choice for two reasons. First, we want the interpretation function to map functions to functions, i.e. algorithms, rather than functional relations, i.e. specifications. Second, with functional relations as morphisms it does not seem possible to keep propositionally equal elements distinct intensionally[6], so that we lose the decidable semantic equality and also the choice principle. Of course, setoids with functional relations as morphisms have their own merits, but for this project we found it interesting to look at the other alternative. Some more thoughts on the issue of functions *vs.* functional relations and unique choice can be found in Section 6.5.1. We also remark that if we express totality of a functional relation using Σ-types instead of existential quantification then there is no difference between functions and functional relations. This is for example the case in the type theory suggested in Remark 5.1.11.

Bénabou also considers a categorical version of empirical sets called *théorie d'appartenance et égalité formelle* in which a "setoid" is simply a span in a category along with morphisms witnessing reflexivity, symmetry, and transitivity. These morphisms are to obey equational laws similar to the groupoid equations in Section 5.2. Morphisms between these objects are again (suitably defined) functional relations. Closely related to this is Carboni's explicit description of the exact completion of a left exact category [14]. This category has as objects

[5] Due to the set-theoretic nature of Bénabou's setting, unique choice is present even if one takes functions instead of functional relations.

[6] We have no precise proof of this because there is more than one way of setting up a type-theoretic setoid model with functional relations as morphisms, but in the obvious one one has to make this identification for example in order to define substitution in families.

spans with structural morphisms witnessing reflexivity, symmetry, and transitivity; and morphisms are morphisms in the underlying category respecting the relations.

It deserves mention that in none of these categorical approaches is type dependency a primitive. It can be encoded using pullbacks or subsets of cartesian products, but only in an extensional setting where definitional and propositional equality are identified.

Jacobs [57] and Mendler [82] have looked at categorical re-formulations of the syntax of quotient types in the context of simple type theory and extensional type theory, respectively. Jacobs identifies (extensional) quotient types as left adjoint to equality, whereas Mendler proves that they correspond to co-equalisers. For our purposes the syntax is not really important, we feel free to add or remove rules as long as this is sound with respect to the setoid semantics and corresponds to (constructive) mathematical practice.

6. Applications

In this chapter we describe some small applications of the extended type theories studied in this thesis. Most of the applications only use the extensional constructs *qua* constants; the additional definitional equalities gained by the interpretation in the various syntactic models are not used. This allows us to carry out the examples within the existing Lego system by merely working in a context containing assumptions for the various extensional constructs (see Appendix Appendix A). Most of the examples have been carried out in Lego, but we prefer to stick to the notation used so far and not give actual Lego syntax here. The main purpose of the examples is to explain the usefulness of extensional constructs as intermediate "macro language" rather than working with setoids or deliverables directly.

This becomes particularly clear in Section 6.1 where subset types are heavily used and in Section 6.2 which explores an encoding of streams using functional extensionality. The example of category theory in type theory (Section 6.3) shows how the obvious encoding of categories in type theories with extensional constructs gets translated in the various setoid models. The resulting encodings of category theory in pure intensional type theory are quite complex, which we see as evidence for the structuring and taxonomic aspect of setoid models.

The explicit unfolding of the interpretation in the setoid models is pursued further in Section 6.4 where we show how a familiar encoding of coproduct types in extensional type theory gets interpreted in pure intensional type theory. On the one hand, this constitutes an interesting application of functional extensionality, on the other hand it exemplifies the elimination of this construct by way of the interpretation in the setoid models.

Section 6.5 aims at a comparison between quotient types and the usual set-theoretic encoding of quotienting. This also sheds some light on the difference between setoid models and toposes and the role of "unique choice". The last example (Section 6.6) provides evidence for the usefulness of the conservativity of equality reflection over intensional type theory with extensional constructs (Theorem 3.2.5).

In order to prove the practical usefulness of intensional type theory with extensional constructs, larger case studies taken from Computer Science and constructive mathematics would be required. However, we felt that subsequent research in this field would benefit more from the present short examples, each of which highlights a specific topic or problem raised in this work.

6.1 Tarski's fixpoint theorem

Our aim is to give a formalisation of the following general form of Tarski's fixpoint theorem [105]. Our formalisation is based on a previous one by Randy Pollack [94] which uses the Extended Calculus of Constructions [70] without extensional constructs added.

Definition 6.1.1. A poset $(T, \leq)$ is a *complete lattice* if every subset $P \subset T$ has a (necessarily unique) least upper bound (lub) $\bigvee P$ in T.

Theorem 6.1.2 (Tarski). *Let $(T, \leq)$ be a complete lattice and $f : T \to T$ be monotone. The sub-poset of fixpoints of f is then a complete lattice, too.*

"Informal" proof. We first note that if $P \subset T$ then P has also a greatest lower bound in T given by $\bigwedge P := \bigvee\{x|x \leq P\}$. This implies that the dual of a complete lattice is a complete lattice again.

Now we show that the lub of the set of "post-fixpoints" of f, i.e. $\bigvee\{x|x \leq f(x)\}$, is a fixpoint of f and that it is actually the greatest fixpoint, i.e. that

$$\bigvee\{x|x \leq f(x)\} = \bigvee\{x|x = f(x)\}$$

Both properties follow by expanding the definition of "least upper bound". By duality we also obtain the existence of least fixpoints.

Next we observe that for any $x \in T$ the sub-poset $\uparrow x = \{y|x \leq y\}$ of elements above x is a complete lattice. The lub of $P \subset \uparrow x$ in $\uparrow x$ is given by the lub of $P \cup \{x\}$ in T. The verification is by case distinction on whether or not P is empty.

Now given a subset $P \subset \{x|x = f(x)\}$ of the fixpoints of f let l be its lub in T. First, we show that f restricts to $\uparrow l$. Suppose that $l \leq x$. We want to show that $l \leq f(x)$. Since l is a lub this amounts to showing that $P \leq f(x)$, but this follows from the assumption and the fact that P contains fixpoints only. Now let w be the least fixpoint of f viewed as a function on $\uparrow l$. This fixpoint exists since by the above $\uparrow l$ is a complete lattice. We claim that w is the desired lub of P in $\{x|x = f(x)\}$. Obviously, w is an upper bound of P since $w \geq l \geq P$. On the other hand, if $P \leq w'$ and w' is a fixpoint of f then $l \leq w'$ by definition of l and thus—since w is the least fixpoint of f above l—we have $w \leq w'$ as required.

The structure of this proof is somewhat intriguing since the facts about completeness and fixpoints are first proved for T and later on in the proof instantiated with other partial orders obtained from T by restriction. If we are to formalise this proof in some dependent type theory we therefore have to prove these preliminary facts for arbitrary posets and later on instantiate them. Now if we define a poset as a type together with a binary *Prop*-valued relation for example, then (without subset types) there is no way of identifying the up-closure $\uparrow x$ of an element x as a poset again.

In the formal proof by Pollack this difficulty is overcome by representing the underlying set of a poset not merely as a type, but as a type together with a *Prop*-valued predicate[1] which singles out the intended elements. The poset axioms are then relativised to the intended elements and so are further notions such as lubs and fixpoints. This is where the deliverables model from Chapter 4 comes in. Since in this model ordinary types are in fact modelled as types with *Prop*-valued predicates, we can get away with the previous naive definition of a poset as having merely a type as underlying set. The interpretation of the development in the model should then roughly give rise to Pollack's proof, but we don't even need to carry out the translation if we are merely interested in the existence of a proof. The formal proof obtained thus is substantially shorter than Pollack's original proof and moreover follows much closer the informal proof given above. The reason is that we do not need to verify each time we apply an operation or a law that all the arguments satisfy the relativising predicate. A certain drawback of the use of subset types is the need for explicit coercion functions from subsets to their supersets. In Pollack's proof coercions arise only at the level of the predicates, but not at the level of the terms themselves.

Let us now describe the details of the proof. We work in the internal language of the deliverables model, i.e. we assume the additional term and type formers introduced there, notably subset types and proof irrelevance. We also require a universe (U, El) to define a type of posets which makes the presentation more compact, but could in principle be avoided by using type variables. To increase readability we do not explicitly write down the *El*- and *Prf*-operators and we use the abbreviations for logical connectives and quantifiers introduced in Chapter 4. We also allow infix notation for the ordering $\leq$.

In the context $T\colon U, \leq\colon T \to T \to \mathit{Prop}$ we define a proposition *PO* stating that $(T, \leq)$ forms a partial order and *CL* stating that $(T, \leq)$ is actually a complete lattice. The definition of *CL* uses four auxiliary predicates expressing that an element of T is a lower, upper, least upper bound or greatest lower bound of some property $P : T \to \mathit{Prop}$.

$$
\begin{aligned}
&\mathit{lower}[P\colon T \to \mathit{Prop}, x\colon T] = \forall z\colon T.(P\ z) \Rightarrow x \leq z\\
&\mathit{upper}[P\colon T \to \mathit{Prop}, x\colon T] = \forall z\colon T.(P\ z) \Rightarrow z \leq x\\
&\mathit{lub}[P\colon T \to \mathit{Prop}, x\colon T] = \mathit{upper}[P, x] \wedge \mathit{lower}[\lambda y\colon T.\mathit{upper}[P, y]\ , x]\\
&\mathit{glb}[P\colon T \to \mathit{Prop}, x\colon T] = \mathit{lower}[P, x] \wedge \mathit{upper}[\lambda y\colon T.\mathit{lower}[P, y]\ , x]\\
&\mathit{CL} = \mathit{PO} \wedge \forall P\colon T \to \mathit{Prop}.\exists x\colon T.\mathit{lub}[P, x]
\end{aligned}
$$

Furthermore, if $f\colon T \to T$ we define a predicate $\mathit{mon}[f]$ stating that f is monotone with respect to $\leq$ and if in addition $P\colon T \to \mathit{Prop}$ we define a predicate $\mathit{closed}[f, P]$ expressing that P is preserved by f. We also define predicates $\mathit{fix}[f, x] := (x \stackrel{L}{=} f\ x)$ and $\mathit{prefix}[f, x] := (x \leq f\ x)$. Notice that all the definitions made so far depend on T and $\leq$ and that these can be instantiated using substitutions. For example, the proposition $\mathit{PO}[T, \lambda x, y\colon T.y \leq x]$ expresses that the dual of $(T, \leq)$ is a poset.

[1] Actually, Pollack uses partial equivalence relations rather than just predicates, but since the present proof does not make use of quotienting, the extra generality offered by using PERs is not needed.

We keep the variables $T, \leq, f$ and assume proofs of the propositions $CL[T, \leq]$ and $mon[T, \leq, f]$. We may then construct the following proofs:

$$fpu[f] : \forall u{:}\, T.lub[\lambda x{:}\, T.prefix[f, x]\ ,\ u] \Rightarrow fixpt[f, u]$$

$$lubfpsu[f] : \forall u \in T.lub[\lambda x{:}\, T.prefix[f, x]\ ,\ u] \Rightarrow lub[\lambda x{:}\, T.fix[f, x]\ ,\ u]$$

establishing that the lub of the set of pre-fixpoints is a fixpoint and actually the greatest fixpoint. We also prove that complete lattices have glbs.

$$Glb : \forall P{:}\, T \to Prop.\exists g{:}\, T.glb[P, g]$$

by constructing g as the *lub* of the lower bounds.

Now while keeping T, $\leq$, and f we instantiate the definitions made so far with different arguments. Whenever $P : T \to Prop$ we may form the subset type $\{x{:}\, T|P\ x\}$ and define an order $\leq'$ on this type by pre-composing the relation $\leq$ by the injection $wit(-)$.

$$\leq' = \lambda x, y{:}\, \{x{:}\, T|P\ x\}.wit(x) \leq wit(y)$$

Henceforth we write $\leq$ instead of $\leq'$. Now we prove the proposition

$$PO[\{x{:}\, T|Px\}, \leq]$$

i.e. that a subset of a poset is a poset again and use this to establish the main lemma, namely that the up-closure of some element of a complete lattice is a complete lattice. The up-closure is represented as $\uparrow x := \{y{:}\, T|x \leq y\}$.

$$upsCL : \forall x{:}\, T.CL[\uparrow x, \leq]$$

Roughly, this proof *upsCL* is constructed as follows. Assuming $x{:}\, T$ and $P : \{y{:}\, T|x \leq y\} \to Prop$ we construct a predicate P' on T as

$$\lambda y{:}\, T.\exists p{:}\, (x \leq y).(P(y)_p) \vee x \stackrel{L}{=} y$$

following the set-theoretic definition in the informal proof (Recall that $(y)_p : \uparrow x$ is an instance of the introduction operator for subset types.). The existential quantification over proofs that y lies in the subset is required to lift the predicate P to the whole of T. Another way of doing so would be to use the *extend*-operator from Section 4.6.2.

We may now use proof irrelevance to deduce that if $y{:}\, T$ and $p{:}\, x \leq y$ then $P'\ y$ iff $(P\ (y)_p) \vee x \stackrel{L}{=} y$. (Had we used the *extend*-operator this would be an instance of {}-ELIM-NONPROP-COMP.) We then follow the informal proof.

Now we prove that f is increasing on lubs of sets of fixpoints.

$$\forall P{:}\, T \to Prop.(\forall a{:}\, T.(P\ a) \Rightarrow fixpt[f, a]) \Rightarrow \forall l{:}\, T.lub[P, l] \Rightarrow l \leq (f\ l)$$

We then form the subset of fixpoints $Fix := \{x{:}\, T|x \stackrel{L}{=} f\ x\}$ and focus on the main goal

$$CL[Fix, \leq]$$

where again $\leq$ refers to the restriction of the original ordering to *Fix*. It is trivial to show that this is again a partial order. Given $P : Fix \to Prop$ we use existential elimination on $\exists l{:}T.lub[\lambda a{:}T.\exists p{:}fixpt[f,a].P\ (a)_p, l]$ giving a variable $l{:}T$ and a proof that l is the lub in T of the P "lifted" to T using existential quantification over proofs as above. Next we use existential elimination with

$$Glb[\uparrow l, \leq, upsCL\ l]\ \lambda c{:}\uparrow l.f\ \ wit(c) \leq wit(c)$$

giving $w{:}\uparrow l$ and a proof that it is the greatest lower bound of the post-fixpoints in $\uparrow l$. Here we use f still as a function on T. We then need to show that $l \leq f(wit(c))$ if $c{:}\uparrow l$ in order to get that w is actually a fixpoint of f. Another possibility would be to explicitly construct an $f'{:}\uparrow l \to \uparrow l$ and to take the least fixpoint of this function. One concludes as in the informal proof.

6.1.1 Discussion

The pattern of the formal proof follows quite closely the informal one except for the coercion functions between subsets and their supersets. B. Reus, who has been using a similar type theory to develop synthetic domain theory in Lego, has reported that these coercions are extremely cumbersome. In general, the coercions seem unavoidable, unless one gives up decidability and explicit proof objects; one could, however, imagine a system in which coercions from a subset type to a *non-propositional* supertype in the sense of Section 4.6.2 may be elided. This ties in with Hayashi's ATTT [43] where one has implicit coercions between refinement types and basic types. The details of this in the context of the deliverables model remain to be worked out, though.

Apart from its simplicity, the present formalisation using subset types also has the advantage of being more convincing than the proof in the pure Calculus of Constructions by Pollack, since it is closer to the intended set-theoretic understanding of posets and lattices.

Remark. The present proof follows closely Tarski's development in [105]. After seeing it, A. Pitts has come up with a direct formula for the computation of least upper bounds in the sub-poset of fixpoints. Namely, if $P \subset \{x \mid x = f(x)\}$ then $w := \bigwedge\{x \mid f(x) \leq x \wedge x \geq P\}$ is the lub of P in $\{x \mid x = f(x)\}$ as can be shown by direct verification.

6.2 Streams in type theory

We assume a type theory with functional extensionality, e.g. the internal language of the model $\mathbf{S}_0$ or $\mathbf{S}_1$. For definiteness we assume $\mathbf{S}_0$ and thus functional extensionality with respect to Leibniz equality. Given a type σ (not necessarily in the empty context) we form a type of infinite lists ("streams") over σ as

$$Str(\sigma) := \mathbf{N} \to \sigma$$

If $S : Str(\sigma)$ we put

$$Hd_\sigma(S) := S\ 0 : \sigma$$

and

$$Tl_\sigma(S) := \lambda n{:}\,\mathbf{N}.S\ (Suc(n))$$

Now we define a *bisimulation* as a relation $s, s'{:}\,Str(\sigma) \vdash R : Prop$ with the property that

$$\forall s, s'{:}\,Str(\sigma).R[s, s'] \Rightarrow Hd(s) \stackrel{L}{=} Hd(s')$$

and

$$\forall s, s'{:}\,Str(\sigma).R[s, s'] \Rightarrow R[Tl(s), Tl(s')]$$

In this case we can establish the *coinduction principle*

$$s, s'{:}\,Str(\sigma) \vdash R[s, s'] \Rightarrow s \stackrel{L}{=} s'\ true$$

In other words, bisimilar streams may be substituted for one another in an arbitrary context. The (straightforward) proof of this uses functional extensionality to reduce equality on streams to equality on σ and induction ($R^{\mathbf{N}}$).

Streams can be constructed using co-iteration [64, 114] as follows. If $out : \tau \rightarrow \sigma$ and $step : \tau \rightarrow \tau$ then for $init : \tau$ we put

$$CoIt_{\sigma,\tau}(out, step, init) = \lambda n{:}\,\mathbf{N}.out\ (step^n\ init) : Str(\sigma)$$

where $-^n$ is the n-th iteration of a function defined using $R^{\mathbf{N}}$. Now by the rules for definitional equality we have

$$Hd(CoIt_{\sigma,\tau}(out, step, init)) = out\ init : \sigma \tag{6.1}$$

$$Tl(CoIt_{\sigma,\tau}(out, step, init)) = CoIt_{\sigma,\tau}(out, step, step\ init) : Str(\sigma) \tag{6.2}$$

For example if $M : \sigma$ we may define a constant stream

$$Const(M) = CoIt_{\sigma,\mathbf{1}}(\lambda x{:}\,\mathbf{1}.M, \lambda x{:}\,\mathbf{1}.x, \star)$$

It satisfies $Hd(Const(M)) = M$ and $Tl(Const(M)) \stackrel{L}{=} Const(M)$ by coinduction.

More generally, using coinduction we can show *terminality* of $Str(\sigma)$, i.e. that any function $f : \tau \rightarrow Str(\sigma)$ which satisfies the two equations 6.1 and 6.2 above up to propositional equality must be propositionally equal to $\lambda x{:}\,\tau.CoIt_{\sigma,\tau}(out, step, x)$.

To be independent of the "implementation" of streams as functions one also needs a prefixing[2] operation: if $M{:}\,\sigma$ and $S{:}\,Str(\sigma)$ then

$$Pref(M, S) = \lambda n{:}\,\mathbf{N}.R^{\mathbf{N}}_{[x:\mathbf{N}]\sigma}(M, \lambda x{:}\,\mathbf{N}.\lambda s{:}\,\sigma.S\ x\ , n)$$

satisfying $Hd(Pref(M, S)) = M$ and $Tl(Pref(M, S)) = S$. Prefixing is also primitive in Coquand's approach to infinite objects [19] and infinite lists in

[2] In fact, prefixing is definable in terms of co-iteration using function types, but it seems more economic to define it directly as above. Prefixing can also be defined in terms of a generalised version of co-recursion, as is shown in [64].

Martin-Löf's non-standard type theory [76, 77]. Using coinduction we can prove that $Pref(Hd(S), Tl(S)) \stackrel{L}{=} S$ for $S : Str(S)$.

Possibly terminating streams, i.e. solutions to domain equations of the form

$$\Theta \cong (\sigma \otimes \Theta)_\perp$$

have been proposed as a model of *dataflow networks* [60]—these are networks of agents communicating via channels with unbounded capacity playing a role in protocol verification [101, 30].

If the infinite streams described above are used instead of possibly terminating ones, one has the advantage that *liveness*, i.e. the absence of deadlock modelled by a terminating stream, is built into the typing. A certain difficulty arises from the ubiquitous use of the fixpoint combinator in the definition of particular domain-theoretic streams. (For example, the behaviour of a dataflow network is defined as the least solution to a system of fixpoint equations obtained from the specification of the agents and the geometry of the network.) Since for the type $Str(\sigma)$ obviously no fixpoint combinator can be available (for instance $\lambda s\colon Str(\mathbf{N}).Const(Suc(Hd(s)))$ has no fixpoint and $\lambda s\colon Str(\sigma).Tl(s)$ has no unique fixpoint which could be obtained by an iterative process[3]), we cannot immediately compute the solution of such equations. Not surprisingly, this is the point where we have to supply a proof of "liveness", i.e. a proof that the domain-theoretic least fixpoint of a function on streams is indeed infinite. Fortunately, due to the simplicity of the above domain equation we can always compute a candidate for this least fixpoint which coincides with the least fixpoint if the latter happens to be infinite.

Suppose that $F : Str(\sigma) \to Str(\sigma)$ and that σ is inhabited, e.g. by $M : \sigma$. If the least fixpoint of F[4] is total then $Hd(F\ s)$ must be independent of s so a "guess" for it is $M_0 := Hd(F\ (Const(M)))$. Continuing with the iteration, we obtain the tail of the fixpoint as the fixpoint of the function

$$\lambda s\colon Str(\sigma).Tl(F\ (Pref(M_0, s)))$$

This suggests the following definition for the candidate for fixpoints.

$$\begin{aligned}
&Fix(F, M) = CoIt_{\sigma, Str(\sigma)\to Str(\sigma)}(\\
&\quad \lambda F\colon Str(\sigma) \to Str(\sigma).Hd(F\ (Const(M)))\ ,\\
&\quad \lambda F\colon Str(\sigma) \to Str(\sigma).\lambda x\colon Str(\sigma).Tl(F\ (Pref(Hd(F\ (Const(M))), x)))\ ,\\
&\quad F)
\end{aligned}$$

Admittedly, this is not a very efficient implementation, but if so desired one may always translate instances of *Fix* into general recursive programs and prove once and for all that the thus extracted programs never deadlock. In [91] a similar approach has been adopted to deal with well-founded recursion efficiently.

[3] The domain-theoretic least fixpoint of these functions is in both cases the undefined stream.

[4] "Least fixpoint" is understood informally here via some again informal embedding of $Str(\sigma)$ into Θ—the domain of possibly terminating streams.

Along with this candidate we define a totality predicate on "stream transformers" as a greatest fixpoint:

$$\begin{array}{l} \mathit{Total}(F : \mathit{Str}(\sigma) \rightarrow \mathit{Str}(\sigma)) = \\ \quad \exists P\colon (\mathit{Str}(\sigma) \rightarrow \mathit{Str}(\sigma)) \rightarrow \mathit{Prop}.(P\ F) \wedge (\forall F.(P\ F) \Rightarrow \quad \star \\ \quad\quad \exists x\colon \sigma.(\forall s\colon \mathit{Str}(\sigma).\mathit{Hd}(F\ s) \stackrel{L}{=} x) \wedge \\ \quad\quad\quad P\ (\lambda s\colon \mathit{Str}(\sigma).\mathit{Tl}(F\ (\mathit{Pref}(x, s))))) \end{array}$$

The first line of the definiens is a higher-order encoding of the greatest fixpoint of the remainder when abstracted from P. Using coinduction it is possible to prove the following proposition:

$$\begin{array}{l} \forall F\colon \mathit{Str}(\sigma) \rightarrow \mathit{Str}(\sigma).\forall m\colon \sigma. \\ \quad \mathit{Total}(F) \Rightarrow F\ (\mathit{Fix}(F, x)) \stackrel{L}{=} \mathit{Fix}(F, x) \end{array}$$

and also that every other fixpoint of F equals $\mathit{Fix}(F, x)$. This proof (which we have carried out in Lego) makes heavy use of the property that every function and predicate respects bisimilarity. It would highly complicate the proof if one did not have coinduction and instead used a defined "book-equality" on streams. If, for conceptual reasons, one is interested in such a development, one may always translate the present proof using the interpretation in one of the setoid models.

It is possible to justify the choice of *Total* using an intuitive domain-theoretic argument. It turns out that whenever $F : \Theta \rightarrow \Theta$ has an infinite least fixpoint in the domain-theoretic sense and the domain S of elements of the streams is flat, then F must satisfy the predicate *Total*. Indeed, if the least fixpoint of F is infinite then $F\ \bot$ must be different from $\bot$ and therefore $m_0 := \mathit{Hd}(F\ \bot)$ must be different from $\bot$. Thus, since S is flat, $\lambda s.\mathit{Hd}(F\ \bot)$ must be the constant function $\lambda s.m_0$. Now recall the "dinaturality principle" [93] for the least fixpoint $\mathbf{Y}(f \circ g) = f(\mathbf{Y}(g \circ f))$. Consider $\mathit{Tl}(\mathbf{Y}(F))$. We have:

$$\begin{array}{lll} & \mathit{Tl}(\mathbf{Y}(F)) & \\ = & \mathit{Tl}(\mathbf{Y}(\lambda s.\mathit{Pref}(\mathit{Hd}(F\ s), \mathit{Tl}(F\ s)))) & \text{since } F\ s \neq \bot \\ = & \mathit{Tl}(\mathbf{Y}(\lambda s.\mathit{Pref}(m_0, \mathit{Tl}(F\ s)))) & \\ = & \mathbf{Y}(\lambda s.\mathit{Tl}(F\ (\mathit{Pref}(m_0, s)))) & \text{by dinaturality.} \end{array}$$

Thus $\mathit{Total}(F)$ is satisfied for the predicate

$$P := \lambda F.\text{“}\mathbf{Y}(F) \text{ is infinite”}$$

A simple application of *Fix* and *Total* is the recursive definition of $\mathit{Const}(M)$ as the fixpoint of $F = \lambda s\colon \mathit{Str}(\sigma).\mathit{Pref}(M, s)$. This satisfies *Total* with the obvious choice $P = \lambda f.f \stackrel{L}{=} F$ because $\mathit{Hd}(F\ s) = M$ and $\mathit{Tl}(F(\mathit{Pref}(M, s))) = F$.

In a Lego development it has been possible to prove "*Total*" for the central loop of the alternating bit protocol in the setup of Dybjer and Sander [30] and thus to redo their domain-theoretic verification in the setting of (total) type theory $\mathbf{S}_0$.

6.3 Category theory in type theory

Formulating category theory within type theory is a nice application of dependent types and propositional equality. Several authors have attempted to formalise bits of category theory in type-theoretic systems with varying results [33, 51, 4]. In fact, Luo [69] reports that giving a sound foundation for category theory was one of Martin-Löf's original motivations for introducing his type theories [73].

Our own experience suggests that the simplest approach is to use propositional equality for both objects and morphisms. In this way we have managed to give a formal proof of the Yoneda Lemma in Lego—a nontrivial task because it necessitates the formalisation of presheaf categories which in turn mix up equality of objects and of morphisms. Our formalisation makes use of functional extensionality in various places.

Here we will not reproduce this fairly obvious formalisation, but rather explain how the definition of a category using propositional equality gets translated into setoids using the models $\mathbf{S}_0$ and $\mathbf{S}_1$. The resulting structures bear resemblance with Hornung's and Huet's formalisations [33, 51] which, however, lack object equality and therefore do not seem to permit the definition of functor categories. (In [33] this *is* possible using intensional equality for objects, however more complicated constructions mixing equality of objects and arrows do not seem to go through anymore.) This application shows how the setoid models can be used as a guideline for formalisations if one does insist on using (partial) equivalence relations rather than propositional equality.

Let us first recall the definition of a category using propositional equality.

Definition 6.3.1. In type theory with extensional constructs a *category* is given by a type $\vdash Ob$ of objects, a family of types

$$x, y{:}\ Ob \vdash Mor[x, y]$$

of *morphisms*, an operation

$$x{:}\ Ob \vdash id[x] : Mor[x, x]$$

an operation

$$x, y, z{:}\ Ob\ ,\ f{:}\ Mor[x, y], g{:}\ Mor[y, z] \vdash comp[x, y, z, g, f] : Mor[x, z]$$

and the following operations corresponding to the usual axioms where $comp[x, y, z, g, f]$ is abbreviated by $g \circ f$:

$$x, y{:}\ Ob\ ,\ f{:}\ Mor[x, y] \vdash idL[f] : Id_{Mor[x,y]}(f \circ id[x], f)$$

$$x, y{:}\ Ob\ ,\ f{:}\ Mor[x, y] \vdash idR[f] : Id_{Mor[x,y]}(id[y] \circ f, f)$$

$$x, y, z, w{:}\ Ob\ ,\ f{:}\ Mor[x, y], g{:}\ Mor[y, z], h{:}\ Mor[z, w] \vdash$$
$$Ass[f, g, h] : Id_{Mor[x,w]}(h \circ (g \circ f), (h \circ g) \circ f)$$

It should be clear that this definition is stable under constructions like functor category, arrow category, slice category, opposite category etc. and has—whenever $x{:}\,\sigma \vdash \tau[x]$ is a family of types—a particular instance based on $Ob = \sigma$ and $Mor[x,y] = \tau[x] \to \tau[y]$, called $\mathit{Full}[\sigma,\tau]$[5]. Of special interest is the category $\mathit{Full}[U, El]$ for a universe (U, El), which can be compared to the category of small sets. The use of propositional equality on objects (Id_{Ob}) is needed for instance in the definition of slice categories.

The typing of composition may suggest that composable morphisms must have definitionally rather than propositionally equal domain and codomain, respectively, and that thus we cannot use propositional equality on objects. However, using *Subst* we can adapt the domain and codomain of a morphism. For instance, if $X, X', Y : Ob$ and $P : Id_{Ob}(X, X')$ and $F : Mor[X, Y]$ then we have

$$F' := \mathit{Subst}_{Ob,[x:Ob]Mor[x,Y]}(P, F) : Mor[X', Y]$$

and using Id-Elim-J we can establish properties of F'. For example, if F is an isomorphism then so is F'.

6.3.1 Categories in $\mathbf{S}_0$

In the model $\mathbf{S}_0$, a category according to the above definition (in the empty context, for simplicity) expands to the following. The objects are represented by a type $\vdash Ob$ and a partial equivalence relation $x, y{:}\, Ob \vdash x = y : Prop$. (We use infix notation here to increase readability. This *Prop*-valued equivalence on objects is not to be confounded with definitional equality which is denoted $x = y : Ob$.) Morphisms are represented by a *single* type $\vdash Mor$ of morphisms and a relation

$$x, y{:}\, Ob\ ,\ f, g{:}\, Mor \vdash f =_{x,y} g : Prop$$

stating that f and g are equal *qua* morphisms from x to y. These data are subject to the following axioms where we have used conjunction ($\wedge$) and implication ($\Rightarrow$) instead of context extensions:

$$x, y{:}\, Ob\ ,\ f, g{:}\, Mor \vdash (x = x) \wedge (y = y) \wedge f =_{x,y} g) \Rightarrow g =_{x,y} f\ \mathit{true}$$

and

$$x, y{:}\, Ob\ ,\ f, g, h{:}\, Mor \vdash (x = x) \wedge (y = y) \wedge \\ (f =_{x,y} g) \wedge (g =_{x,y} h) \Rightarrow (f =_{x,y} h)\ \mathit{true}$$

and

$$x, y, x', y'{:}\, Ob\ ,\ f, g{:}\, Mor \vdash (x = x') \wedge (y = y') \wedge (f =_{x,y} g) \Rightarrow f =_{x',y'} g\ \mathit{true}$$

The identity operation now takes the form

$$x{:}\, Ob \vdash id[x] : Mor$$

[5] This corresponds to the category-theoretic notion of "full internal subcategory" defined for example in [58, Theorem 4.5.7], where it is attributed to Pénon.

with

$$x, x' : Ob \vdash (x = x') \Rightarrow id[x] =_{x,x} id[x'] \; true$$

and composition takes the form

$$x, y, z : Ob \; , \; f, g : Mor \vdash comp[x, y, z, g, f] : Mor$$

with

$$\begin{array}{l} x, y, z, x', y', z' : Ob \; , \; f, g, f', g' : Mor \vdash \\ \quad x = x' \wedge y = y' \wedge z = z' \wedge \\ \quad\quad f =_{x,y} f' \wedge g =_{y,z} g' \Rightarrow \\ \quad\quad comp[x, y, z, g, f] =_{x,z} comp[x', y', z', g', f'] \; true \end{array}$$

In view of Lemma 5.1.6, the laws are then expressed in terms of $- =_{-,-} -$, e.g. *idL* becomes

$$x, y : Ob \; , \; f : Mor \vdash x = x \wedge y = y \wedge f =_{x,y} f \Rightarrow comp[x, x, y, f, id[x]] =_{x,y} f \; true$$

The salient feature of this encoding is the single type of morphisms independent of the respective domains and codomains. This makes it easy to replace domain and codomain of a morphism by propositionally equal ones, but it precludes the particular instance of a category based on a universe ($Full[U, El]$), since such a universe is not available in $\mathbf{S}_0$. In the extension of $\mathbf{S}_0$ by universes described in Section 5.1.8 the situation is different. Here we can of course construct a category with U as its set of objects, but then we do not have a substitutive equality on objects (at least not an equality with extensional constructs). Let us sketch how a category looks in this extension under the assumption that Ob has trivial $-_{set}$ and $-_{rel}$-components, and that Mor has trivial $-_{type}$-component. A category is then given by a type Ob of objects (for instance $Ob = U$) and a family of types $x, y : Ob \vdash Mor[x, y]$ such that each of the types $Mor[x, y]$ is endowed with a partial equivalence relation:

$$x, y : Ob \; , \; f, g : Mor[x, y] \vdash f =_{x,y} g : Prop$$

$$x, y : Ob \; , \; f, g : Mor[x, y] \vdash f =_{x,y} g \Rightarrow g =_{x,y} f \; true$$

$$x, y : Ob \; , \; f, g, h : Mor[x, y] \vdash f =_{x,y} g \wedge g =_{x,y} h \Rightarrow f =_{x,y} h \; true$$

The composition operation has the usual typing:

$$x, y, z : Ob \; , \; f : Mor[x, y], g : Mor[y, z] \vdash comp[x, y, z, g, f] : Mor[x, z]$$

and the congruence laws and the other components are as one expects them to be. This encoding of categories is quite similar to the ones offered in [51, 33]:

Of course, this notion of category is not stable under constructions such as functor categories because of the lack of object equality. But in the extension of $\mathbf{S}_0$ with universes both notions of category coexist (and even mixed forms where Ob has nontrivial $-_{set}$- *and* $-_{type}$-components), and it makes sense to call a category "small" if it has a "set of objects", i.e. if in the source type theory $Ob = El(X)$ for some $X : U$ (or if in the target type theory it has trivial $-_{type}$-component). One would then allow the formation of a functor category $\mathbf{C} \rightarrow \mathbf{D}$ only if $\mathbf{C}$ is "small". The same restrictions on functor categories apply in set theory, but for size reasons, not for lack of substitutive equality. We remark that such a distinction between categories which have object equality and those which don't also is made by Bénabou in [7], albeit for different reasons.

6.3.2 Categories in $\mathbf{S}_1$

In $\mathbf{S}_1$ we have both dependency of morphisms on objects and a substitutive object equality. This is achieved by a "reindexing" function on morphisms. Again we only treat the case of a category in the empty context and use infix "=" (not to be confused with definitional equality) for the relations on objects and morphisms. A category then consists of a type $\vdash Ob$ and a partial equivalence relation $x, y: Ob \vdash x = y$. Unlike in $\mathbf{S}_0$, symmetry and transitivity must now be witnessed by terms which are part of the category structure.

$$x, y: Ob\ ,\ p: x = y \vdash Ob_{sym}[p] : y = x$$

$$x, y, z: Ob\ ,\ p: x = y\ ,\ q: y = z \vdash Ob_{trans}[p, q] : x = z$$

Next we have a set of morphisms indexed over "existing" objects

$$x, y: Ob\ ,\ \bar{x}: x = x\ ,\ \bar{y}: y = y \vdash Mor[\bar{x}, \bar{y}]$$

Notice that we do not indicate the dependency on x and y since these can be inferred from $\bar{x}$ and $\bar{y}$. The equality relation is defined for arbitrary morphisms not necessarily with common domain and codomain.

$$\begin{aligned}&x, y, x', y': Ob\ ,\ \bar{x}: x = x\ ,\ \bar{y}: y = y\ ,\ \bar{x}': x' = x'\ ,\ \bar{y}': y' = y'\ ,\\ &\quad f: Mor[\bar{x}, \bar{y}]\ ,\ g: Mor[\bar{x}', \bar{y}'] \vdash\\ &\quad f =_{\bar{x},\bar{y},\bar{x}',\bar{y}'} g\end{aligned}$$

This equality is again symmetric and transitive with the restrictions made in Definition 5.3.4. We do not explicitly state these. More interesting is the "reindexer" for morphisms over related objects:

$$\begin{aligned}&x, y, x', y': Ob\ ,\ \bar{x}: x = x\ ,\ \bar{y}: y = y\ ,\ \bar{x}': x' = x'\ ,\ \bar{y}': y' = y'\ ,\\ &p: x = x', q: y = y'\ ,\ f: Mor[\bar{x}, \bar{y}]\ ,\ \bar{f}: f =_{\bar{x},\bar{y},\bar{x},\bar{y}} f\\ &\vdash\\ &reindex[p, q, \bar{f}] : Mor[\bar{x}', \bar{y}']\end{aligned}$$

and in addition the axiom on reindexing

$$\Gamma \vdash ax[p, q, \bar{f}] : f =_{\bar{x},\bar{y},\bar{x}',\bar{y}'} reindex[p, q, \bar{f}]$$

where Γ is the context in the judgement for "*reindex*". Using symmetry and transitivity we can construct a proof that $reindex[p, q, \bar{f}]$ is related to itself. The identities take the form

$$\begin{aligned}&x: Ob, \bar{x}: x = x \vdash id[\bar{x}] : Mor[\bar{x}, \bar{x}]\\ &x, x': Ob, \bar{x}: x = x, \bar{x}': x' = x', p: x = x' \vdash id\_resp[p] : id[x] =_{\bar{x},\bar{x},\bar{x}',\bar{x}'} id[x']\end{aligned}$$

A special case of *id_resp* is

$$x: Ob, \bar{x}: x = x \vdash id\_resp[\bar{x}] : id[\bar{x}] =_{\bar{x},\bar{x},\bar{x},\bar{x}} id[\bar{x}]$$

We abbreviate this term by $\overline{id}[\bar{x}]$. Composition is defined similarly:

$$\begin{array}{l} x,y,z\colon Ob\ ,\ \bar{x}\colon x=x, \bar{y}\colon y=y, \bar{z}\colon z=z\ , \\ f\colon Mor[\bar{x},\bar{y}], g\colon Mor[\bar{y},\bar{z}]\ ,\ \bar{f}\colon f=f, \bar{g}\colon g=g \\ \vdash comp[\bar{g},\bar{f}] : Mor[\bar{x},\bar{z}] \end{array}$$

$$\begin{array}{l} x,y,z\colon Ob\ ,\ \bar{x}\colon x=x, \bar{y}\colon y=y, \bar{z}\colon z=z\ , \\ f\colon Mor[\bar{x},\bar{y}], g\colon Mor[\bar{y},\bar{z}]\ ,\ \bar{f}\colon f=f, \bar{g}\colon g=g \\ x',y',z'\colon Ob\ ,\ \bar{x}'\colon x'=x', \bar{y}'\colon y'=y', \bar{z}'\colon z'=z'\ , \\ f'\colon Mor[\bar{x}',\bar{y}'], g'\colon Mor[\bar{y}',\bar{z}']\ ,\ \bar{f}'\colon f'=f', \bar{g}'\colon g'=g' \\ p\colon x=x', q\colon y=y', r\colon z=z', u\colon f =_{\bar{x},\bar{y},\bar{x}',\bar{y}'} f', v\colon g =_{\bar{y},\bar{z},\bar{y}',\bar{z}'} g' \\ \vdash comp\_resp[p,q,r,u,v] : comp[\bar{g},\bar{f}] =_{x,y,x',z'} comp[\bar{g}',\bar{f}'] \end{array}$$

Again we can define a term $\overline{comp}[\bar{g},\bar{f}]$ witnessing "existence" of $comp[\bar{g},\bar{f}]$. As an example of an equational law we give again the left identity law:

$$\begin{array}{l} x,y\colon Ob\ ,\ \bar{x}\colon x=x\ ,\ \bar{y}\colon y=y\ ,\ f\colon Mor[\bar{x},\bar{y}], \bar{f} : f =_{\bar{x},\bar{y},\bar{x},\bar{y}} f \\ \vdash idL[\bar{f}] : comp[\bar{f},\overline{id}[\bar{x}]] =_{\bar{x},\bar{y},\bar{x},\bar{y}} f \end{array}$$

It follows from the fact that $\mathbf{S}_1$ forms a syntactic category with attributes closed under Π-types, Σ-types, and identity types that the above encoding of categories is stable under all constructions on categories which can be carried out with the encoding based on propositional equality using these type formers. This includes functor categories, arrow categories, slice categories, the instances $Full(\sigma,\tau)$, and many more. Of course, many of these instances do not fully exhaust the generality of the encoding, as for example the possible dependency of composition on the "existence proofs" $\bar{f}$. Nevertheless, this generality has proven necessary in order to obtain a syntactic category with attributes and so one can expect counterexamples to every simplification of the above encoding. It is, however, difficult to construct such counterexamples explicitly because even with machine support the encoding becomes unmanageable as soon as one goes beyond the simplest cases.

6.3.3 Discussion

We have seen how some reasonable formulations of category theory in type theory arise as translations of the obvious encoding based on propositional equality with extensional constructs added. We consider such an encoding useful for the following two reasons.

First, a development of category theory in constructive type theory permits to directly explore the constructive content of many category-theoretic arguments, which gets hidden if tools like set-theoretic quotienting and choice from equivalence classes are being used. A case in point is Lafont's proof [21, Section 4.10] that the addition of function types to an algebraic theory constitutes a conservative extension. Second, type theory as a basis of category theory is philosophically appealing as the consistency of type-theoretic systems like TT, $\mathbf{S}_0$, $\mathbf{S}_1$ even with universes can be proved within ordinary mathematics (probably even in primitive recursive arithmetic), whereas the relative consistency of a set-theoretic foundation of category theory like Mac-Lane's universes [72, I.6] goes beyond ZFC.

6.4 Encoding of the coproduct type

In the literature on extensional type theory various encodings of type formers from simpler or more general ones have been proposed. In particular, we have the encoding of a sum (coproduct) type using natural numbers and functions by Troelstra [109, Example 11.4.7] and the encoding of inductive types like the natural numbers in terms of Martin-Löf's well-ordering type [88, 39]. By the conservativity theorem 3.2.5 these encodings can also be carried out within TT_I and therefore give rise to counterparts in the various setoid models. Since we have not looked at interpretations of the well-ordering type in the setoid models, we only work out the example with coproducts here and leave the study of well-ordering types to future work.

By *coproduct type* we mean a type former which to any two types σ and τ associates a type $\sigma + \tau$ together with the following term formers: if $M : \sigma$ then $inl_{\sigma,\tau} : \sigma + \tau$ and if $M : \tau$ then $inr_{\sigma,\tau}(M) : \sigma + \tau$. If $z{:}\,\sigma + \tau \vdash \rho[z]$ and $x{:}\,\sigma \vdash L[x]{:}\,\rho[inl_{\sigma,\tau}(x)]$ and $y{:}\,\tau \vdash R[y]{:}\,\rho[inr_{\sigma,\tau}(y)]$ then whenever $M : \sigma + \tau$ we have $R^{\sigma+\tau}_\rho(L, R, M) : \rho[M]$ and the definitional equalities $R^{\sigma+\tau}_\rho(L, R, inl(M)) = L[M]$ and $R^{\sigma+\tau}_\rho(L, R, inr(M)) = R[M]$.

The abovementioned encoding of coproducts in TT_I requires a type (called $\mathbf{N}_2$ in [75]) with two elements (called 0 and 1) and corresponding elimination rule which in the situation $z{:}\,\mathbf{N}_2 \vdash \sigma[x]$, $L : \sigma[0]$, $R : \sigma[1]$, and $M : \mathbf{N}_2$ gives an element $R^{\mathbf{N}_2}_\sigma(L, R, M) : \sigma$ in such a way that $R^{\mathbf{N}_2}_\sigma(L, R, 0) = L$ and $R^{\mathbf{N}_2}_\sigma(L, R, 1) = R$. Moreover, for each type $\Gamma \vdash \sigma$ a term $\bot_\sigma : Id_{\mathbf{N}_2}(0,1) \to \sigma$ is required. The type $\mathbf{N}_2$ may be defined as

$$\mathbf{N}_2 := \Sigma m{:}\,\mathbf{N}.\Sigma n{:}\,\mathbf{N}.Id_{\mathbf{N}}(m + n, Suc(0))$$

where $+$ is the obvious encoding of addition in terms of $R^{\mathbf{N}}$; the terms $\bot_\sigma$ may be defined by induction on the structure of σ [109, p. 592, Proposition 4.1]. The interesting case is when $\sigma = Id_\tau(M, N)$. Here one defines a function from $\mathbf{N}_2$ to τ using $R^{\mathbf{N}_2}$ which takes on the values M, N on 0, 1, respectively, and then obtains $\bot_{Id_\tau(M,N)}$ using *Resp* . Unlike in the extensional theory studied in *loc. cit.*, the term $\bot_\sigma$ may thus contain a variable of type $Id_{\mathbf{N}_2}(0,1)$. We remark that this definition is not possible in the presence of an empty type, but becomes again possible if the empty type is contained in a universe [100]. Troelstra and van Dalen now offer the following encoding of the coproduct type $\sigma + \tau$ for types σ, τ:

$$\sigma + \tau := \Sigma n{:}\,\mathbf{N}_2.(Id_{\mathbf{N}_2}(n, 0) \to \sigma) \times (Id_{\mathbf{N}_2}(n, 1) \to \tau)$$

Indeed, if $M : \sigma$ then we have

$$inl(M) := (0, \lambda p{:}\,Id(0,0).M, \lambda p{:}\,Id(0,1).\bot_\tau\ p) : \sigma + \tau$$

and analogously we define the right injection using $\bot_\sigma$. Conversely, assume $z{:}\,\sigma + \tau \vdash \rho[z]$, $x{:}\,\sigma \vdash L : \rho[inl(x)]$ and $y{:}\,\tau \vdash R : \rho[inr(x)]$ and $M : \sigma + \tau$. Using R^Σ and $R^{\mathbf{N}_2}$ the task of finding an element of $\rho[M]$ may be reduced to finding elements of the following two types:

$$u\colon Id(0,0) \to \sigma, v\colon Id(0,1) \to \tau \vdash \rho[(0,u,v)]$$

$$u\colon Id(1,0) \to \sigma, v\colon Id(1,1) \to \tau \vdash \rho[(1,u,v)]$$

These elements in turn can be constructed using *Subst* and L, R, provided we can show that the variables u and v are propositionally equal to the second and third components of some instance of *inl* and *inr*, respectively. Let us focus on the first type. If $u\colon Id(0,0) \to \sigma$ and $v\colon Id(0,1) \to \tau$ then using functional extensionality and uniqueness of identity we find that $Id_{Id(0,0)\to\sigma}(u, \lambda p\colon Id(0,0).u\ (\mathit{Refl}(0)))$ is inhabited, and using functional extensionality and an instance of $\bot$, we find that $Id_{Id(0,1)\to\tau}(v, \lambda p\colon Id(0,1).\bot_\tau\ p)$ is inhabited. Therefore, the triple $(0,u,v)$ is propositionally equal to the object $inl(u\ (\mathit{Refl}(0)))$. Similarly for the second type.

The two equalities associated with the coproduct type can only be shown propositionally. The reason is that in the above construction *IdUni* gets applied to a variable and thus no reduction is possible.

6.4.1 Development in the setoid models

The above argument goes through immediately in the model $\mathbf{S}_1$ because all the constructions used are supported by this model. It also goes through in $\mathbf{S}_0$ when we replace the identity type by Leibniz equality (which is substitutive by Proposition 5.1.8), but due to the presence of an impredicative universe the definition of the terms $\bot_\sigma$ requires some explanation. The terms $\bot_\sigma$: $Prf(0 \stackrel{L}{=} 1) \to \sigma$ are again defined by induction on the structure of σ. If $\sigma = Prop$ we choose some arbitrary proposition; if $\sigma = Prf(P)$ then we define a function $F : \mathbf{N}_2 \to Prop$ with $F\ 0 = tt$ and $F\ 1 = P$ and define $\bot_{Prf(P)}$ as

$$\lambda p\colon Prf(0 \stackrel{L}{=} 1).p\ (\lambda x\colon \mathbf{N}_2.Prf(F\ x))\ I$$

where $I = \lambda x\colon Prop.\lambda p\colon Prf(x).p$ is the canonical proof of tt.

Notice that this implies that each of the types $[\![\Gamma \mid \sigma]\!]_{set}$, which all "live" in the empty context, is inhabited by some element $\bot'_\sigma$. This in turn is not surprising since these types are generated from $\mathbf{1}$, *Prop*, $\mathbf{N}$ by $\times$ and $\to$.

For simplicity we assume that $\mathbf{N}_2$ is available in the target type theory; if we were to carry out its definition in terms of $\mathbf{N}$ and the identity type then its underlying type would be $\mathbf{N} \times \mathbf{N}$ and the relation would be Leibniz equality restricted to pairs with sum equal to 1.

Now let σ and τ be families of setoids which arise as values of the interpretation function (so that $\bot$ is available). The underlying type of (the interpretation of) $\sigma + \tau$ becomes

$$(\sigma + \tau)_{set} := \mathbf{N}_2 \times \sigma_{set} \times \tau_{set}$$

Notice here that the underlying type of a Leibniz equality is the extensional unit type which disappears at the left of a function space.

The injections *inl* and *inr* are defined using the default elements $\bot'$ in the obvious (pointwise) way. The relation on this type singles out the elements which lie in the image of either injection and on these elements uses the relations on σ and τ.

The elimination rule applies to a family of setoids indexed over $\sigma + \tau$, but by definition of families its underlying type (call it ρ) does not depend on $\sigma + \tau$. The input to the elimination rule thus consists of functions $\sigma_{set} \to \rho$ and $\tau_{set} \to \rho$ and an element M of $\sigma + \tau$. The conclusion is then obtained by case distinction on the first component of M and application of either function to the second and third component, respectively. We omit the verifications.

Explicitly carrying out the construction in $\mathbf{S}_1$ is quite complicated and would probably not clarify much. The key step, however, where the elimination of functional extensionality takes place, is quite instructive and relatively easy to explain. The need for functional extensionality in the definition of the eliminator $R^{\sigma+\tau}$ came from the need for constructing an element of $\rho[(0, u, v)]$ from an element of $\rho[(0, u', v')]$ (for some u' and v') under the assumption that u, u' and v, v' are point-wise equal functions. In TT_I one first concludes that these functions are propositionally equal and then uses *Subst*; within the model $\mathbf{S}_1$, the family ρ comes with a "reindexer" which allows to pass between fibres over related elements, so that in order to pass between the two instances of ρ one only needs relatedness of $(0, u, v)$ and $(0, u', v')$ which is defined pointwise.

6.5 Some basic constructions with quotient types

In a series of books a collective of French mathematicians has under the pseudonym "N. Bourbaki" attempted to put all of mathematics, in particular geometry and algebra, on a solid axiomatic foundation based on set theory. They emphasise abstraction and modularisation: the set-theoretic encoding of a concept is used as little as possible and derived universal properties are employed instead. Their first book [10] is on basic set-theoretic constructions and contains a section on quotient sets where some universal constructions with them are given. As an application of the quotient type formers we shall in this section try to redo these constructions in a type-theoretic setting. The emphasis on abstraction in the Bourbaki approach makes it particularly apt for a development in type theory.

As in various places higher-order logic is required, we work in the impredicative type theory $\mathbf{S}_0$. Parts of the development can also be carried out using the quotient types for first-order Martin-Löf type theory given in Section 3.2.6.1.

6.5.1 Canonical factorisation of a function

A function $x{:}\sigma \vdash f[x] : \tau$ is a *surjection* if $\forall y{:}\tau.\exists x{:}\sigma.f[x] \stackrel{L}{=} y$ holds; f is an *injection* if $\forall x, x'{:}\sigma.f[x] \stackrel{L}{=} f[x'] \Rightarrow x \stackrel{L}{=} x'$ holds. Notice that due to the lack of the axiom of choice it is not the case in type theory that every surjection has a right inverse. It is also not the case that if a function is both surjective and injective then it has an inverse. This would be equivalent to the axiom of unique choice (AC!) which is not available in $\mathbf{S}_0$. (We come back to this point later.) If $y{:}\,\tau \vdash g[y] : \rho$ is another function then we write $g \circ f$ for the function $x{:}\,\sigma \vdash g[f[x]] : \rho$. We extend Leibniz equality to functions in the obvious way.

Now let $x{:}\,\sigma \vdash f[x] : \tau$ be any function and consider the equivalence relation $R[x, x'{:}\,\sigma] := (f[x]) \stackrel{L}{=} (f[x'])$. By definition of R the function f lifts to the quotient type σ/R, that is we have

$$x{:}\,\sigma/R \vdash g[x] := plug_R\ x\ in\ f\ using\ ([x, x'{:}\,\sigma, p{:}\,R[x, x']]p) : \tau$$

and by Q-COMP we get $x{:}\,\sigma \vdash f[x] = g[[x]_R]$. Using *Qind* we readily establish that g is injective and that $h[x{:}\,\sigma] := [x]_R$ is surjective. The thus obtained factorisation of f into a surjection and an injection is initial among all such. This means that if $x{:}\,\sigma \vdash s[x] : \rho$ is surjective and $y{:}\,\rho \vdash i[y] : \tau$ is injective and $f \stackrel{L}{=} i \circ s$ then there exists a unique (up to $\stackrel{L}{=}$) function $x{:}\,\sigma/R \vdash u[x] : \rho$ such that $u \circ h \stackrel{L}{=} s$ and $i \circ h \stackrel{L}{=} g$. Indeed, we obtain u as the lifting of s.

Bourbaki now points out that the quotient σ/R is in bijective correspondence with the image of σ under f, i.e. the subset of τ consisting of those elements which have an inverse image in σ. In fact they factor f as a three-fold composition of the class-map, this bijection, and the injection from the image into τ. If we code the image of f in type theory as[6]

$$Im(f) := \Sigma y{:}\,\tau.Prf(\exists x{:}\,\sigma.(f\ x) \stackrel{L}{=} y)$$

then we obtain another factorisation of f into a surjection and an injection, this time through $Im(f)$. It is easy to see that this factorisation is terminal (up to propositional equality). Using either initiality or terminality we obtain a function $l : \sigma/R \to Im(f)$.

Since we lack unique choice, we are, however, unable to define an inverse to this. If we had AC! in the sense of Section 5.4 we could construct an inverse as follows. Given $u : Im(f)$ we form the type $\Sigma z{:}\,\sigma/R.l[x] \stackrel{L}{=} u$ and prove using $\exists$-elimination that this type contains a unique element. If P is a proof term for this then we can form $\lambda u{:}\,Im(f).AC!(P)$ providing the desired inverse to l.

Let us note that in the model $\mathbf{S}_0$ the above mapping l is the function f itself, whereas an inverse to l would have to be a mapping in the other direction which need not necessarily exist, for instance, if σ_{set} is empty.

We remark that if we had unique choice then we could *define* quotient types in the usual way as the subset of $\sigma \to Prop$ consisting of "equivalence classes". We also remark that if we use the Σ-type instead of $\exists$ to define images then we have of course "choice", but the projection from $Im(f)$ to τ would not be injective.

Bourbaki also explains that any equivalence relation (in the sense of Section 5.1.10) R on σ can be seen as being of the above form using the function f which sends an element $x{:}\,\sigma$ to the set of elements related to x. In the impredicative type theory $\mathbf{S}_0$ this can be mimicked by putting $f[x] := \lambda x'{:}\,\sigma.R[x, x'] : \sigma \to Prop$. We have

[6] We could also use a subset type former, but since we haven't explicitly described it in $\mathbf{S}_0$ we use Σ-types instead.

$$x, y{:}\,\sigma \vdash (f[x] \overset{L}{=} f[y]) \overset{L}{=} R[x, y]\ \mathit{true}$$

by propositional extensionality. In first-order type theory ($\mathbf{S}_1$) this is, however, not possible.

6.5.2 Some categorical properties of $\mathbf{S}_0$

One may look at the issue of factorisation from a more modern category-theoretic point of view as follows. A *regular epi* is a surjection which can be obtained as the coequaliser of two functions. Conversely, a *regular mono* is an equaliser of a parallel pair. Let $\mathbf{S}_0/\overset{L}{=}$ be the category obtained from $\mathbf{S}_0$ by taking closed types as objects and functions up to propositional equality as morphisms. The regular epis in $\mathbf{S}_0/\overset{L}{=}$ are precisely those which can be obtained from class maps by pre- and post-composing with isomorphisms (notice that $x{:}\,\sigma \vdash [x]_R : \sigma/R$ is the coequaliser of the two projections from $\Sigma x{:}\,\sigma.\Sigma y{:}\,\sigma.\mathit{Prf}(R[x, x'])$ to σ. Similarly, the regular monos are precisely those which can be obtained in this way from projections out of Σ-types with propositional second component (= subset types). Here propositional extensionality is needed. In $\mathbf{S}_0/\overset{L}{=}$ the regular epis together with injections and the surjections with regular monos each form a *factorisation system* [5, p. 190]. The content of unique choice in the framework of $\mathbf{S}_0$ is precisely that all surjections are regular epis and that all injections are regular monos. We also remark without proof that $\mathbf{S}_0/\overset{L}{=}$ is *regular* [5, p. 187], i.e. has finite limits[7] and coequalisers (given by quotient types) which are stable under pullback (the verification of the latter property requires the dependent elimination rule for quotients Q-I-ELIM). In addition this category has the property that the regular monos are "classified" by the object *Prop*. It is also very easy to see that the category with the same objects, but with *functional relations* as morphisms (again up to propositional equality) is a *topos*. One only has to check that $\sigma \to \mathit{Prop}$ is a "power object" in the sense of [5]. We have not pursued this category-theoretic analysis any further in this thesis because it glosses over the distinction between definitional and propositional equality, which we consider as crucial in type theory.

6.5.3 Subsets and quotients

Let σ be a type and $x, x'{:}\,\sigma \vdash R[x, x'] : \mathit{Prop}$ an equivalence relation. If $x{:}\,\sigma \vdash P[x] : \mathit{Prop}$ we may form the "subset" $\tau := \Sigma x{:}\,\sigma.\mathit{Prf}(P[x])$ and denote by $i[x{:}\,\tau] := x.1 : \sigma$ the first projection which is injective by PR-IR. In first-order type theory, e.g. $\mathbf{S}_1$, we must restrict to a predicate P which is single-valued, i.e. any two proofs of which are propositionally equal, or use the squash type former.

On τ we have the induced equivalence relation $R_\tau[x, x'] := R[i[x], i[x']]$. Now we can define a bijective correspondence between the quotient of τ by R_τ

[7] In order to have a canonical choice for these finite limits it is advisable to identify objects with equal $-_{\mathit{set}}$ component and extensionally equal $-_{\mathit{rel}}$-components.

$$\rho := \tau/R_\tau$$

and the image of τ under the class map from σ to σ/R, i.e. the type

$$\varphi := \Sigma z\colon \sigma/R.\exists x\colon \tau.z \stackrel{L}{=} [i[x]]_R$$

In one direction (from ρ to φ) we use the lifting of the obvious map from τ to φ obtained by restricting the codomain of $x\colon\tau \vdash [i[x]]_R : \sigma/\rho$. The other direction is more complicated. In set theory, i.e. in the development of [10], one uses a right inverse to the class map from τ to τ/R_τ. In type theory we use the (derived) dependent elimination rule Q-I-ELIM from Sections 3.2.6.1 and 5.1.5.1. We define

$$\zeta[z\colon \sigma/R] := \Pi x\colon \tau.Prf(z \stackrel{L}{=} [i[x]]_R) \to \tau/R_\tau$$

By $\exists$-elimination a term of type $x\colon \sigma/R \vdash \zeta[x]$ gives rise to a function from φ to ρ, and such term can be defined using the dependent elimination rule for quotients provided we can find a term M with

$$x\colon\sigma \vdash M[x] : \zeta[[x]_R]$$

together with a proof of

$$x, x'\colon \sigma\ ,\ p\colon Prf(R[x,x']) \vdash Subst\,(Qax_R(p), M[x]) \stackrel{L}{=} M[x']$$

We define

$$M[z\colon\sigma] := \lambda x\colon \tau.\lambda p\colon Prf(z \stackrel{L}{=} [i[x]]_R).[x]_{R_\tau}$$

and the proviso is readily established using functional extensionality and the definitions of τ and R_τ. The verification that the function obtained thus is the required inverse is straightforward.

In $\mathbf{S}_0$ this inverse would have been obtained more easily by working in the model directly rather than by using the "internal language" given by the syntax for quotient types.

6.5.4 Saturated subsets

The predicate P from above is called *saturated* if for each x in P the whole equivalence class associated to x is contained in P, more precisely if

$$\forall x, y\colon \sigma.P[x] \Rightarrow R[x,y] \Rightarrow P[y]$$

In [10] it is noted that a subset is saturated iff it is the inverse image under the class map of some subset of the quotient set σ/R. This is true in type theory as well. Indeed, if P is saturated then we define $P'[z\colon\sigma/R] := \exists x\colon\sigma.P[x] \wedge [x]_R \stackrel{L}{=} z$ and we have that

$$x\colon\sigma \vdash P[x] \stackrel{L}{=} P'[[x]_R]$$

using propositional extensionality and effectiveness of the quotient type as defined in Section 5.1.6.4. Without effectiveness this example does not go through even if we would replace equality of predicates by bi-implication. We could also define P' using lifting (*plug*). That both definitions agree requires again effectiveness.

The other direction, i.e. that an inverse image along the class map is saturated, is straightforward and only requires Qax.

Bourbaki also defines the saturation of a subset P of σ as the inverse image of the image of P under the class map. We can show that this saturation is equal to $x{:}\sigma \vdash \exists x'{:}\sigma.R[x,x'] \wedge P[x']$, but again this requires effectiveness.

In *loc. cit.* saturatedness is extended to predicates with parameters, i.e. relations, and then called compatibility. Of particular interest is the case of functional relations where the image along the class map defines a set-theoretic analogue to lifting (*plug*). Since (due to the lack of unique choice) functions and functional relations are different in our type theory, we cannot sensibly compare the two definitions.

In *loc. cit.* a more general notion of lifting is also defined which takes a function $x{:}\sigma \vdash h[x] : \tau$ with $x,x'{:}\sigma \vdash R[x,x'] \Rightarrow Q[h[x],h[x']]$ for equivalence relations R and Q on σ and τ, respectively into a function from σ/R to τ/Q. In type theory we do this by lifting the composition of h with the class map for τ/Q.

6.5.5 Iterated quotients

Let σ, R be as above and $z,z'{:}\sigma/R \vdash S[z,z'] : Prop$ be an equivalence relation on the quotient of σ by R. On σ we define an equivalence relation T by $T[x,x'] := S[[x]_R,[x']_R]$. It is argued in [10] that the quotient types $(\sigma/R)/S$ and σ/T are in bijective correspondence by relating elements which come from one and the same element in σ by taking equivalence classes. In type theory we can define the bijection by appropriately lifting the class maps. In the model $\mathbf{S}_0$ the thus obtained bijection is simply the identity on σ_{set} in accordance with the set-theoretic definition.

Conversely, if T is any equivalence relation on σ which has the property that $R[x,x']$ entails $T[x,x']$ for $x,x'{:}\sigma$, then (since T is saturated in both x and x') we can by the analysis in Section 6.5.4 define an equivalence relation S on σ/R and find that T is equal to the relation defined by pre-composing S with the class map for σ/R. Bourbaki calls the thus obtained relation S the *quotient* of T by R denoted T/R. One thus obtains a bijective correspondence between $(\sigma/R)/(T/R)$ and σ/T.

That the above line of reasoning goes through both in the set-theoretic framework of [10] and in type theory is a nice example for the emphasis on abstraction and modularity in the Bourbaki approach.

6.5.6 Quotients and products

Let R and S be equivalence relations on σ and τ, respectively. An equivalence relation $R \times S$ on the product $\sigma \times \tau$ is defined component-wise. One has a bijective correspondence between $(\sigma/R) \times (\tau/S)$ and $(\sigma \times \tau)/(R \times S)$. One direction is obtained by lifting the obvious map from $\sigma\times\tau$ to $\sigma/R\times\tau/S$. For the other direction we consider the function $x{:}\,\sigma, y{:}\,\tau \vdash [(x,y)]_{R\times S} : (\sigma\times\tau)/(R\times S)$ and lift along y to obtain

$$w{:}\,\tau/S, x{:}\,\sigma \vdash \mathit{plug}_{R\times S}\ w\ \mathit{in}\ [y{:}\,\tau][(x,y)]_{R\times S}\ \mathit{using}\ H : \sigma\times\tau/R\times S$$

where H uses reflexivity of R, the definition of $R\times S$ and Qax. Now in order to lift this function along x we must use *Qind* to replace w by an equivalence class and the lifted function by its definition. This method is of use when we want to define a binary function on a quotient type. By giving such a function on representatives and showing that it respects the relations component-wise, one has, strictly speaking, defined a function on $(\sigma\times\tau)/(R\times S)$. Pre-composition with the said bijection yields the desired binary function.

This concludes the type-theoretic analysis of Bourbaki's constructions with quotients.

6.5.7 Quotients and function spaces

Motivated by the last construction we may ask whether quotienting is compatible with function spaces, e.g. whether the type $\sigma \to \tau/S$ is isomorphic to the quotient of $\sigma\to\tau$ by the equivalence relation $\sigma\to S := [u,v{:}\,\sigma\to\tau]\forall x{:}\,\sigma.S[u\ x, v\ x]$. In the model $\mathbf{S}_0$ and also in set theory this is the case, but apparently this is not derivable from our syntax for quotient types. The reason lies in the fact that in topos logic, where every surjection is isomorphic to a class map, this property implies the internal axiom of choice (IAC) from Section 5.1.4.2 which is in general not valid. (Recall that IAC means the proposition $(\forall x{:}\,\sigma.\exists y{:}\,\tau.P[x,y]) \to (\exists f{:}\,\sigma\to\tau.\forall x{:}\,\sigma.P[x,f\ x])$ for all types σ,τ.) Parts of the following development also appear in a category-theoretic guise in [5, p. 251 ff.].

IAC is equivalent to saying that all surjections (in the sense of Section 6.5.1) split internally, i.e.

$$\forall s:\varphi\to\psi.(\forall y{:}\,\psi\exists x{:}\,\varphi.(s\ x)\stackrel{L}{=}y) \Rightarrow \exists t{:}\,\psi\to\varphi.\forall y{:}\,\psi.s(t\ y)\stackrel{L}{=}y$$

In order to obtain IAC from this one starts from P with $\forall x{:}\,\sigma.\exists y{:}\,\tau.P[x,y]$ and instantiates with s being the first projection from $\varphi := \Sigma x{:}\,\sigma.\Sigma y{:}\,\tau.\mathit{Prf}(P[x,y])$ to $\psi := \sigma$. Surjectivity of this function is precisely the assumed premise to IAC, a splitting of this surjection on the other hand gives the conclusion of IAC.

That all surjections split internally is in turn equivalent to saying that if $s:\varphi\to\psi$ is surjective then so is $\rho\to s := \lambda u{:}\,\rho\to\varphi.s\circ u$ for all types ρ. Let us call (following [5]) a surjection with this property *powerful*. To obtain an internal splitting of powerful s we instantiate with $\rho=\psi$ and apply surjectivity of $\rho\to s$ to the identity on ψ. The other direction is easy.

Now under the assumption that for every equivalence relation S on τ the types $\sigma\to\tau/S$ and $(\sigma\to\tau)/(\sigma\to S)$ are isomorphic, we find that every "class map" $\lambda x{:}\,\tau.[x]_S$ is powerful. This is the case in $\mathbf{S}_0$. Unlike in a topos it is, however, not the case in $\mathbf{S}_0$ that every surjection is isomorphic to a class map. For example the surjection onto $Im(f)$ described above (Section 6.5.1) does not have this property. Thus we obtain that compatibility of quotienting with function spaces is not true in a topos, unless it satisfies IAC and thus is Boolean, and this can not be derived from our syntax for quotient types which (except for the choice operator) can be interpreted in any topos. The choice operator, on the other hand, is of no use here because it only works for non-quotiented context and we want compatibility of quotients with function spaces even if one of the involved types contains free variables of quotient type.

In view of this one may consider including the isomorphism between $\sigma \to \tau/S$ and $(\sigma \to \tau)/(\sigma \to S)$ as a primitive in the syntax of quotient types.

6.6 Σ is co-continuous—intensionally

In [80], Mendler shows that in a locally cartesian closed category **C** every "Σ-functor" $\Sigma_f : \mathbf{C}/A \to \mathbf{C}/B$ sending $u : X \to A$ to $f \circ u : X \to B$ preserves directed limits and in particular limits of ω-chains. In view of the correspondence between locally cartesian closed categories and extensional type theory [99, 47], this result carries over to TT_E and by Theorem 3.2.5 also to TT_I. The result has applications to encodings of coinductive types in TT_I (in fact the encoding of streams as $\mathbf{N} \to \sigma$ is a special case) and to the intensional formulation of bisimilarity avoiding the use of impredicativity, i.e. the quantification over all bisimulations [63]. We sketch the set-theoretic instance of a special case of Mendler's result and give a translation into TT_E. The crux of this example is to demonstrate the power of the conservativity theorem 3.2.5 because a direct proof in TT_I seems to be rather complicated.

6.6.1 Parametrised limits of ω-chains

As in Section 5.2 we use informal Martin-Löf type theory to refer to sets and families of sets and functions between them.

Consider an ω chain of sets and functions

$$\cdots \xrightarrow{b_{i+1}} B_{i+1} \xrightarrow{b_i} B_i \xrightarrow{b_{i-1}} \cdots \xrightarrow{b_1} B_1 \xrightarrow{b_0} B_0$$

and let L be its limit with projections $\pi_i : L \to B_i$. This means that $b_i \circ \pi_{i+1} = \pi_i$ for $i \in \omega$ and that for every "compatible family" $(x_i)_{i\in\omega}$ with $x_i \in B_i$ and $b_i(x_{i+1}) = x_i$ there exists a unique element $\langle x_i | i \in \omega\rangle \in L$ with $\pi_i(\langle x_i | i \in \omega\rangle) = x_i$.

This defines the limit up to isomorphism. Recall that such limit L always exists and that it may be explicitly constructed as the set of compatible families by

$$L := \{f \in \Pi i \in \omega.B_i \mid \forall i \in \omega.b_i(f_{i+1}) = f(i)\}$$

The projections are then defined by $\pi_i(f \in L) = f(i)$ and if $(x_i)_{i\in\omega}$ is a compatible family as above then we define the unique associated element of L as

$$\langle x_i \mid i \in \omega\rangle = \lambda i \in \omega.x_i$$

The verification is routine.

Now suppose that the whole development so far depended upon some parameter $\vartheta \in \Theta$ and that we form the disjoint union over all possible values of this parameter. This means that we consider the chain

$$\cdots \xrightarrow{b'_{i+1}} \Sigma\vartheta \in \Theta.B^{\vartheta}_{i+1} \xrightarrow{b'_i} \Sigma\vartheta \in \Theta.B^{\vartheta}_i \xrightarrow{b'_{i-1}} \cdots$$
$$\cdots \xrightarrow{b'_1} \Sigma\vartheta \in \Theta.B^{\vartheta}_1 \xrightarrow{b'_0} \Sigma\vartheta \in \Theta.B^{\vartheta}_0$$

Here the notation B^{ϑ}_i indicates the dependency of B_i on ϑ and may subsequently be used to denote substitution. This applies to the other meta-variables (L, b_i,...) as well. The functions b'_i are given by

$$b'_i(x \in \Sigma\vartheta \in \Theta.B^{\vartheta}_{i+1}) := (x.1, b^{\vartheta}_i(x.2)) \in \Sigma\vartheta \in \Theta.B^{\vartheta}_i$$

Now we have:

Proposition 6.6.1 (Mendler). *The set $\Sigma\vartheta \in \Theta.L^{\vartheta}$ with projections*

$$\pi'_i((\vartheta, f) \in \Sigma\vartheta \in \Theta.L) = (\vartheta, \pi_i(f)) : \Sigma\vartheta \in \Theta.B^{\vartheta}_i$$

is a limit of the chain of the b'_i.

Proof. Let $x_i \in \Sigma\vartheta \in \Theta.B^{\vartheta}_i$ with

$$b'_i(x_{i+1}) = x_i \qquad \text{for } i \in \omega \tag{6.3}$$

be a compatible family. By applying the first projection to this equation and induction on i we find that all the first components $x_i.1$ are equal (to $\vartheta_0 := x_0.1$ for instance). Therefore we have $x_i.2 \in B^{\vartheta_0}_i$ and $b^{\vartheta_0}_i(x_{i+1}.2) = x_i.2$ by applying the second projection to Equation 6.3. This allows us to form

$$(\vartheta_0, \langle x_i.2 \mid i \in \omega\rangle) \in \Sigma\vartheta \in \Theta.L^{\vartheta}$$

It is routine to check that this satisfies the commutation requirement and that it is unique among those.

We remark that the "proof-relevant" character of Σ is crucial here. If we replace it by "existential quantification", i.e. $\exists\vartheta \in \Theta.B^{\vartheta}_i := \emptyset$ if $B^{\vartheta}_i = \emptyset$ for all ϑ and $\{\star\}$ otherwise, then $\exists\vartheta \in \Theta.L^{\vartheta}$ is *not* a limit of the sequence of the $\exists\vartheta \in \Theta.B^{\vartheta}_i$. For example, if $\Theta = \omega$ and $B^{\vartheta}_i = \{\star\}$ iff $\vartheta \geq i$ then $x_i := \star$ is a compatible family, but $\exists\vartheta \in \Theta.L^{\vartheta}$ is empty. This phenomenon is the gist of Hallnäs' definition of bisimulation as an ω-limit by replacing $\exists$ by Σ.

6.6.2 Development in $\mathbf{TT}_E$

The above proof can be formalised in TT_E—the extensional type theory introduced in Section 3.2.3.1—almost without changes. We assume a chain depending on $\vartheta{:}\,\Theta$:

- $\vartheta{:}\,\Theta\ ,\ i{:}\,\mathbf{N} \vdash B[\vartheta, i]$
- $\vartheta{:}\,\Theta\ ,\ i{:}\,\mathbf{N} \vdash b[\vartheta, i] : B[\vartheta, Suc(i)] \to B[\vartheta, i]$

and define the type of compatible families by

$$CompFam[\vartheta{:}\,\Theta] := \Sigma f{:}\,\Pi i{:}\,\mathbf{N}.B[\vartheta, i]\,.\,\Pi i{:}\,\mathbf{N}.Id(b[\vartheta, i]\ (f\ (Suc(i)))\ ,\ f\ i)$$

For simplicity we identify the limit L with this type. Now we make the following definitions:

- $B'[i{:}\,\mathbf{N}] := \Sigma\vartheta{:}\,\Theta.B[\vartheta, i]$
- $b'[i{:}\,\mathbf{N}] := \lambda x{:}\,B'[Suc(i)].(x.1, b[x.1, i]\ x.2)$
- $CompFam' := \Sigma f{:}\,\Pi i{:}\,\mathbf{N}.B'[i]\,.\,\Pi i{:}\,\mathbf{N}.Id(b'[i]\ (f\ (Suc(i)))\ ,\ f\ i)$
- If $x{:}\,\sigma \vdash \tau[x]$ then $\Sigma!x{:}\,\sigma.\tau = \Sigma x{:}\,\sigma.(\tau \times \Pi y{:}\,\sigma.\tau[y] \to Id_\sigma(x, y))$

and establish in TT_E

$$\begin{aligned}\vdash_E\ &\Pi x{:}\,CompFam'.\Sigma!z{:}\,(\Sigma\vartheta{:}\,\Theta.CompFam[\vartheta]).\\ &\Pi i{:}\,\mathbf{N}.Id_{B'[i]}((z.1, z.2.1\ i)\ ,\ x.1\ i)\ \mathit{true}\end{aligned}$$

i.e. that $\Sigma\vartheta{:}\,\Theta.CompFam[\vartheta]$ is the limit of the $B'[i]$. Notice that if $x{:}\,CompFam'$ and $i{:}\,\mathbf{N}$ then $x.1\ i : B'[i]$ because $x.2$ is the proof of compatibility, and similarly for elements of *CompFam*.

6.6.3 Development in $\mathbf{TT}_I$

Now from the conservativity theorem 3.2.5 it follows immediately that the same judgement holds in intensional type theory with functional extensionality and uniqueness of identity (TT_I), provided the assumptions on B and b are valid in TT_I. However, giving a direct derivation of this in TT_I is quite difficult, because the "candidate" $z{:}\,\Sigma\vartheta{:}\,\Theta.CompFam[\vartheta]$ already contains instances of *Subst* so for the verification we have to reason about terms containing instances of *Subst*. More precisely, we first construct a term

$$x{:}\,CompFam', i{:}\,\mathbf{N} \vdash_I Uniform : Id_\Theta(\underbrace{(x.1\ 0).1}_{\text{“}\vartheta_0\text{”}}\ ,\ (x.1\ i).1)$$

using $x.2$ (the proof that $x.1$ is compatible) and $R^{\mathbf{N}}$. Then we construct an element of

$$x{:}\,CompFam' \vdash_I CompFam[(x.1\ 0).1]$$

the first component of which (of type $\Pi i{:}\,\mathbf{N}.B[(x.1\ 0).1, i]$) is given by

$$x{:}\,CompFam' \vdash_I \lambda i{:}\,\mathbf{N}.Subst_{\Theta,[\vartheta:\Theta]B[\vartheta,i]}(Uniform[x, i]\ ,\ (x.1\ i).2)$$

The verification now uses *IdUni* (and the derived induction principle J) and *Ext*. The author has spend considerable time with Lego trying to formalise this completely without succeeding. The strategy is of course clear and is indeed dictated by the proof of Theorem 3.2.5, but carrying out all the necessary dependent substitutions in Lego turns out to be extremely cumbersome. The lesson is that machine-supported reasoning in TT_I, in order to be efficient, must be supported by tactics, which could very well be based on the conservativity theorem.

One could e.g. imagine that one would conduct parts of a development in TT_E. This has the advantage that certain terms which can be proven propositionally equal are syntactically identified, for instance $\mathit{Subst}_{\sigma,\tau}(P, M)$ and $\mathit{Subst}_{\sigma,\tau}(P', M)$ for some $M : \tau[N_1]$ and $P, P' : \mathit{Id}_\sigma(N_1, N_2)$.

7. Conclusions and further work

We have compared extensional and intensional formulations of type theory and argued for the need for adding "extensional constructs" to the latter. These extensional constructs have been justified using syntactic models based on pure intensional type theory.

A natural next step would be to develop an implementation of the syntactic models along the lines of Section 1.4.1 so as to gain more experience with them. We think (although this may turn out to be wrong) that this implementation task is essentially straightforward and merely requires a lot of time. It is worth mentioning that in a certain sense such an implementation already exists in the form of the combinators for the syntactic models defined in Lego (Section 4.4 & 5.1.2), but the variable-free syntax they support is of very limited practical use.

More difficult, but certainly feasible, would be an implementation of a theorem proving tactic based on the conservativity result Theorem 3.2.5 as sketched in the previous section.

A more tentative task would consist of an implementation of the groupoid model. This model as we have presented it is based on classical extensional set theory and as such is not amenable to a direct implementation. However, it seems that the fragment of the groupoid model which is actually taken on by the interpretation function from the syntax does admit such an effective description. The intended application of such an endeavour would be to have a formal framework for interpreting propositional equality on types as isomorphism (Section 5.2.4).

All of these implementation tasks could influence further development of the setoid interpretation, for instance suggest the right formulation of universes, cf. Section 5.2.2.6.

Another important question which has only been marginally addressed in this thesis is the expressiveness of the "source type theories" with extensional constructs with respect to the setoid models in which they are interpreted. It seems that for every setoid in the model there exists an isomorphic one which is taken on by the interpretation function and that every section of definable families is definable. This would work by exhibiting a given setoid as the quotient of its $-_{set}$ component by its $-_{rel}$-component. The resulting relation on the quotient type will, however, only be equivalent (not equal) to the original one, hence the isomorphism. Such a completeness result (if it holds) would not be all too useful since we still do not know which type and term formers can be interpreted. So what one would really like to have is an overview over the principles and rules (like countable choice or compatibility of quotients with exponentiation) which are valid in the setoid models.

More specially, the meta-theory of quotient types should be studied more thoroughly. We have made some attempts at that in the course of the examples in Section 6.5, but these were far from exhaustive. For instance, it would be interesting to shed some light on the role of *effectiveness* of quotient types in applications.

The interpretability of type formers in the setoid models ought to be studied from a more general perspective. For instance one would like to know whether the difficulties with interpreting inductive types and universes in $\mathbf{S}_1$ are inherent or stem from a flawed definition of the model. After two years of experience with setoid interpretation the author tends towards the first alternative, but this is of course only a personal impression.

Appendix A. Lego context approximating $\mathbf{S}_0$

Below is a sequence of Lego declarations which simulates quotient types, proof irrelevance, and functional and propositional extensionality as supported by $\mathbf{S}_0$, cf. Appendix Appendix B. This code is slightly more general than $\mathbf{S}_0$ as Lego supports an infinite hierarchy of universes. It should, however, be possible to interpret the code using the method described in Section 5.1.8. To be absolutely safe, one can of course avoid the use of Lego universes when working with quotient types. The code does not contain subset types as those can be simulated by Σ-types and it does not support the non-standard operations *extend* and *choice* as the syntactic restrictions for those cannot be stated in Lego. A version of this code extended by proofs of certain properties such as effectiveness of quotient types is part of the Lego distribution. Notice that `Q` refers to Leibniz equality.

A.1 Extensionality axioms

```
$[Bi_Imp:{X,Y:Prop}(X->Y)->(Y->X)->Q X Y];
```
propositional extensionality

```
$[Pr_Ir:{X:Prop}{x,y:X}Q x y];
```
proof irrelevance

```
$[Ext:{A:Type(0)}{B:A->Type(0)}{U,V:{a:A}B a}
       ({a:A}Q (F a)(G a))->Q F G];
```
functional extensionality

A.2 Quotient types

```
[A|Type(0)][R:A->A->Prop];
```
all declarations are relative to `A` and `R`

```
$[QU : Set // R];
```
Q-FORM. `//R` indicates the dependency of `QU` on `R`

```
$[QU_class:A->QU//R];
```
Q-INTRO

```
$[QU_cor:{a1,a2:A}(R a1 a2) -> Q (QU_class a1) (QU_class a2)];
```
Q-AX

```
$[QU_it:{C|Set}{f:A->C}
       ({a1,a2:A}(R a1 a2) -> (Q (f a1) (f a2))) -> QU -> C];
```
Q-ELIM

```
$[QU_ind:{P:QU->Prop}({a:A}P(QU_class a)) -> {q:QU}P q];              Q-IND

[[C|Set][f:A->C][H:{a1,a2:A}(R a1 a2) -> Q (f a1) (f a2)][a:A]        Q-COMP
        QU_it f H (QU_class a) ==> f a];
Discharge A;                          relativising the declarations to A and R
```

After the last `Discharge` command the constant `QU` has the type

`{A|Type(0)}{R:A->A->Prop}Type(0)`

thus it associates a type `QU R` to each type `A` and relation `R:A->A->Prop`. The other constants from `QU_class` to `QU_ind` are similarly generalised.

A.3 Further axioms

One may add axioms approximating the substitutivity property of Leibniz equality (Proposition 5.1.8) and the special case of Proposition 4.6.7 where $\sigma = \mathbf{N}$. The corresponding definitional equalities can, however, only be approximated by propositional equalities, which hampers their efficient use.

```
$[Q_Subst:{A|Type(0)}{B:A->Type(0)}{a,a'|A}(Q a a')->(B a)->(B a')];
$[Q_Subst_Comp:{A|Type(0)}{B:A->Type(0)}{a:A}{b:B a}
        Q (Q_Subst B (Q_refl a) b) b];

$[extend : {P:Prop}(P -> Nat) -> Nat];
$[extend_Comp : {P:Prop}{f:P->Nat}{p:P}Q (f p) (extend f)];
```

Appendix B. Syntax

For the convenience of the reader all of the rules appearing throughout this thesis are summarised in this appendix. Notice that some of the rules are incompatible with each other, e.g. ID-UNI-I and UNIV-ID. Notice also that since the rules are literally the same as in the text, some appear in abbreviated form and others do not.

Rules for TT

$$\frac{}{\diamond \vdash}\ \text{EMPTY} \qquad \frac{\Gamma \vdash \sigma}{\Gamma, x:\sigma \vdash}\ \text{COMPR}$$

$$\frac{\Gamma, x:\sigma \vdash \tau}{\Gamma \vdash \Pi x:\sigma.\tau}\ \Pi\text{-FORM} \qquad \frac{\Gamma \vdash M:\sigma \quad \Gamma \vdash N:\sigma}{\Gamma \vdash Id_\sigma(M,N)}\ \text{ID-FORM} \qquad \frac{\Gamma \vdash}{\Gamma \vdash \mathbf{N}}\ \text{N-FORM}$$

$$\frac{\Gamma, x:\sigma, \Delta \vdash}{\Gamma, x:\sigma, \Delta \vdash x:\sigma}\ \text{VAR} \qquad \frac{\Gamma \vdash \sigma = \tau \quad \Gamma \vdash M:\sigma}{\Gamma \vdash M:\tau}\ \text{CONV}$$

$$\frac{\Gamma, x:\sigma \vdash M:\tau}{\Gamma \vdash \lambda x:\sigma.M^\tau : \Pi x:\sigma.\tau}\ \Pi\text{-INTRO} \qquad \frac{\Gamma \vdash M:\Pi x:\sigma.\tau \qquad \Gamma \vdash N:\sigma}{\Gamma \vdash App_{\sigma,\tau}(M,N):\tau[x:=N]}\ \Pi\text{-ELIM}$$

$$\frac{\Gamma \vdash}{\Gamma \vdash 0:\mathbf{N}}\ \text{N-INTRO-0} \qquad \frac{\Gamma \vdash M:\mathbf{N}}{\Gamma \vdash Suc(M):\mathbf{N}}\ \text{N-INTRO-SUC}$$

$$\frac{\Gamma, x:\mathbf{N} \vdash \sigma \quad \Gamma \vdash M_z:\sigma[x:=0] \quad \Gamma, x:\mathbf{N}, p:\sigma \vdash M_s:\sigma[x:=Suc(x)] \quad \Gamma \vdash N:\mathbf{N}}{\Gamma \vdash R^{\mathbf{N}}_\sigma(M_z, M_s, N):\sigma[x:=N]}\ \text{N-ELIM}$$

$$\frac{\Gamma \vdash M:\sigma}{\Gamma \vdash Refl_\sigma(M):Id_\sigma(M,M)}\ \text{ID-INTRO}$$

$$\frac{\Gamma, x:\sigma, y:\sigma, p:Id_\sigma(x,y) \vdash \tau \quad \Gamma, x:\sigma \vdash M:\tau[x:=x][y:=x][p:=Refl_\sigma(x)] \quad \Gamma \vdash N_1:\sigma \quad \Gamma \vdash N_2:\sigma \quad \Gamma \vdash P:Id_\sigma(N_1,N_2)}{\Gamma \vdash J_{\sigma,\tau}(M,N_1,N_2,P):\tau[x:=N_1][y:=N_2][p:=P]}\ \text{ID-ELIM-J}$$

Equality rules—a selection

$$\frac{\Gamma \vdash \sigma}{\Gamma \vdash \sigma = \sigma}\ \text{T-REFL} \qquad \frac{\Gamma \vdash \sigma = \tau}{\Gamma \vdash \tau = \sigma}\ \text{T-SYM} \qquad \frac{\Gamma \vdash \sigma = \tau \qquad \Gamma \vdash \tau = \rho}{\Gamma \vdash \sigma = \rho}\ \text{T-TRANS}$$

$$\frac{\Gamma \vdash M = M' : \mathbf{N}}{\Gamma \vdash Suc(M) = Suc(M') : \mathbf{N}}\ \text{C-Suc} \qquad \frac{\Gamma \vdash \sigma = \sigma' \qquad \Gamma, x{:}\sigma \vdash \tau = \tau'}{\Gamma \vdash \Pi x{:}\sigma.\tau = \Pi x{:}\sigma'.\tau'}\ \text{C-}Pi$$

$$\frac{\vdash \Gamma = \Delta \qquad \Gamma \vdash \sigma = \tau}{\vdash \Gamma, x{:}\sigma = \Delta, y{:}\tau}\ \text{C-Compr}$$

$$\frac{\Gamma \vdash \Pi x{:}\sigma.\tau = \Pi x{:}\sigma'.\tau' \qquad \Gamma, x{:}\sigma \vdash M = M' : \tau}{\Gamma \vdash \lambda x{:}\sigma.M^{\tau} = \lambda x{:}\sigma'.M'^{\tau'} : \Pi x{:}\sigma.\tau}\ \text{C-Abstr}$$

Rules for semantically defined definitional equality (Section 4.7)

$$\frac{[\![\Gamma \mid M]\!] = [\![\Gamma \mid N]\!] \in Sect([\![\Gamma \mid \sigma]\!])}{\Gamma \vdash M = N : \sigma}\ \text{Eq-Term}$$

$$\frac{[\![\Gamma \mid \sigma]\!] = [\![\Gamma \mid \tau]\!] \in Fam([\![\Gamma]\!])}{\Gamma \vdash \sigma = \tau}\ \text{Eq-Type}$$

$$\frac{[\![\Gamma]\!] = [\![\Delta]\!] \in Ob(\mathbf{C})}{\vdash \Gamma = \Delta}\ \text{Eq-Cont}$$

Computation rules These are always understood under the premise that left and right hand sides both have the indicated type.

$$\frac{}{\Gamma \vdash App_{\sigma,\tau}(\lambda x{:}\sigma.M^{\tau}, N) = M[x := N] : \tau[x := N]}\ \Pi\text{-Beta}$$

$$\frac{}{\Gamma \vdash R^{\mathbf{N}}_{\sigma}(M_z, M_s, 0) = M_z : \sigma[x := 0]}\ \text{Nat-Comp-Zero}$$

$$\frac{}{\begin{array}{l}\Gamma \vdash R^{\mathbf{N}}_{\sigma}(M_z, M_s, Suc(N)) = \\ \qquad M_s[x := N][p := R^{\mathbf{N}}_{\sigma}(M_z, M_s, N)] : \sigma[x := Suc(N)]\end{array}}\ \text{Nat-Comp-Suc}$$

$$\frac{}{\begin{array}{l}\Gamma \vdash J_{\sigma,\tau}(M, N, N, Refl_{\sigma}(N)) = \\ \qquad M[x := N] : \tau[x := N][y := N][p := Refl_{\sigma}(N)]\end{array}}\ \text{Id-Comp}$$

Rules for Σ-types and unit type

$$\frac{}{\vdash \mathbf{1}}\ \text{Unit-Form} \qquad \frac{}{\star : \mathbf{1}}\ \text{Unit-Intro}$$

$$\frac{x{:}\mathbf{1} \vdash \sigma[x] \qquad \vdash M : \sigma[\star] \qquad \vdash N : \mathbf{1}}{\vdash R^{\mathbf{1}}_{\sigma}(M, N) : \sigma[N] \qquad \vdash R^{\mathbf{1}}_{\sigma}(M, \star) = M}\ \text{Unit-Elim/Comp}$$

$$\frac{\vdash \sigma \qquad x{:}\sigma \vdash \tau[x]}{\vdash \Sigma x{:}\sigma.\tau[x]}\ \Sigma\text{-Form} \qquad \frac{\vdash M : \sigma \qquad N : \tau[M]}{\vdash pair_{\sigma,\tau}(M, N) : \Sigma x{:}\sigma.\tau[x]}\ \Sigma\text{-Intro}$$

$$\frac{\begin{array}{l}p : \Sigma x{:}\sigma.\tau[x] \vdash \rho[p] \\ x{:}\sigma, y{:}\tau[x] \vdash M[x,y] : \rho[pair_{\sigma,\tau}(x,y)] \qquad \vdash P : \Sigma x{:}\sigma.\tau[x]\end{array}}{\vdash R^{\Sigma}_{\sigma,\tau,\rho}(M, P) : \rho[P]}\ \Sigma\text{-Elim}$$

$$\frac{}{R^{\Sigma}_{\sigma,\tau,\rho}(M, pair_{\sigma,\tau}(N, O)) = M[N, O] : \rho[pair_{\sigma,\tau}(N, O)]}\ \Sigma\text{-Comp}$$

Rules for the extensional unit type

$$\frac{\Gamma \vdash}{\Gamma \vdash \mathbf{1}_E}\ \text{Unit-Form} \qquad \frac{\Gamma \vdash}{\Gamma \vdash \star : \mathbf{1}_E}\ \text{Unit-Intro}$$

$$\frac{\Gamma \vdash M : \mathbf{1}_E}{\Gamma \vdash M = \star : \mathbf{1}_E} \quad \text{UNIT-EQ}$$

Rules for the Calculus of Constructions

$$\frac{}{\vdash Prop} \quad \text{PROP-FORM} \qquad \frac{\vdash S : Prop}{\vdash Prf(S)} \quad \text{PRF-FORM}$$

$$\frac{x{:}\,\sigma \vdash S(x) : Prop}{\vdash \forall x{:}\,\sigma.S[x] : Prop} \quad \text{PROP-INTRO}$$

$$\frac{}{\vdash Prf(\forall x{:}\,\sigma.S[x]) = \Pi x{:}\,\sigma.Prf(S[x])} \quad \text{PROP-EQ}$$

Universes

$$\frac{}{\vdash U} \quad U\text{-FORM} \qquad \frac{\vdash M : U}{\vdash El(M)} \quad El\text{-FORM}$$

$$\frac{\vdash S : U \qquad x : El(S) \vdash T : U}{\vdash \hat{\Pi}(S,T) : U} \quad U\text{-INTRO-}\Pi$$

$$\frac{}{El(\hat{\Pi}(S,T)) = \Pi x{:}\, El(S).El(T)} \quad U\text{-EQ-}\Pi$$

$$\frac{}{\vdash \hat{\mathbf{N}} : U} \quad U\text{-INTRO-N} \qquad \frac{}{\vdash El(\hat{\mathbf{N}}) = \mathbf{N}} \quad U\text{-EQ-N}$$

Peano's fourth axiom

$$\frac{\Gamma \vdash M : Id_{\mathbf{N}}(0, Suc(0)) \qquad \Gamma \vdash \sigma}{\Gamma \vdash Peano_\sigma(M) : \sigma} \quad \text{PEANO-4}$$

Further rules for the identity type

$$\frac{\begin{array}{l}\Gamma\,,\, x, y : \sigma\,,\, p : Id_\sigma(x,y) \vdash \tau[x,y,p] \\ \Gamma \vdash N_1 : \sigma \qquad \Gamma \vdash N_2 : \sigma \\ \Gamma \vdash M : \tau[N_1, N_1, Refl_\sigma(N_1)] \\ \Gamma \vdash P : Id_\sigma(N_1, N_2)\end{array}}{\Gamma \vdash J'_{\sigma,\tau}(M, N_1, N_2, P) : \tau[N_1, N_2, P]} \quad \text{ID-ELIM-J}'$$

$$\frac{\Gamma \vdash \sigma \qquad \Gamma \vdash M : \sigma \qquad \Gamma \vdash P : Id_\sigma(M,M)}{\Gamma \vdash IdUni_\sigma(M,P) : Id_{\mathrm{Id}_\sigma(M,M)}(P, Refl(M))} \quad \text{ID-UNI-I}$$

$$\frac{\Gamma \vdash \sigma \qquad \Gamma \vdash M : \sigma}{\Gamma \vdash IdUni_\sigma(M, Refl(M)) = Refl(Refl(M)) : Id_\sigma(M,M)} \quad \text{ID-UNI-COMP}$$

$$\frac{\begin{array}{l}\Gamma \vdash U, V : \Pi x{:}\,\sigma.\tau \\ \Gamma, x{:}\,\sigma \vdash P : Id_\tau(U\ x, V\ x)\end{array}}{\Gamma \vdash Ext_{\sigma,\tau}(U,V,P) : Id_{\Pi x:\sigma.\tau}(U,V)} \quad \text{EXT-FORM}$$

$$\frac{\begin{array}{l}\Gamma, x{:}\,\sigma \vdash \tau[x] \\ \Gamma \vdash M_1, M_2 : \sigma \\ \Gamma \vdash P : Id_\sigma(M_1, M_2) \\ \Gamma \vdash N : \tau[M_1]\end{array}}{\Gamma \vdash Subst_{\sigma,\tau}(M_1, M_2, P, N) : \tau[M_2]} \quad \text{LEIBNIZ}$$

$$\frac{}{\Gamma \vdash Subst_{\sigma,\tau}(M, M, Refl_\sigma(M), N) = N : \tau[M]} \quad \text{LEIBNIZ-COMP}$$

Rules for extensional type theory

$$\frac{\Gamma\vdash\sigma \qquad \Gamma\vdash M:\sigma \qquad \Gamma\vdash N:\sigma \qquad \Gamma\vdash P: Id_\sigma(M,N)}{\Gamma\vdash M=N:\sigma}\ \textsc{Id-Defeq}$$

$$\frac{\Gamma\vdash\sigma \qquad \Gamma\vdash M:\sigma \qquad \Gamma\vdash N:\sigma \qquad \Gamma\vdash P: Id_\sigma(M,N)}{\Gamma\vdash P=Refl_\sigma(M): Id_\sigma(M,N)}\ \textsc{Id-Uni}$$

$$\frac{\Gamma\vdash M:\Pi x{:}\sigma.\tau}{\Gamma\vdash M=\lambda x{:}\sigma.M\ x:\Pi x{:}\sigma.\tau}\ \Pi\textsc{-Eta}$$

Rules for $\mathbf{TT}_{\Re}$

$$\frac{\Gamma\vdash_I M,N:\sigma \qquad \Gamma\vdash_I P: Id_\sigma(M,N)}{\Gamma\vdash_I \Re_\sigma(M,N,P):\sigma}\ \Re\textsc{-Form}$$

$$\frac{\Gamma\vdash_I \Re_\sigma(M,N,P):\sigma}{\Gamma\vdash_I \Re_\sigma(M,N,P)=N:\sigma}\ \Re\textsc{-Eq}$$

Rules for quotient types in $\mathbf{TT}_E$

$$\frac{\Gamma\vdash\sigma \qquad \Gamma,x,x'{:}\sigma\vdash\rho[x,x']}{\Gamma\vdash\sigma/\rho}\ \textsc{Q-E-Form}$$

$$\frac{\Gamma\vdash M{:}\sigma \qquad \Gamma\vdash\sigma/\rho}{\Gamma\vdash[M]_\rho{:}\sigma/\rho}\ \textsc{Q-E-Intro}$$

$$\frac{\Gamma\vdash M,N:\sigma \qquad \Gamma\vdash H:\rho[M,N]}{\Gamma\vdash[M]_\rho=[N]_\rho:\sigma/\rho}\ \textsc{Q-E-Eq}$$

$$\frac{\begin{array}{l}\Gamma,x{:}\sigma/\rho\vdash\tau[x] \qquad \Gamma,x{:}\sigma\vdash M[x]:\tau[[x]_\rho]\\ \Gamma,x,x'{:}\sigma,p{:}\rho[x,x']\vdash M[x]=M[x']:\tau[[x']_\rho]\\ \Gamma\vdash N:\sigma/\rho\end{array}}{\Gamma\vdash plug_\rho\,N\ in\ M:\tau[N]}\ \textsc{Q-E-Elim}$$

$$\frac{}{\Gamma\vdash plug_\rho\,[N]_\rho\ in\ M=M[N]:\tau[[N]_\rho]}\ \textsc{Q-E-Comp}$$

Rules for quotient types in $\mathbf{TT}_I$

$$\frac{\Gamma\vdash\sigma \qquad \Gamma,x,x'{:}\sigma\vdash\rho[x,x']}{\Gamma\vdash\sigma/\rho}\ \textsc{Q-I-Form} \qquad \frac{\Gamma\vdash M{:}\sigma \qquad \Gamma\vdash\sigma/\rho}{\Gamma\vdash[M]_\rho{:}\sigma/\rho}\ \textsc{Q-I-Intro}$$

$$\frac{\Gamma\vdash M,N:\sigma \qquad \Gamma\vdash H:\rho[M,N]}{\Gamma\vdash Qax_\rho(H): Id_{\sigma/\rho}([M]_\rho,[N]_\rho)}\ \textsc{Q-I-Ax}$$

$$\frac{\begin{array}{l}\Gamma,x{:}\sigma/\rho\vdash\tau[x]\\ \Gamma,x{:}\sigma\vdash M[x]:\tau[[x]_\rho]\\ \Gamma,x,x'{:}\sigma,p{:}\rho[x,x']\vdash\\ \qquad H: Id_{\tau[[x']_\rho]}(Subst_{\sigma/\rho,\tau}(Qax_\rho(p),M[x]),\ M[x'])\\ \Gamma\vdash N:\sigma/\rho\end{array}}{\Gamma\vdash plug_\rho\,N\ in\ M\ using\ H:\tau[N]}\ \textsc{Q-I-Elim}$$

$$\frac{}{\Gamma \vdash \mathit{plug}_\rho\, [N]_\rho \; \mathit{in}\; M \;\mathit{using}\; H = M[N] : \tau[[N]_\rho]} \quad \text{Q-I-COMP}$$

Proof irrelevance

$$\frac{\Gamma \vdash A : \mathit{Prop} \qquad \Gamma \vdash M, N : \mathit{Prf}(A)}{\Gamma \vdash \mathit{Pr\_Ir}(A, M, N) : \mathit{Prf}(M \stackrel{L}{=} N)} \quad \text{PR-IR}$$

Rules for subset types

$$\frac{\vdash \sigma \qquad , x{:}\sigma \vdash P : \mathit{Prop}}{\vdash \{x{:}\sigma \mid P\}} \quad \{\}\text{-FORM}$$

$$\frac{\vdash M : \sigma \qquad \vdash H : \mathit{Prf}(P[M])}{\vdash (M)_H : \{x{:}\sigma \mid P\}} \quad \{\}\text{-INTRO}$$

$$\frac{\vdash M : \{x{:}\sigma \mid P\}}{\vdash \mathit{wit}(M) : \sigma} \quad \{\}\text{-WIT}$$

$$\frac{\vdash M : \{x{:}\sigma \mid P\}}{\vdash \mathit{cor}(M) : \mathit{Prf}(P[\mathit{wit}(M)])} \quad \{\}\text{-COR}$$

$$\frac{}{\vdash \mathit{wit}((M)_H) = M \in \sigma \qquad \vdash \mathit{cor}((M)_H) = H \in \mathit{Prf}(P[M])} \quad \{\}\text{-BETA}$$

$$\frac{\begin{array}{l}\Gamma, x{:}\sigma \vdash \tau[x]\ \mathit{Nprop} \\ \Gamma, x{:}\sigma \vdash P[x] : \mathit{Prop} \\ \Gamma, p{:}\{x{:}\sigma \mid P[x]\} \vdash M[p] {:}\, \tau[\mathit{wit}(p)] \\ \Gamma \vdash N{:}\sigma\end{array}}{\Gamma \vdash \mathit{extend}_{\sigma,P,\tau}(M, N) : \tau[N]} \quad \{\}\text{-ELIM-NONPROP}$$

Same premises as {}-ELIM-NONPROP and

$$\frac{\Gamma \vdash H : \mathit{Prf}(P[M])}{\Gamma \vdash \mathit{extend}_{\sigma,P,\tau}(M, N) = M[(N)_H] : \tau[N]} \quad \{\}\text{-ELIM-NONPROP-COMP}$$

Extensional constructs in S_0

$$\frac{\Gamma \vdash A : \mathit{Prop} \qquad \Gamma \vdash M, N : \mathit{Prf}(A)}{\Gamma \vdash M = N : \mathit{Prf}(A)} \quad \text{PR-IR}$$

$$\frac{\Gamma \vdash P : \mathit{Prop} \qquad \Gamma \vdash Q : \mathit{Prop} \qquad \Gamma \vdash H : \mathit{Prf}(P \Leftrightarrow Q)}{\Gamma \vdash \mathit{Bi\_Imp}(P, Q, H) : \mathit{Prf}(P \stackrel{L}{=} Q)} \quad \text{BI-IMP}$$

$$\frac{\begin{array}{l}\Gamma \vdash U, V : \Pi x{:}\sigma.\tau \\ \Gamma, x{:}\sigma \vdash H : \mathit{Prf}(U\ x \stackrel{L}{=} V\ x)\end{array}}{\Gamma \vdash \mathit{Ext}(H) : \mathit{Prf}(U \stackrel{L}{=} V)} \quad \text{EXT}$$

Quotient types in S_0

$$\frac{\Gamma \vdash \sigma \qquad \Gamma\ , s, s'{:}\sigma \vdash R[s, s'] : \mathit{Prop}}{\Gamma \vdash \sigma/R} \quad \text{Q-FORM}$$

$$\frac{\Gamma \vdash M\!:\sigma}{\Gamma \vdash [M]_R\!:\sigma/R}\ \textsc{Q-Intro}$$

$$\frac{\Gamma \vdash \tau \qquad \Gamma\,,\, s\!:\sigma \vdash M[s] : \tau \qquad \Gamma \vdash N : \sigma/R \qquad \Gamma\,,\, s, s'\!:\sigma\,,\, p\!: Prf(R[s,s']) \vdash H : Prf(M[s] \stackrel{L}{=} M[s'])}{\Gamma \vdash plug_R\, N \; in\; M \; using\; H : \tau}\ \textsc{Q-Elim}$$

$$\frac{}{\Gamma \vdash plug_R\, [N]_R \; in\; M \; using\; H = M[N] : \tau}\ \textsc{Q-Comp}$$

$$\frac{\Gamma \vdash M, N : \sigma \qquad \Gamma \vdash H : Prf(R[M,N])}{\Gamma \vdash Qax_R(H) : Prf([M]_R \stackrel{L}{=} [N]_R)}\ \textsc{Q-Ax}$$

$$\frac{\Gamma\,,\, x\!:\sigma/R \vdash P[x] : Prop \qquad \Gamma\,,\, s\!:\sigma \vdash H : Prf(P[[s]_R]) \qquad \Gamma \vdash M : \sigma/R}{\Gamma \vdash Qind_R(H, M) : Prf(P[M])}\ \textsc{Q-Ind}$$

Rules for the choice operator

$$\frac{\Gamma \vdash M : \sigma/R \qquad \begin{array}{l}\text{There exists a non-quotiented context } \Delta \text{ and}\\ \text{a type } \tau \text{ and a term } N \text{ with } \Delta \vdash N : \tau \text{ and a}\\ \text{syntactic context morphism } \Gamma \vdash f \Rightarrow \Delta \text{ such}\\ \text{that } M \equiv N[f] \text{ and } \sigma/R \equiv \tau[f].\end{array}}{\Gamma \vdash choice(M) : \sigma}\ \textsc{Q-Choice}$$

$$\frac{\Gamma \vdash choice([M]_R) : \sigma \qquad \Gamma \vdash M : \sigma}{\Gamma \vdash choice([M]_R) = M : \sigma}\ \textsc{Q-Choice-Comp}$$

$$\frac{\Gamma \vdash M : \sigma/R \qquad \Gamma \vdash choice(M) : \sigma}{\Gamma \vdash [choice(M)]_R = M : \sigma/R}\ \textsc{Q-Choice-Ax}$$

Rules for the identity type on universes

$$\frac{\Gamma \vdash X, Y : U \qquad \Gamma \vdash F : Iso[X,Y]}{\Gamma \vdash UnivId(X, Y, F) : Id_U(X, Y)}\ \textsc{Univ-Id}$$

$$\frac{\Gamma \vdash F : Iso[X,Y] \qquad \Gamma \vdash M : El(X)}{\Gamma \vdash Subst_{U,El}(X, Y, UnivId(F), M) = F\; M : El(Y)}\ \textsc{Univ-Id-Eq}$$

$$\frac{}{\Gamma \vdash UnivId(X, X, id[X]) = Refl_U(X) : Id_U(X, X)}\ \textsc{Univ-Id-Eq'}$$

Rules for squash types

$$\frac{\Gamma \vdash \sigma}{\Gamma \vdash \Box\sigma}\ \Box\text{-}\textsc{Form} \qquad \frac{\Gamma \vdash M : \sigma}{\Gamma \vdash \Box M : \Box\sigma}\ \Box\text{-}\textsc{Intro}$$

$$\frac{\Gamma \vdash M, N : \Box\sigma}{\Gamma \vdash \Box\_ax(M,N) : Id_{\Box\sigma}(M,N)}\ \Box\text{-}\textsc{Ax}$$

$$\frac{\Gamma \vdash \sigma, \tau \qquad \Gamma, x{:}\sigma \vdash M[x] : \tau \qquad \Gamma \vdash N : \Box\sigma \qquad \Gamma\ , x, x'{:}\sigma \vdash H : Id_\tau(M[x], M[x'])}{\Gamma \vdash plug_{\Box}\ N\ in\ M\ using\ H : \tau}\ \Box\text{-}\textsc{Elim}$$

$$\frac{\Gamma \vdash plug_{\Box}\ \Box N\ in\ M\ using\ H\ : \tau}{\Gamma \vdash \Box\_Eq(M,N,H) : plug_{\Box}\ \Box N\ in\ M\ using\ H = M[N] : \tau}\ \Box\text{-}\textsc{Eq}$$

Axioms of Choice and Unique Choice

$$\frac{\Gamma \vdash \sigma \qquad \Gamma \vdash \tau}{\Gamma \vdash (\forall x{:}\,\sigma.\exists y{:}\,\tau.R[x,y]) \to (\exists f{:}\,\sigma \to \tau.\forall x{:}\,\sigma.R[x, f\ x])\ true}\ \text{IAC}$$

$$\frac{\Gamma \vdash \sigma \qquad \Gamma \vdash P : Prf(\exists x{:}\,\sigma.\forall y{:}\,\sigma.x \stackrel{L}{=} y)}{\Gamma \vdash AC!(P) : \sigma}\ \text{AC!}$$

Appendix C. A glossary of type theories

We use various type theories sometimes in parallel, so that the following summary, which gives the name or symbol of each relevant type theory together with a brief description and a reference to its first appearance in the text, may be helpful.

TT, core type theory. Dependent type theory with dependent products, sums natural numbers, unit type, and identity types. (p. 22)

Calculus of Constructions. A part of TT together with an impredicative universe. In the C.o.C. propositional equality may be defined as Leibniz equality (see 2.3.4). It is generally weaker than the identity type; in $\mathbf{S}_0$ (see below) the two are, however, equivalent. (p. 22)

TT_I. The type theory TT together with the extensional constructs of functional extensionality and uniqueness of identity. The elimination operator J is replaced by the Leibniz principle *Subst* for convenience. The extensional constructs are merely assumed as constants. (p. 64)

TT_E. The type theory TT together with the equality reflection rule which identifies definitional and propositional equality. Also called "extensional type theory". (p. 64)

$TT_{\mathfrak{R}}$. Extension of TT_I with an intensional counterpart to equality reflection ("$\mathfrak{R}$"). For every derivable judgement in TT_E there exists a decoration of this judgement with *Subst* and $\mathfrak{R}$ derivable in $\mathrm{TT}_{\mathfrak{R}}$. (p. 66)

D. Type theory defined by the deliverables model—a kind of subset interpretation of the Calculus of Constructions. **D** supports all type and term formers of the C.o.C. and in addition proof irrelevance and subset types. The symbol **D** is also used for the deliverables model itself. (p. 94)

$\mathbf{S}_0$. Type theory defined by the "non-dependent" setoid model in Section 5.1 (also called $\mathbf{S}_0$). It includes the C.o.C. and supports proof irrelevance, subset types, functional and propositional extensionality, and quotient types. The extensional constructs in $\mathbf{S}_0$ are based on Leibniz equality which in $\mathbf{S}_0$ is equivalent in strength to the identity type. $\mathbf{S}_0$ does not support universes, except for the impredicative universe of propositions. There exists an extension of $\mathbf{S}_0$ which makes universes available, but extensional constructs are then restricted to types inside the universe, see Section 5.1.8. (p. 116)

$\mathbf{S}_1$. Type theory defined by the "dependent" setoid model in Section 5.3. It supports all type and term formers of TT_I including functional extensionality based on the identity type. Probably quotient types are also supported. Unlike $\mathbf{S}_0$ the type theory $\mathbf{S}_1$ caters for proper type dependency, however, it is not clear whether a universe is supported. (p. 153)

Source type theory. Type theory defined by a syntactic model, e.g. $\mathbf{S}_1$ or $\mathbf{D}$.

Target type theory. Type theory used to construct a syntactic model. Usually TT or the C.o.C.

Appendix D. Index of symbols

The following summarises the mathematical symbols used throughout the thesis together with a brief explanation and a reference to their first appearance in the text.

$-^{op}$	opposite category	31
Sets	category of sets and functions	31
Subst	substitution operator for identity types	55
Id	identity type, propositional equality	15
Refl	reflexivity operator for id. types (canonical element)	56
Sym	symmetry operator for id. types	56
Trans	transitivity operator for id. types	56
Resp	compatibility of id. types with function application	56
J	elimination operator for id. types	16
IdUni	uniqueness of identity proofs	57
$\vdash$	judgement relation in type theory	14
$\Gamma \vdash \sigma$ *true*	judgement expressing that σ is non-empty	17
$\circ$	composition (of morphisms, substitutions, and terms)	20
id	identity morphism	26
$\mathbf{N}$	type of natural numbers	15
$0, Suc$	canonical elements of $\mathbf{N}$, zero and successor	15
$R^{\diamond}$	elimination operator for type former $\diamond$	15
$\equiv$	syntactic identity	18
Int	type of positive and negative integers	125
Real	type of real numbers	131
$Str(\sigma)$	type of streams over σ	167
$Hd_{\sigma}(-)$	first element of a stream	167
$Tl(-)$	tail of a stream	168
CoIt	coiterator for streams	168
$\diamond$	empty context	14
$\mathbf{C}$	a category, category of contexts	25
$\top$	terminal object, semantic equivalent to the empty context	25

References

1. Stuart Allen. A non-type-theoretic account of Martin-Löf's types. *Symposium on Logic in Computer Science*, 1987.
2. Thorsten Altenkirch. *Constructions, Normalization, and Inductive Types*. PhD thesis, University of Edinburgh, 1994.
3. Thorsten Altenkirch. Proving strong normalization of CC by modifying realizability semantics. Henk Barendregt and Tobias Nipkow, editors, *Types for Proofs and Programs, Springer LNCS Vol. 806*, pages 3–18, 1994.
4. Thorsten Altenkirch, Zhaohui Luo, et al. Equality for categories. E-mail, Apr.–Sep. 1990. Unpublished.
5. Michael Barr and Charles Wells. *Toposes, Triples and Theories*. Springer, 1985.
6. Michael Beeson. *Foundations of Constructive Mathematics*. Springer, 1985.
7. J. Bénabou. Fibred categories and the foundations of naive category theory. *Journal of Symbolic Logic*, 50:10–37, 1985.
8. J. Bénabou. Théorie des ensembles empiriques. Cahiers de poétique comparée, MEZURA, No. 17 & 21, Institut National des Langues et Civilisations Orientales, 2, rue de Lille, 75007 Paris, 1990.
9. Errett Bishop and Douglas Bridges. *Constructive Analysis*. Springer, 1985.
10. N. Bourbaki. *Eléments de mathématique, Livre I, Théorie des ensembles—fascicule de résultats*. Hermann, Paris, 1964.
11. Rod Burstall. Programming with Modules as Typed Functional Programming. *Proc. Inter. Conf. on Fifth Generation Computer Systems, Tokyo*, 1984.
12. Rod Burstall and Butler Lampson. Pebble, a kernel language for modules and abstract data types. *Information and Computation*, 76:278–346, 1988.
13. Rod Burstall and James McKinna. Deliverables: An approach to program semantics in constructions. *Proc. MFCS '93, Springer LNCS, Vol. 711*, 1993. Also as LFCS technical report ECS-LFCS-91-133.
14. A. Carboni. Some free constructions in realizability and proof theory. *Journal of Pure and Applied Algebra*, to appear.
15. J. Cartmell. *Generalized Algebraic Theories and Contextual Categories*. PhD thesis, Univ. Oxford, 1978.
16. Robert Constable et al. *Implementing Mathematics with the Nuprl Development System*. Prentice-Hall, 1986.
17. Thierry Coquand. Metamathematical investigations of a calculus of constructions. Piergiorgio Odifreddi, editor, *Logic and Computer Science*, pages 91–118. Academic Press Ltd., 1990.
18. Thierry Coquand. Pattern matching with dependent types. *Workshop on Logical Frameworks, Båstad*, 1992. Preliminary Proceedings.
19. Thierry Coquand. Infinite objects in type theory. H. Barendregt and T. Nipkow, editors, *Types for Proofs and Programs*. Springer, 1994. LNCS 806.
20. Thierry Coquand and Gérard Huet. The Calculus of Constructions. *Information and Computation*, 76:95–120, 1988.
21. R. Crole. *Categories for Types*. Cambridge University Press, 1993.

22. Pierre-Louis Curien. *Categorical Combinators, Sequential Algorithms and Functional Programming*. Pitman, 1986.
23. Pierre-Louis Curien. Alpha-conversion, conditions on variables and categorical logic. *Studia Logica*, 3:318–360, 1989.
24. Pierre-Louis Curien. Substitution up to isomorphism. *Fundamenta Informaticae*, 19:51–86, 1993.
25. N. J. Cutland. *An Introduction to Recursive Function Theory*. Cambridge University Press, 1980.
26. N. G. de Bruijn. A survey of the project Automath. J. P. Seldin and J. R. Hindley, editors, *To H. B. Curry: Essays on Combinatory Logic, Lambda Calculus, and Formalism*, pages 589–606. Academic Press, London, 1980.
27. N. G. de Bruijn. Telescopic mappings in typed lambda calculus. *Information and Computation*, 91(2):189–204, April 1991.
28. Gilles Dowek et al. The COQ proof assistant user's guide, V5.6. Rapport technique 134, INRIA Rocquencourt, Dec. 1991.
29. Peter Dybjer. Inductive sets and families in Martin-Löf's type theory and their set-theoretic semantics. G. Huet and G. Plotkin, editors, *Logical Frameworks*, pages 280–306, 1991.
30. Peter Dybjer and Herbert Sander. A functional programming approach to the specification and verification of concurrent systems. *Specification and Verification of Concurrent Systems*, pages 331–343. Springer, 1990. Workshops in Computing.
31. Thomas Ehrhard. *Une sémantique catégorique des types dépendants. Application au Calcul des Constructions*. PhD thesis, Univ. Paris VII, 1988.
32. Solomon Feferman. Theories of finite type. Jon Barwise, editor, *Handbook of Mathematical Logic*, chapter D.4. North-Holland, 1977.
33. G. Huet and A. Saïbi. Constructive category theory. Draft, presented at the Joint CLICS-TYPES Workshop on Categories and Type Theory, Gothenburg, Jan 8–10, 1995.
34. D.M. Gabbay and R.J.G.B. de Queiroz. Equality in labelled deductive systems and the functional interpretation of propositional equality. P. Dekker and M. Stokhof, editors, *Proceedings of the Ninth Amsterdam Colloquium*, pages 547–565, 1994.
35. Robin Gandy. On the axiom of extensionality I. *Journal of Symbolic Logic*, 21:36–48, 1956.
36. Robin Gandy. On the axiom of extensionality II. *Journal of Symbolic Logic*, 24:287–300, 1959.
37. Herman Geuvers and Benjamin Werner. On the Church-Rosser property for expressive type systems and its consequences for their metatheoretic study. *Proceedings of the 9th Annual IEEE Symposium on Logic in Computer Science*, 1994.
38. Healfdene Goguen. *A Typed Operational Semantics for Type Theory*. PhD thesis, University of Edinburgh, 1994.
39. Healfdene Goguen and Zhaohui Luo. Inductive data types: Well-ordering types revisited. Technical Report ECS-LFCS-92-209, LFCS, April 1992.
40. Robert Goldblatt. *Topoi: The Categorial Analysis of Logic*. North-Holland, revised edition, 1984.
41. R. Harper and J. C. Mitchell. On the type structure of Standard ML. *ACM Trans. Programming Lang. and Systems*, 15(2):211–252, 1993.
42. R. Harper, J. C. Mitchell, and E. Moggi. Higher-order modules and the phase distinction. *Conference record of the 17th ACM Symposium on Principles of Programming Languages (POPL)*, pages 341–354, San Francisco, CA USA, 1990.
43. Susumu Hayashi. Singleton, union, and intersection types for program extraction. *Information and Computation*, 109(1/2):174–210, 1994.

44. M. Hofmann. Conservativity of equality reflection over intensional type theory. *Proc. BRA TYPES workshop, Torino, June 1995, Springer LNCS*, 1996. To appear.
45. Martin Hofmann. Syntax and semantics of dependent types. P. Dybjer and A. M. Pitts, editors, *Semantics and Logics of Computation*. Cambridge University Press, to appear 1997.
46. Martin Hofmann. Elimination of extensionality for Martin-Löf type theory. H. Barendregt and T. Nipkow, editors, *Types for Proofs and Programs*. Springer, 1994. LNCS 806.
47. Martin Hofmann. On the interpretation of type theory in locally cartesian closed categories. Jerzy Tiuryn and Leszek Pacholski, editors, *Proc. CSL '94, Kazimierz, Poland, Springer LNCS, Vol. 933*, pages 427–442, 1995.
48. Martin Hofmann. A simple model for quotient types. *Proc. TLCA '95, Edinburgh, Springer LNCS Vol. 902*, pages 216–234, 1995.
49. Martin Hofmann and Thomas Streicher. A groupoid model refutes uniqueness of identity proofs. *Proceedings of the 9th Symposium on Logic in Computer Science (LICS), Paris*, 1994.
50. Martin Hofmann and Thomas Streicher. The groupoid interpretation of type theory. Submitted to the Proceedings of the meeting *Twenty-five years of constructive type theory*, Venice, October 1995.
51. Wolfgang Hornung. *Entwicklung eines LEGO-Kontexts für die Verifikation kategorieller Aussagen*. Universität Erlangen, 1992.
52. Gérard Huet. Cartesian closed categories and λ-calculus. Guy Cousineau et al., editors, *Combinators and Functional Programming Languages*, volume 242 of *LNCS*, pages 123–135. Springer, Berlin, Heidelberg, New York, 1985.
53. J. M. E. Hyland. The effective topos. A. S. Troelstra and D. van Dalen, editors, *The L. E. J. Brouwer Centenary Symposium*, pages 165–216. North-Holland, 1982.
54. J. M. E. Hyland, P. T. Johnstone, and A. M. Pitts. Tripos theory. *Mathematical Proceedings of the Cambridge Philosophical Society*, 88:205–232, 1980.
55. Martin Hyland and Andrew Pitts. The Theory of Constructions: Categorical Semantics and Topos-Theoretic Models. *Categories in Computer Science and Logic*. AMS, 1989.
56. B. Jacobs, E. Moggi, and Th. Streicher. Relating Models of Impredicative Type Theories. D. Pitt *et al.*, editor, *Proc. Conf. Category Theory and Computer Science, Paris, France*, pages 197–218. Springer, LNCS vol. 530, 1991.
57. Bart Jacobs. Quotients in simple type theory. manuscript.
58. Bart Jacobs. *Categorical Type Theory*. PhD thesis, University of Nijmegen, 1991.
59. Bart Jacobs. Comprehension categories and the semantics of type theory. *Theoretical Computer Science*, 107:169–207, 1993.
60. Gilles Kahn. The semantics of a simple language for parallel programming. *Information Processing*, 74:471–475, 1974.
61. François Lamarche. A Proposal about Foundations I. Manuscript, 1991(?).
62. Joachim Lambek and Philip Scott. *Introduction to Higher-Order Categorical Logic*. Cambridge University Press, 1985.
63. Lars Hallnäs. An intensional characterisation of the largest bisimulation. *Theoretical Computer Science*, 53, 1987.
64. François Leclerc and Christine Paulin-Mohring. Programming with Streams in COQ. A Case Study: The Sieve of Eratosthenes. H. Barendregt and T. Nipkow, editors, *Types for Proofs and Programs*. Springer, 1994. LNCS 806.
65. A. Levy. *Basic Set Theory*. Springer-Verlag, 1979.
66. Horst Luckhardt. *Extensional Gödel Functional Interpretation. A Consistency Proof of Classical Analysis*, volume 306 of *Lecture Notes in Mathematics*. Springer, Berlin, 1973.

67. Z. Luo. Program specification and data refinement in type theory. S. Abramsky and T. S E. Maibaum, editors, *Proc. TAPSOFT '91, Springer LNCS, Vol. 493*, pages 142–168, 1991.
68. Zhaohui Luo. *An Extended Calculus of Constructions*. PhD thesis, University of Edinburgh, July 1990.
69. Zhaohui Luo. Type Theory, Logic, and Computer Science. Course notes for the LFCS Postgraduate Course on type theory, January 1991.
70. Zhaohui Luo. *Computation and Reasoning*. Oxford University Press, 1994.
71. Zhaohui Luo and Robert Pollack. LEGO proof development system: User's manual. Technical Report ECS-LFCS-92-211, LFCS, Computer Science Dept., University of Edinburgh, 1992.
72. Saunders Mac Lane. *Categories for the Working Mathematician*. Springer, 1971.
73. Per Martin-Löf. A theory of types. Manuscript, 1971.
74. Per Martin-Löf. An intuitionistic theory of types: Predicative part. H. E. Rose and J. C. Sheperdson, editors, *Logic Colloquium 1973*, pages 73–118. North-Holland, 1975.
75. Per Martin-Löf. *Intuitionistic Type Theory*. Bibliopolis·Napoli, 1984.
76. Per Martin-Löf. Mathematics of infinity. P. Martin-Löf and G. Mints, editors, *Proc. of COLOG '88, Tallinn, USSR*, pages 146–197. Springer LNCS vol. 417, 1988.
77. Per Martin-Löf. Nondeterministic definitions in type theory. Lecture notes by Peter Dybjer, March 1991.
78. James McKinna. *Deliverables: A Categorical Approach to Program Development in Type Theory*. PhD thesis, University of Edinburgh, 1992.
79. Paul-André Mellies. Typed λ-calculi with explicit substitutions may not terminate. *Proceedings of TLCA '95, Edinburgh, LNCS*, 1995. to appear.
80. N. P. Mendler. Σ is cocontinuous. Manuscript.
81. N. P. Mendler. *Inductive Definition in Type Theory*. PhD thesis, Cornell University, 1988.
82. Nax P. Mendler. Quotient types via coequalisers in Martin-Löf's type theory. in the informal proceedings of the workshop on Logical Frameworks, Antibes, May 1990.
83. Robin Milner and Mads Tofte. *Commentary on Standard ML*. MIT Press, 1991.
84. John C. Mitchell and Gordon D. Plotkin. Abstract types have existential type. *ACM Transactions on Programming Languages and Systems*, 10(3):470–502, July 1988.
85. Ieke Moerdijk and Saunders Mac Lane. *Sheaves in Geometry and Logic. A First Introduction to Topos Theory*. Springer, 1992.
86. Eugenio Moggi. Computational lambda calculus and monads. *4th Symposium on Logic in Computer Science*. IEEE, 1989.
87. Eugenio Moggi. A category-theoretic account of program modules. *Math. Struct. in Comp. Sci.*, 1(1):103–139, 1991.
88. B. Nordström, K. Petersson, and J. M. Smith. *Programming in Martin-Löf's Type Theory, An Introduction*. Clarendon Press, Oxford, 1990.
89. Adam Obtułowicz. Categorical and algebraic aspects of Martin-Löf type theory. *Studia Logica*, 3:299–317, 1989.
90. Christine Paulin-Mohring. Extracting F_ω's programs from proofs in the calculus of constructions. *Principles of Programming Languages (POPL)*, pages 1–17. ACM, 1989.
91. Christine Paulin-Mohring. *Extraction de programmes dans le Calcul des Constructions*. PhD thesis, Université Paris VII, 1989.
92. Andrew Pitts. Categorical logic. *Handbook of Logic in Computer Science (Vol. VI)*. Oxford University Press, 1997. To appear.
93. Gordon Plotkin. Generic notes on domains. Notes for lectures from several terms at the University of Edinburgh, 1989-90.

94. Randy Pollack. A formalisation of Tarski's fixpoint theorem in Lego. Part of the Lego-distribution and personal communication.
95. Eike Ritter. *Categorical Abstract Machines for Higher-Order Typed Lambda Calculi*. PhD thesis, University of Cambridge, 1992.
96. A. Salvesen and J. M. Smith. The strength of the subset type in Martin-Löf's type theory. *IEEE Symposium on Logic in Computer Science*, 1988.
97. D. Sannella. Formal program development in Extended ML for the working programmer. *Proc. 3rd BCS/FACS Workshop on Refinement, Hursley Park, 1990*, pages 99–130. Springer Workshops in Computing, 1991.
98. D. S. Scott. Identity and existence in intuitionistic logic. M. P. Fourman, C. J. Mulvey, and D. S. Scott, editors, *Applications of Sheaves*, pages 660–696. Springer-Verlag, 1977.
99. Robert A. G. Seely. Locally cartesian closed categories and type theory. *Mathematical Proceedings of the Cambridge Philosophical Society*, 95:33–48, 1984.
100. Jan Smith. The independence of Peano's fourth axiom from Martin-Löf's type theory without universes. *Journal of Symbolic Logic*, 53(3), 1988.
101. Thomas Streicher. A verification method for finite dataflow networks with constraints applied to the verification of the alternating bit protocol. Technical Report MIP 8706, Universität Passau, 1987.
102. Thomas Streicher. *Correctness and Completeness of a Categorical Semantics of the Calculus of Constructions*. PhD thesis, Universität Passau, 1989.
103. Thomas Streicher. *Semantics of Type Theory*. Birkhäuser, 1991.
104. Thomas Streicher. *Semantical Investigations into Intensional Type Theory*. Habilitationsschrift, LMU München, 1993.
105. Alfred Tarski. A lattice-theoretical fixpoint theorem and its applications. *Pacific Journal of Mathematics*, 5:285–309, 1955.
106. Paul Taylor. *Recursive Domains, Indexed Category Theory, and Polymorphism*. PhD thesis, University of Cambridge, 1986.
107. A.S. Troelstra. Aspects of constructive mathematics. Jon Barwise, editor, *Handbook of Mathematical Logic*, chapter D.5, pages 973–1052. North-Holland, 1977.
108. A.S. Troelstra. On the syntax of Martin-Löf's type theories. *Theoretical Computer Science*, 51, 1987.
109. A.S. Troelstra and D. van Dalen. *Constructivism in Mathematics, An Introduction*, volume II. North-Holland, 1988.
110. A.S. Troelstra and D. van Dalen. *Constructivism in Mathematics, An Introduction*, volume I. North-Holland, 1988.
111. David Turner. A new formulation of constructive type theory. P. Dybjer, editor, *Proceedings of the Workshop on Programming Logic*, pages 258–294. Programming Methodology Group, Univ. of Göteborg, May 1989.
112. F.W. von Henke, A. Dold, M. Grosse, H. Rueß, D. Schwier, and M. Strecker. Construction and deduction methods for the formal development of software. *Arbeitsberichte des* KORSO-*Projekts*. Springer LNCS, to appear 1995.
113. P. L. Wadler. Comprehending monads. *Mathematical Structures in Computer Science*, 2:461–493, 1992.
114. Gavin C. Wraith. A note on categorical datatypes. *Proc. Conf. Category Theory and Computer Science, Manchester, UK*, pages 213–223. Springer LNCS vol. 389, 1989.

Index